Introductory Heterocyclic Chemistry

Introductory Heterocyclic Chemistry

Peter A. Jacobi
Dartmouth College
New Hampshire, United States

Registered Office(s)
John Wiley & Sons, Inc., 111 River Street, Hoboken, NJ 07030, USA
John Wiley & Sons Ltd, The Atrium, Southern Gate, Chichester, West Sussex, PO19 8SQ, UK

Editorial Office
The Atrium, Southern Gate, Chichester, West Sussex, PO19 8SQ, UK

For details of our global editorial offices, customer services, and more information about Wiley products visit us at www.wiley.com.

Wiley also publishes its books in a variety of electronic formats and by print-on-demand. Some content that appears in standard print versions of this book may not be available in other formats.

Library of Congress Cataloging-in-Publication Data applied for
Paperback : 9781119417590

Cover design by Wiley
Cover Images: © iyadsm/iStockphoto
Courtesy of Peter A. Jacobi

Set in 10/12pt WarnockPro by SPi Global, Chennai, India

Printed in the UK

In tribute to Ted, whose fascination with heterocyclic chemistry was boundless—a treasured friend, and an inspiring mentor to generations of heterocyclic chemists

Contents

Preface

Why heterocycles? Why has Nature chosen these ring systems as a foundation for so many of her life processes? And why have organic chemists focused so much effort on understanding and synthesizing these materials? One reason is that heterocyclic rings (i.e. those containing atoms other than carbon) are present in the majority of known natural products, contributing to enormous structural diversity. In addition, they often possess significant biological activity. Medicinal chemists have embraced this last property in designing most of the small molecule drugs in use today. Indeed, a 2014 study found that nearly 60% of all FDA-approved drugs in this category incorporate a nitrogen heterocycle (totaling some 640!).* Oxygen and sulfur heterocycles are also well represented.

The chemistry of these substances is quite different from that encountered in carbocyclic systems, and an appreciation of these differences is important for understanding biochemistry and molecular biology at a fundamental level. Also, it is not unusual for heterocyclic species to undergo rearrangements and transformations having no parallel in carbocyclic chemistry, which provides a good training ground for sharpening mechanistic skills.

This leads to the question of at what level should the study of heterocyclic chemistry begin. It is fair to say that most courses in this area are geared toward graduate education, at least in the U.S. But this need not be the case, and the author has many years of experience teaching this material at the junior and senior undergraduate level. This text builds upon that experience, and it should be appropriate for introducing this topic to aspiring heterocyclic chemists at various stages of their careers. It should also find use as a resource for chemists in related fields who simply wish to "test the waters" of heterocyclic chemistry.

---

*Vitaku, E.; Smith, D. T.; Njardarson, J.T. *J. Med. Chem.* 2014, 57, 10257–10274.

Acknowledgments

The author is indebted to my good friend and colleague, Professor Gordon W. Gribble, for reading and commenting on the entire text. A special thanks is also due to Ms. Lora C. Leligdon, of Kresge Library, Dartmouth College, for her invaluable aid in tracking down innumerable references. And finally, I am grateful to my dear wife Lee Ann, who, although not a chemist persevered through the entire project and served as proofreader extraordinaire.

1

Some Biologically Important Heterocycles of Nature

Heterocyclic rings come in many sizes and shapes, and they may be either aromatic or non-aromatic, fused or non-fused. This chapter provides a brief survey of some of the most biologically important heterocycles found in nature.

Let us begin our discussion with the three common amino acids tryptophan, proline, and histidine (Figure 1.1). Tryptophan contains an indole skeleton (blue), which is aromatic by virtue of having 10 π-electrons in a cyclic conjugated array, two of which are donated by the ring nitrogen. Proline, on the other hand, is clearly non-aromatic, since the pyrrolidine ring has only the free electron pair on nitrogen. But what about histidine, the distinguishing feature of which is an imidazole core with a total of eight electrons? Does this ring system satisfy Hückel's rule? The answer is yes, since one of the electron pairs resides in an orthogonal sp^2-orbital, leaving 6 π-electrons to constitute the aromatic sextet. Lastly, while not an amino acid itself, we include in this introduction serotonin, a product of catabolism of tryptophan [1a]. The chief function of this indole alkaloid is as a neurotransmitter in the brain, and as such, it plays a key role in regulating mood. Drugs that alter the concentration of serotonin in the brain have found use in treating depression and anxiety disorders.

As to their own biological role, amino acids are ubiquitous as the molecular building blocks of peptides, proteins, and enzymes. In addition to the three heterocyclic amino acids just described, 17 others make up the class of 20 naturally occurring amino acids, 9 of which are considered "essential" (i.e., they cannot be biosynthesized by humans, and must be provided by diet). In this context, one might wonder how nature functions at such a complex level with such a limited "tool chest." Part of the answer is given by the simple equation $N=20^n$, where N equals the number of possible peptides/proteins, 20 equals the number of monomer building blocks, and n equals the number of amino acids in the chain. The results can be staggering! For example, the number of structurally unique dipeptides would total 400; for n=5 this number jumps to 3,200,000; and for n=100, which is still quite a small protein, the possible combinations would be, well, astronomical (many orders of magnitude greater than the estimated number of atoms in the universe) [1b].

Introductory Heterocyclic Chemistry, First Edition. Peter A. Jacobi.
© 2019 John Wiley & Sons Ltd. Published 2019 by John Wiley & Sons Ltd.

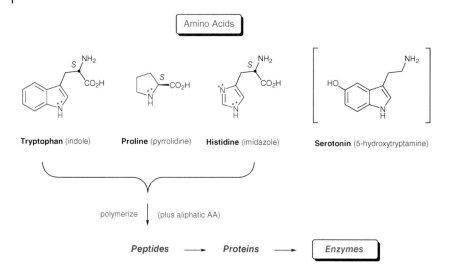

Figure 1.1 Some naturally occurring heterocycles.

Proteins are one example of a class of compounds known as informational macromolecules, which control crucial life processes. We saw above how great complexity can be generated from a relatively small group of amino acid building blocks, even considering just the primary structure of the derived proteins. However, proteins are not unique in this capability, and it has been estimated that over 90% of all of the organic material found in living organisms, including many thousands of macromolecules, can be generated from about three dozen monomeric species [1a]. Of these, 20 constitute the naturally occurring amino acids. Another five are the DNA and RNA bases adenine (abbreviated A), thymine (T), guanine (G), and cytosine (C), all found in DNA, and uracil (U), which replaces thymine in RNA (Figure 1.2). A and G are examples of purine heterocycles, while T, C, and U are pyrimidines.

DNA is *formally* derived from A, T, G, and C by initial condensation of the NH groups shown in red with 2-deoxyribose (an example of a furanose heterocycle, and another of nature's basic building blocks). The resultant deoxyribonucleosides then undergo phosphorylation at the primary hydroxyl group to afford the corresponding deoxyribonucleotides, which on polymerization lead to single stranded DNA. But this is not the end of the story. Most readers will be aware of the double helical nature of DNA, wherein base pairing through hydrogen bonding assures that identical concentrations of A and T will always be present. The same holds true for G and C. This observation was one of the keys to unraveling the structure of DNA, and it provided a molecular basis for the process of replication. RNA, produced from DNA by transcription, is a

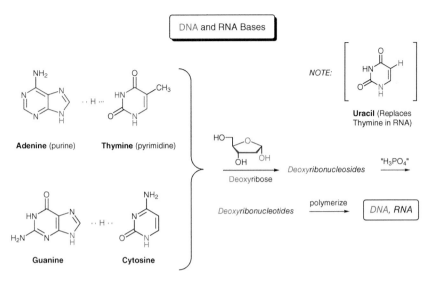

Figure 1.2 Naturally occurring heterocycles.

single stranded polynucleotide with uracil substituting for thymine, and ribose replacing deoxyribose. In the third step for processing genetic information, the message encoded in RNA is translated by ribosomes to that required for synthesizing specific proteins.

1.1 Vitamins

Who among us has not at some point been concerned with "getting enough vitamins?" Although generally required in only trace quantities, these micro-nutrients are one of five essential components of a healthy diet (the others being carbohydrates, fats, proteins, and certain mineral elements). Vitamins perform a myriad of biological functions, and they are generally classified as being either fat soluble or water soluble. All of the water-soluble vitamins are heterocycles, a sampling of which are described below (Figure 1.3) [1].

Pyridoxine is one of a group of three closely related pyridine heterocycles that constitute the vitamin B_6 group of vitamins (Figure 1.3). These species are best known for their role in transamination reactions, wherein an amino group from one α-amino acid is reversibly transferred to the α-carbon of an α-keto acid. This process is coupled with the interconversion of pyridoxal and pyridoxamine, according to the overall equation:

$$\text{pyridoxal} + R\left(CHNH_2\right) - CO_2H \rightleftharpoons R\left(C = O\right) - CO_2H + \text{pyridoxamine}$$

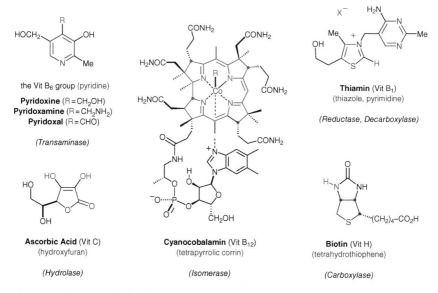

Figure 1.3 Heterocyclic vitamins.

While the details for this enzyme catalyzed transformation are complex, we shall see later that each step finds precedent in common bond forming reactions employed in heterocycle synthesis.

Ascorbic acid (vitamin C) is best known as a preventive and curative agent for the debilitating disease scurvy, which is characterized by lethargy, brown spots on the skin and gums, open sores, and if left untreated, death from bleeding. Human beings are among the few vertebrates who cannot biosynthesize this substance from glucose. Up until the early 1800s scurvy was common among sailors and other adventurers lacking access to fresh citrus fruit, an excellent source of vitamin C. The British Navy is credited with making the observation that a daily ration of citrus juice, mixed of course with grog, prevented scurvy (hence the term "limey," originally intended as derogatory slang for sailors in the Royal Navy). In this capacity, vitamin C serves as a co-factor in the enzymatic hydroxylation of proline to afford 4-hydroxyproline, a major component of connective tissue:

$$\alpha\text{-ketoglutarate} + \underset{\text{Proline}}{\boxed{\quad}}\text{—CO}_2\text{H} \xrightarrow[\substack{\text{prolyl hydroxylase} \\ O_2}]{\text{ascorbate}} \underset{\text{4-Hydroxyproline}}{\text{HO}\,\boxed{\quad}\text{—CO}_2\text{H}} + \text{succinate}$$

It is, of course, also well known for its antioxidant properties by virtue of its enediol functionality (shown in red in Figure 1.3).

As ascorbic acid is to scurvy, so vitamin B_{12} is to pernicious anemia, an auto-immune disorder in which the body fails to produce sufficient healthy red blood cells. A major breakthrough in treating this disease was made in 1926, when it was found that a diet rich in partially cooked liver effected a cure. It remained until the late 1940s for the anti-pernicious anemia factor to be iso-lated in the crystalline state, in the form of cyanocobalamin (R=CN in Figure 1.3; in the physiologically active forms, R=5'-deoxyadenosyl or CH_3). Another 10 years were required to determine its very complex structure, con-taining a cobalt atom complexed in a corrin nucleus. In humans, vitamin B_{12} serves as a coenzyme in two important enzymatic processes. In the methylco-balamin form it functions as a methyltransferase, in which a methyl group transfers between two molecules. Alternatively, in the adenosylcobalamin form it functions as an isomerase, in which a hydrogen atom shifts from one carbon atom to an adjacent one, with concomitant exchange of a second group. See, for example, the conversion of glutamic acid to β-methylaspartic acid below, where the group shown in red migrates from C3 to C4, exchanging with the hydrogen shown in blue [1a]:

Glutamic acid   β-Methylaspartic acid

Thiamin (vitamin B_1) contains both a thiazole and a pyrimidine ring and it has played a rich role in the history of vitamin discovery. In fact, the name "vitamine" derives from early studies on this material, when it was recognized as a <u>vit</u>al, <u>amine</u> containing dietary component in preventing beriberi (the "e" was later dropped when it was discovered that not all vitamins are amines). In animal tissues it is present mainly as a pyrophosphate derivative (TPP), in which form it serves as a coenzyme in several cellular processes. One of these involves decarboxylations of α-keto acids, such as pyruvate, where TPP func-tions as a carrier of an intermediate aldehyde:

$$CH_3(CO) - CO_2^- + H_2O + \text{pyruvate decarboxylase} \rightarrow CH_3 - CH = O + HCO_3^-$$

All such conversions exploit the unusually high acidity of the C-2 hydrogen shown in red ($pK_a \approx 18$ in the free state [2a]; thought to be considerably lower in the active site [2b]), and the nucleophilicity of the derived ylide (sometimes referred to as a heterocyclic carbene). Thus, the first step in pyruvate decar-boxylation involves nucleophilic addition of TPP ylide across the ketone car-bonyl group of pyruvate (Scheme 1.1). This is followed by decarboxylation to generate an enol, following a mechanism analogous to that observed with

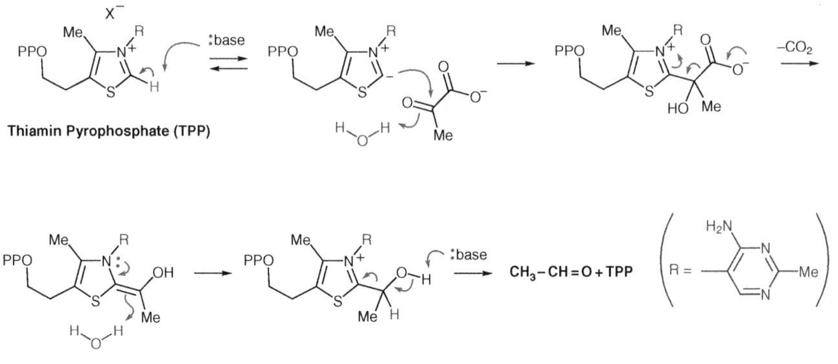

Scheme 1.1

β-ketoacids. Finally, protonation and rapid cleavage of the resultant hydroxyethyl group generates acetaldehyde and returns TPP.

In Scheme 1.1, thiamin pyrophosphate serves as a coenzyme in the decarboxylation of pyruvate to afford acetaldehyde. Biotin (vitamin H), on the other hand, is a coenzyme for carboxylase enzymes, involved in the synthesis of numerous biologically important molecules. As one example, the N-H group highlighted in red (Figure 1.3) serves as a carrier of a carboxy group ($-CO_2^-$) in the carboxylation of pyruvate to oxaloacetate, the biological precursor to glucose. The driving force for this reaction is provided by adenosine triphosphate (ATP), which in the process is converted to ADP:

$$ATP + HCO_3^- + CH_3(CO) - CO_2^- + \text{pyruvate carboxylase} \rightleftharpoons$$
$$ADP + {}^-O_2C - CH_2(CO) - CO_2^- + P_i$$

Historically, vitamin H has sometimes been referred to as the "beauty vitamin," for its purported benefits to hair and skin, or, in Deutsch, "Haar und Haut." Hence the letter "H."

Finally, among the heterocyclic vitamins, folic acid (pteroylglutamic acid) deserves special mention (Figure 1.4). Indeed, it has been stated that "there are few examples in biochemistry where a single compound has provided the key to as many nutritional phenomena as pteroylglutamic acid [3]."

Folic acid itself is found widely distributed in food sources, especially in leafy green vegetables such as spinach. However, as with most vitamins, this species must be converted to an activated form prior to functioning as a coenzyme. The initial step in this process involves enzymatic reduction at the C7-N8 double bond to give 7,8-dihydrofolic acid (FH_2). Further reduction, catalyzed by dihydrofolate reductase, then affords the active coenzyme 5,6,7,8-tetrahydrofolic acid (FH_4), which serves as the carrier of one carbon fragments at the methyl, formaldehyde, and formate oxidation levels. Thus, in the pathway shown, serine hydroxymethyltransferase converts FH_4 to

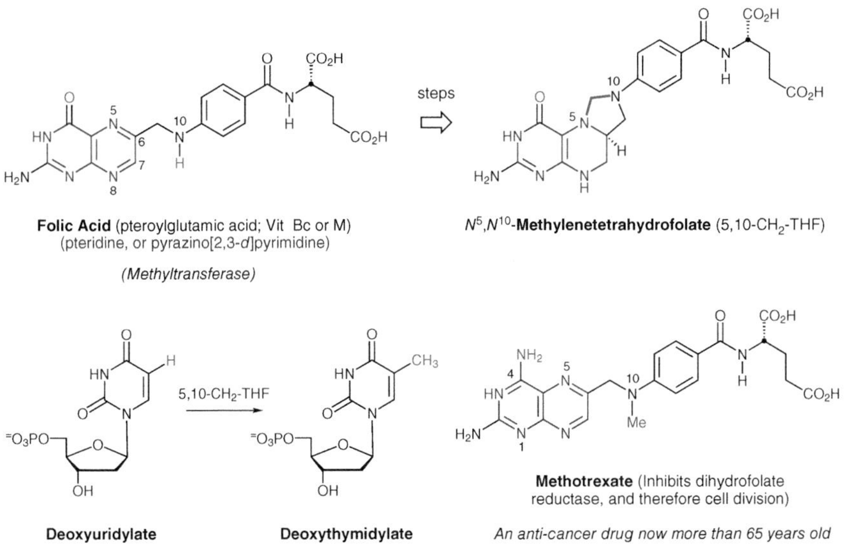

Folic Acid (pteroylglutamic acid; Vit Bc or M)
(pteridine, or pyrazino[2,3-*d*]pyrimidine)

(Methyltransferase)

*N*⁵,*N*¹⁰-**Methylenetetrahydrofolate** (5,10-CH₂-THF)

steps

Deoxyuridylate

5,10-CH₂-THF

Deoxythymidylate

Methotrexate (Inhibits dihydrofolate
reductase, and therefore cell division)

An anti-cancer drug now more than 65 years old

Figure 1.4 Heterocyclic vitamins.

N^5,N^{10}-methylenetetrahydrofolate (5,10-CH$_2$-THF), in which the methylene group to be transferred is highlighted in red. And to where is this group transferred? In a step catalyzed by thymidylate synthase, 5,10-CH$_2$-THF and deoxyuridylate come together to produce deoxythymidylate, by a process involving reductive methylation (the product of oxidation in this step is FH$_2$) [1].

The deciphering of this pathway had enormous implications. Since deoxythymidylate is one of the fundamental building blocks of DNA, which in turn is required of all living cells, it follows that disruption of any of the activating steps described above would lead to cell death. In fact, this strategy has been put to good practice in cancer chemotherapy for over 65 years, focusing both on inhibition of dihydrofolate reductase, as well as thymidylate synthase. Of the near relatives of folic acid, the drug methotrexate binds about 1,000 times stronger to dihydrofolate reductase than does the natural substrate [1b], a result that has been attributed to N-1 protonation in the enzyme pocket, facilitated by the C-4 amino group (Figure 1.4) [4]. Note also that N10 is blocked from activation by a methyl group. In any event, either alone or in combination with other drugs, methotrexate is still employed as a front-line chemotherapy agent in combating a variety of cancers. It is far from a perfect drug, however. The holy grail of chemotherapy would be to find such an agent that was lethal only to cancer cells, which has proven to be an elusive goal. Drugs such as methotrexate owe their effectiveness to the fact that cancer cells replicate at a much greater rate than normal cells, with the consequence that the drug's impact is higher.

1.2 Antibiotics and Tetrapyrroles

No survey of biologically important heterocycles found in nature would be complete without some mention of her antibiotics and tetrapyrroles (Figure 1.5). Penicillin G occupies a special niche in the history of antibiotics, dating back to a chance observation by the Scottish bacteriologist Alexander Fleming in 1928. As the story goes, Fleming left a Petri dish containing the bacterium *Staphylococcus aureus* uncovered while on vacation. When he returned, he was astounded to find that the bacterium had died in some areas of the dish. The causative agent for this demise was eventually traced to a fungus of the genus *Penicillium*, which had apparently drifted in through an open window from a laboratory on the floor below. Fleming postulated that the effect was caused by an antibacterial component present in the fungus, which he named penicillin, but he was unable to further characterize this substance. It remained for Florey and Chain to succeed in purifying penicillin G in 1942, followed by its structure determination in 1945 by Dorothy Hodgkin. For their work, Florey, Chain, and Fleming shared the 1945 Nobel Prize in Medicine (Hodgkin would also eventually win a Nobel, "for her determinations by X-ray techniques of the structures of important biochemical substances," including vitamin B_{12}). For his part, Fleming once alluded to his discovery with the statement "One sometimes finds what one is not looking for." Surely this is one of

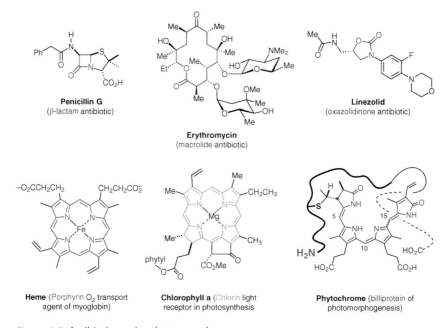

Penicillin G
(β-lactam antibiotic)

Erythromycin
(macrolide antibiotic)

Linezolid
(oxazolidinone antibiotic)

Heme (Porphyrin O_2 transport agent of myoglobin)

Chlorophyll a (Chlorin light receptor in photosynthesis)

Phytochrome (biliprotein of photomorphogenesis)

Figure 1.5 Antibiotics and and tetrapyrroles.

the great understatements of scientific achievement on record. Indeed, were it not for Fleming's observation, the modern age of antibiotics would most likely have been considerably delayed.

Penicillin G was thus the first of the β-lactam antibiotics to be isolated and purified, and compared to other drugs of the era, it had remarkable activity against a wide range of bacteria. This was coupled with low human toxicity. It exerts its activity by specifically inhibiting the transpeptidase that catalyzes the last step in cell wall biosynthesis, which is the cross linking of peptidoglycan. In this capacity the highly reactive β-lactam ring functions as an irreversible acylating agent. Unfortunately, though, through decades of misuse, many bacteria have developed resistance to this class of antibiotics, chiefly through the production of β-lactamases. To overcome this resistance, β-lactam antibiotics are often co-administered with β-lactamase inhibitors.

Erythromycin is an example of a class of antibiotics characterized by incorporating a macrolide ring, and their mode of action is entirely different. In susceptible bacteria, macrolide antibiotics inhibit protein biosynthesis by a mechanism involving disruption of the enzyme peptidyltransferase, thereby interfering with ribosomal translation. Erythromycin and a variety of analogs are particularly useful for treating infections in patients having an allergy for penicillin. Finally, Linezolid is an example of a potent class of antibiotics containing an oxazolidinone ring, which also function by inhibiting the bacterial synthesis of proteins. While more toxic than either penicillin or erythromycin, oxazolidinone antibiotics are oftentimes employed as a "drug of last resort" in hospital settings, due to their effectiveness against gram-positive bacteria that have developed resistance to first line drugs.

This brings us to the topic of tetrapyrroles, of which heme and chlorophyll *a* are the most abundant in nature (Figure 1.5). Heme, of course, is the iron porphyrin component found in hemoglobin, the oxygen binding protein of red blood cells. As such it is responsible for oxygen transport throughout the body, and when fully oxygenated is bright red in color. It is also found in myoglobin, the oxygen carrying and storage protein of muscle. Chlorophyll *a* is one of several closely related bright green pigments found in all higher plants, where it plays a key role as the photoreceptor in photosynthesis. This is the process whereby visible light provides the driving force for the conversion of water and carbon dioxide into glucose and oxygen. The balanced equation for this process is:

$$6\,H_2O + 6\,CO_2 + \text{light} \rightarrow \text{glucose} + 6\,O_2$$

and it is endothermic by 686 kcal/mole. It has been estimated that more than 10^{17} kcal of free energy is stored annually in this manner by the plant world using solar energy [1a]. This far exceeds the per annum amount of energy produced worldwide by the burning of all fossil fuels.

Finally, we close this chapter with a brief discussion of phytochrome (Figure 1.5), which is an example of a billiprotein (i.e., containing both a linear tetrapyrrole chromophore and a covalently bound protein) [5a]. This deep blue pigment is found in only trace amounts in higher plants, but if chlorophyll is the engine of plant life, then phytochrome (Pr) is the driver. In green plants Pr functions as the "on-off" switch for photomorphogenesis, the process by which light transmits growth regulatory information to a cell's genetic apparatus. Information of this type is crucial to the timing of seasonal phenomena, such as seed germination, flowering and fruiting, and chlorophyll production. Mechanistically, Pr functions by undergoing a photoreversible, photochromic interconversion with an activated species known as Pfr. An abundance of evidence points to this change involving photoisomerization about the C15–C16 double bond, with retention of a "semi-extended" conformation [5b]. According to this model, *Z,E*-isomerization induces a change in the tertiary structure of the surrounding protein, providing a molecular basis for transduction of the light signal to the cell's genetic regulatory apparatus. It should be noted that a similar mechanism is involved in the chemistry of vision, wherein 11-cis retinal, the prosthetic group of rhodopsin, undergoes light-induced isomerization to the all trans form, triggering a nerve impulse.

References

1 (a) Lehninger, A. L., *Principles of Biochemistry*, Worth Publishers, Inc., New York, **1982**. (b) Voet, D.; Voet, J. G., *Biochemistry*, 4th edition, John Wiley & Sons, Inc., New York, **2011**.
2 (a) Washabaugh, M. W.; Jencks, W. P. *Biochemistry* **1988**, *27*, 5044–5053. (b) Jordon, F.; Li, H.; Brown, A. *Biochemistry* **1999**, *38*, 6369–6373.
3 Stokstad, E. L. R.; Harris, R. S.; Bethel, F. H., *The Vitamins*, Vol. 3, Sebrell, W. H.; Harris, R. S., Eds., Academic Press, New York, **1954**.
4 Blakley, R. L.; Cocco, L. *Biochemistry* **1985**, *24*, 4704–4709.
5 (a) *Phytochrome and Photoregulation in Plants*, Furuya, M., Ed., Academic Press, New York, **1987**. (b) Rockwell, N. C.; Lagarias, J. C. *ChemPhysChem* **2010**, *11*, 1172–1180. For synthetic aspects, see (c) Jacobi, P. A.; Odeh, I. M. A.; Buddhu, S. C.; Cai, G.; Rajeswari, S.; Fry, D.; Zheng, W.; DeSimone, R. W.; Guo, J.; Coutts, L. D.; Hauck, S. I.; Leung, S. H.; Ghosh, I.; Pippin, D. *Synlett* **2005**, *19*, 2861–2885.

2

Orbitals and Aromaticity; Chemical Reactivity

The same criteria apply for aromatic heterocycles as for their carbocyclic counterparts. Foremost among these is Hückel's rule, which stipulates that for a ring to be aromatic it must contain 4n + 2 π-electrons in a planar, conjugated system, where n = 0 or (in theory) any positive integer. Of course benzene is the best known aromatic species, with 6 π-electrons in 6 p-orbitals contained within a conjugated 6-membered ring (Figure 2.1). Reflecting this aromaticity, benzene is far more stable than predicted for "cyclohexatriene," with a resonance energy (RE) of approximately 36 kcal/mol. It is also neutral and famously resistant to hot acid and base. Pyridine is the closest heterocyclic relative to benzene, wherein one methine group (=CH-) is replaced by a nitrogen atom. It also has 6 π-electrons in 6 p-orbitals, but with an extra electron pair in an orthogonal sp^2-orbital on nitrogen. What are the consequences of this "free" electron pair? Most importantly, pyridine is weakly basic, with a pK_a of 5.2 for the conjugate acid. It also has significant resonance stabilization, although not as high as benzene (RE = 28 kcal/mol versus 36 kcal/mol). It stands to reason that pyridine would have a somewhat lower RE than benzene, since the electronegative nitrogen atom will polarize, and therefore partially localize, the π-cloud toward itself. Nevertheless, pyridine is moderately stable toward hot acid and base.

And what of the five-membered heteroaromatic rings, such as furan (X=O), pyrrole (X=NH) and thiophene (X=S) (Figure 2.1)? Again, each of these species contains 6 π-electrons, but spread over only 5 p-orbitals. Four of these p-orbitals are on carbon atoms, while the fifth is on the heteroatom. Do these ring systems satisfy Hückel's rule? Absolutely! They are fully conjugated, and the electron pair on the p-orbital of the heteroatom is delocalized into the ring. Both furan and thiophene also have a free electron pair in an orthogonal sp^2-orbital, which of course does not contribute to aromaticity. In pyrrole this position is occupied by an sp^2-s σ-bond to hydrogen. Not surprisingly, all are exceedingly weak bases, as exemplified by pyrrole, where protonation would completely destroy aromaticity. As we shall see, their acid stability varies directly with their resonance energy, which ranges from 16–29 kcal/mol.

Introductory Heterocyclic Chemistry, First Edition. Peter A. Jacobi.

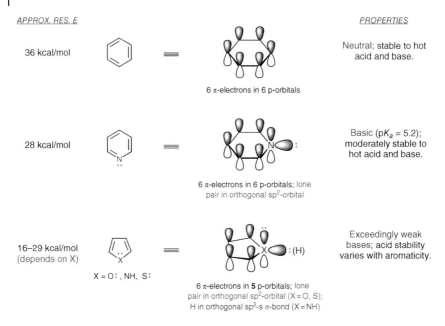

APPROX. RES. E   _PROPERTIES_

36 kcal/mol

6 π-electrons in 6 p-orbitals

Neutral; stable to hot acid and base.

28 kcal/mol

6 π-electrons in 6 p-orbitals; lone pair in orthogonal sp^2-orbital

Basic (pK_a = 5.2); moderately stable to hot acid and base.

16–29 kcal/mol (depends on X)

X = O:, NH, S:

6 π-electrons in **5** p-orbitals; lone pair in orthogonal sp^2-orbital (X = O, S); H in orthogonal sp^2-s σ-bond (X = NH)

Exceedingly weak bases; acid stability varies with aromaticity.

Figure 2.1 Properties of aromatic heterocylics, orbitals.

   Can you predict which of the simple five-membered ring heterocycles will be the most stable? Much in heterocyclic chemistry is intuitive and this is certainly the case with resonance energy. Thus, between thiophene, pyrrole, and furan, one would expect aromaticity to decrease with increasing electronegativity of the heteroatom, due to increasing π-electron localization (Figure 2.2). This is in fact the case, with thiophene (RE = 29 kcal/mol) approaching the aromatic stability of benzene, and furan (RE = 16 kcal/mol) being notoriously acid-labile. For comparison, the calculated RE for cyclopentadienyl anion is in the range of 24–27 kcal/mole, relative to open chain pentadienyl anion [1]. A similar trend is observed in the six-membered ring family of nitrogen heterocycles as one increases the number of nitrogen atoms contained within the ring. With each nitrogen comes increasing π-electron localization and therefore decreasing aromaticity.

   Thus we have now seen that heterocycles can be either aromatic or non-aromatic, and we have a general idea of how to estimate their stability. But what about their chemical reactivity? Among the heteroaromatic species, it is customary to make a further distinction based upon their π-electron density. For example, both pyridine and pyrimidine are classified as π-deficient heterocycles, based upon the fact that most of the ring carbons have a lower electron density than found in benzene (Figure 2.3) [2]. In fact, the highest electron density is found at nitrogen, where the free electron pair(s) occupy orthogonal

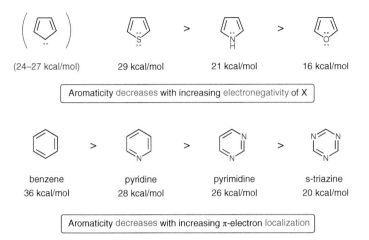

Aromaticity decreases with increasing electronegativity of X

benzene | pyridine | pyrimidine | s-triazine
36 kcal/mol | 28 kcal/mol | 26 kcal/mol | 20 kcal/mol

Aromaticity decreases with increasing π-electron localization

Figure 2.2 Trends in resonance energy.

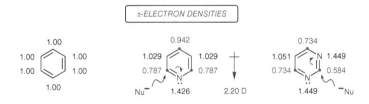

π-ELECTRON DENSITIES

1. Site of highest electron density is the heteroatom.
2. Most ring carbons lower electron density than benzene.
3. No resonance possible involving orthogonal free electron pair.
4. Typical reactions involve nucleophilic addition/substitution at positions α and γ to heteroatom (developing negative charge stabilized).

REACTIVITY COMPARABLE TO NITROBENZENE

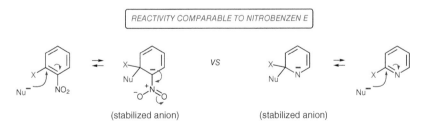

(stabilized anion)   VS   (stabilized anion)

Figure 2.3 π-Deficient heterocycles.

sp^2-orbitals (note that the dipole moment for pyridine passes directly through C4 and the ring nitrogen). As you might expect, then, members of this class are very reluctant partners in electrophilic aromatic substitution. Rather, their typical reactions involve nucleophilic addition/substitution at the positions α and γ to the ring nitrogen(s), where the developing negative charge can be stabilized. In this sense their reactivity is comparable to that found with nitrobenzene.

At the opposite end of the reactivity spectrum are the π-excessive heterocycles, typified by thiophene, pyrrole, and furan (Figure 2.4). These species have higher average π-electron density than found in benzene, a consequence of the resonance structures indicated (six π-electrons on five atoms, as compared to 1.0 π-electron per carbon in benzene) [3]. Note in particular that the dipole moment in pyrrole (1.80 D) actually points away from the ring nitrogen, in contrast to the saturated analog pyrrolidine. Given these properties, it is of no surprise that the typical reactions of π-excessive heterocycles involve electrophilic aromatic substitution, at the positions conjugated to the heteroatom. In this regard, their reactivity is most comparable to anisole, where electrophilic attack is greatly accelerated by resonance donation of the free electron pair on the methoxyl group. Taking furan as an example, electrophilic attack at the 2-position will generate an analogous resonance stabilized cation.

In chapters that follow, we will discuss each of these classes of heterocyclic ring systems in some detail, beginning with simple monocyclic systems and then progressing to more complicated examples. But before proceeding, let us first return to the opening question in the preface to this book: Why heterocycles?

1.046
1.071
1.760

1.090
1.087
1.647   1.80 D

vs

1.57 D

1.067
1.078
1.710

1. Higher average π-electron density than benzene.
2. Resonance involving free electron pair controls reactivity.

3. Typical reactions involve electrophilic aromatic substitution at positions conjugated to electron pair.
4. Reactivity comparable to anisole.

E+   :OMe

E   H   + OMe
(stabilized cation)

vs

H   E   O +
(stabilized cation)

E+

Figure 2.4 π-Excessive heterocycles.

References

1 Bordwell, F. G.; Drucker, G. E; Fried, H. E. *J. Org. Chem.* **1981**, *46*, 632–635.

2 Wiberg, K. B.; Nakaji, D.; Breneman, C. M. *J. Am. Chem. Soc.* **1989**, *111*, 4178–4190.

3 Gilchrist, T. L., *Heterocyclic Chemistry*, 3rd edition, Addison Wesley Longman Limited, Harlow, England, **1997**, p. 13.

.

3

A Prelude to Synthesis

It is true that many of the heterocycles found in nature are complex and have seemingly unusual substitution patterns. However, oftentimes there exists a pattern that may give some clue as to how they were fashioned from relatively simple starting materials. Take, for example, adenine, one of the chief building blocks for DNA (Scheme 3.1). Following the usual tenets of organic synthesis, the assembly of this skeleton might appear to be a daunting task. But look carefully, for the molecular formula of adenine, $C_5H_5N_5$, also corresponds to $(HCN)_5$, or the pentamer of HCN, an abundant material in the earth's prebiotic atmosphere (see the red, numbered bonds in the structure of adenine). Can we write a mechanism leading from HCN to adenine? The answer is not only yes, but it is quite straightforward.

To begin, what is it about HCN that makes it such a versatile starting material? One characteristic is that in the non-ionized state it is highly electrophilic in nature, containing a strongly polarized carbon-nitrogen triple bond. However, HCN is also a weak Brønsted acid, and when dissociated produces a nucleophilic $^-$CN anion. Thus, first one, and then a second $^-$CN can add across HCN to produce the HCN trimer aminomalononitrile (AMN), and at this point we are only two HCN molecules removed from synthesizing adenine! Now is a good time to have the nucleophilic amino group in AMN add across the carbon-nitrogen triple bond of a fourth molecule of HCN (or its equivalent) to produce the amidine $(HCN)_4$. This last material, upon intramolecular cycloaddition and tautomerization, then affords an imidazole ring containing an amino and a cyano group in an *ortho* relationship. This is a functional group arrangement that we will see repeatedly as we analyze further the synthesis of nitrogen heterocycles. Finally, nucleophilic addition to a fifth molecule of HCN, followed in analogous fashion by intramolecular cycloaddition and tautomerization produces adenine. Is this a reasonable process by which adenine might have been produced in the earth's primordial atmosphere? It is at least plausible, since subjecting HCN to conditions thought to exist at that time gives measurable quantities of adenine [1].

Introductory Heterocyclic Chemistry, First Edition. Peter A. Jacobi.
© 2019 John Wiley & Sons Ltd. Published 2019 by John Wiley & Sons Ltd.

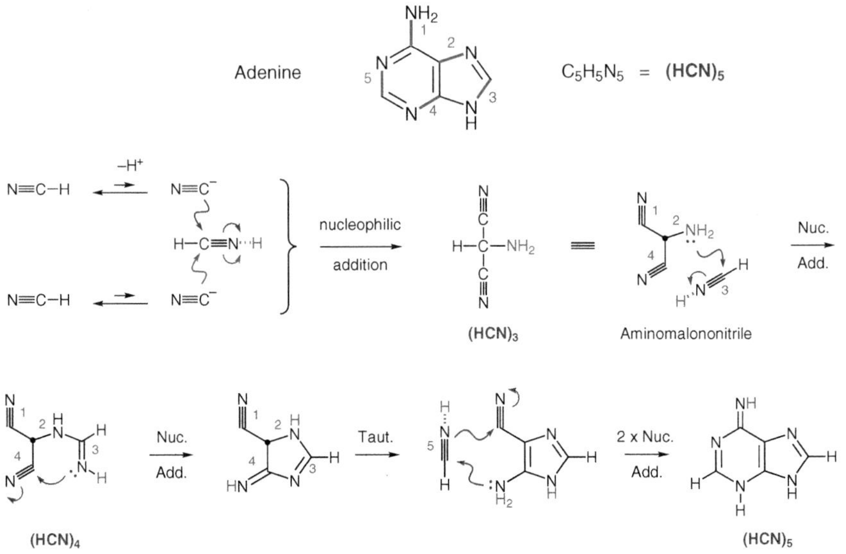

Scheme 3.1 Why heterocycles?

Naturally the synthesis described above, while illustrative, is not practical for laboratory purposes. For this we need a less toxic precursor than HCN. A ready solution is found in the synthesis of aminomalononitrile (AMN) beginning with malononitrile, which due to its acidic character undergoes nitrosation under very mild conditions (Scheme 3.2). The resultant oximinomalononitrile is then reduced either catalytically or with Al(Hg) to give excellent yields of AMN, typically isolated as its tosylate salt. Next, AMN is reacted with the HCN "equivalent" formamidine, which through transamination produces exactly the same intermediate as in step 4 of the mechanism shown in Scheme 3.1. Note that transamination is analogous to nucleophilic acyl substitution, and for intermolecular reactions, it is frequently more efficient than nucleophilic addition of an amine across a nitrile triple bond. Finally, intramolecular cyclization affords the key imidazole precursor described previously (Scheme 3.1), which upon transamination with formamidine and intramolecular cyclization affords adenine [2].

An important "take-home" message from this synthesis is that in every step it is the free electron pair on nitrogen that is functioning as the actual nucleophilic species (see curly arrows). In principle, then, the synthesis of heterocycles should be easier than the corresponding carbocycles, since there is no need to perform activating steps. This is illustrated schematically in Figure 3.1, comparing three of the most useful classes of bond forming reactions. In the

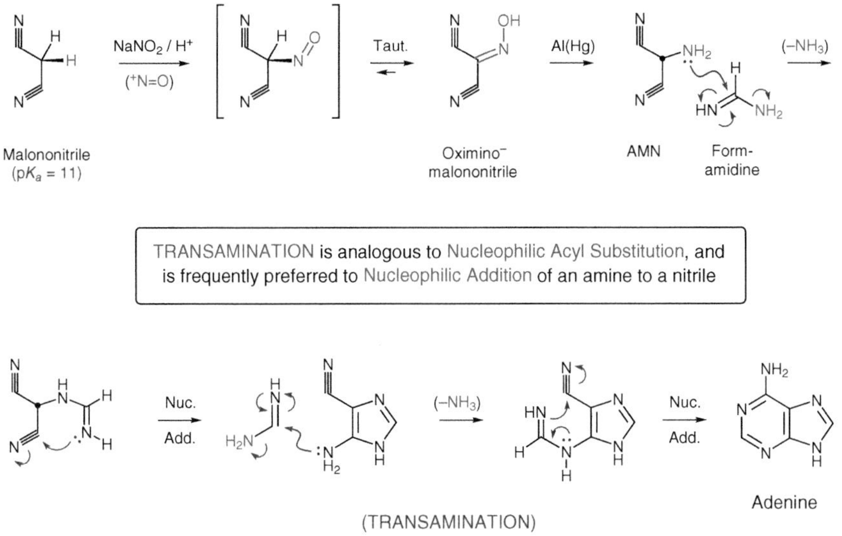

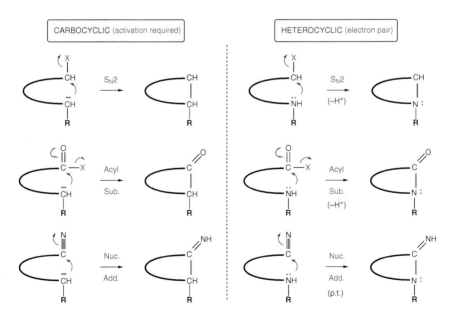

Malononitrile
(pK_a = 11)

Oximino⁻
malononitrile

AMN

Form-
amidine

TRANSAMINATION is analogous to Nucleophilic Acyl Substitution, and
is frequently preferred to Nucleophilic Addition of an amine to a nitrile

Adenine

(TRANSAMINATION)

Scheme 3.2 A laboratory analogy.

| CARBOCYCLIC (activation required) | HETEROCYCLIC (electron pair) |

S$_N$2

Acyl

Sub.

Nuc.

Add.

S$_N$2
(–H⁺)

Acyl

Sub.
(–H⁺)

Nuc.

Add.
(p.t.)

Figure 3.1 Common bond-forming reactions.

carbocyclic series, for example, it is common to employ an intramolecular substitution reaction for ring closure. However, this usually requires the generation of a carbanion, which depending upon the nature of R, might be challenging to accomplish in a regioselective fashion. In contrast, the corresponding reaction in heterocycle synthesis makes use of the free electron pair on the heteroatom (in the example illustrated, nitrogen). Similar contrasts can be drawn between intramolecular acyl substitutions and intramolecular nucleophilic additions for the two classes of molecules. In fact, heterocyclic species, whether they be aromatic or non-aromatic, are in many instances easier to synthesize than the corresponding carbocyclic rings. A word of caution, though: This initial advantage is sometimes compromised by the complicated structural changes which heterocycles might undergo either during preparation, or once in hand.

For example, consider the transformations outlined in Figure 3.2, and ask yourself if you could provide a reasonable mechanism for each observation. The reaction cited in Equation 1 is fairly straightforward, wherein the product of formal S_N2 displacement is contaminated with varying quantities of 2-cyano-5-methylfuran. It is likely this last material is the product of S_N2' substitution, followed by re-aromatization of the furan ring [3a]. Also, regarding Equation 2, many readers will be aware that $CHCl_3$ with strong base leads to the formation of dichlorocarbene, an extremely reactive species that undoubtedly plays a prominent role in the transformation of pyrrole to 3-chloropyridine [3b]. But how can irradiation bring about the conversion of pyrazoles to

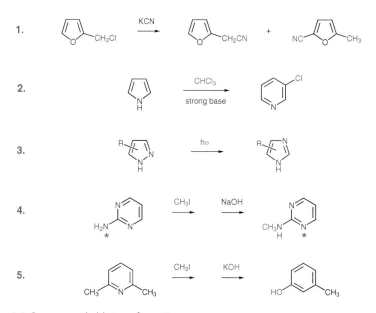

Figure 3.2 **Some remarkable transformations.**

imidazoles shown in Equation 3 [3c]? And even more intriguing, how do we rationalize the labeling experiment depicted in Equation 4? Here the intent was presumably to effect a simple alkylation of the exocyclic 2-amino group with CH_3I, which from most appearances was successful. However, when examined with labeled NH_2 (*), the nitrogen that was exocyclic was transposed into the ring [3d]! And finally, what of the remarkable reaction shown in Equation 5, where 2,6-dimethylpyridine is converted by alkylation and base hydrolysis to *m*-cresol [3e]? Before attempting to rationalize these, and other rearrangements, it is necessary to accumulate more knowledge of the physical properties, synthesis, and reactions of heterocyclic systems, which is the subject of the pages that follow.

References

1 Roy, D.; Najafian, K.; Schleyer, P. V. R. *Proc. Natl. Acad. Sci. USA* **2007**, *104*, 17272–17277.

2 Ferris, J. P.; Orgel, L. E. *J. Am. Chem. Soc.* **1966**, *88*, 3829–3831.

3 (a)Divald, S.; Chun, M. C.; Joullié, M. M. *Tetrahedron Lett.* **1970**, *10*, 777–780. (b)Rice, H. L.; Londergan, T. E. *J. Am. Chem. Soc.* **1955**, *77*, 4678–4679, and references cited therein. (c)Su, M.-D. *J. Phys. Chem. A* **2008**, *112*, 10420–10428, and references cited therein. (d)Goerdeler, J.; Roth, W. *Chem. Ber.* **1963**, *96*, 534–549. See also, Brown, D. J.; Hoerger, E.; Mason, S. F. *J. Chem. Soc.* **1955**, 4035–4040. (e)Lukes, R.; Pergál, M. *Collect. Czech. Chem. Commun.* **1959**, *24*, 36–45.

4

π-Deficient Heterocycles: Some Physical Properties

In general, H_2O-solubility increases with the number of heteroatoms contained within the ring, reflecting the fact that there are more sites available for hydrogen bonding. Thus quinoline, with only one acceptor site, is only slightly soluble in water, while pteridine is essentially infinitely soluble (top Figure 4.1). The effect of ring substituents is less clear. In carbocyclic chemistry, it is a given that hydrophilic substituents increase the H_2O-solubility of aromatic rings, again due to increased hydrogen bonding. Thus, benzene is approximately 50 times less soluble in water than phenol, which in turn is about five times less soluble than resorcinol (bottom Figure 4.1). Do the same trends hold true for heteroaromatic systems? The answer to this question is "it depends," but frequently the opposite effect is observed [1]. Take for example leucopterin, a naturally occurring pteridine found in butterfly wings (middle Figure 4.1). In the tautomeric form shown on the left, one might expect H_2O-solubility to be very high, as is the case with the parent ring system. But in fact leucopterin is for all practical purposes completely insoluble in water. This is because the more favored tri-lactam tautomer is capable of forming exceedingly strong bi-molecular hydrogen bonds of the type illustrated.

We can also make some general predictions regarding basicity. Not surprisingly, π-deficient heterocycles are typically less basic than simple aliphatic amines, since protonation will have a negative impact on aromaticity. For example, pyridine has a pK_a of 5.2, while that of triethylamine is 11.0 (Figure 4.2). Here we must bear in mind that when we discuss amine pK_a's we are usually referring to the degree of dissociation of the conjugate acid (i.e., lower pK_a's correlate with greater dissociation of the conjugate acid, and hence weaker basicity of the free amine). It follows that any factor, which destabilizes the conjugate acid of an amine will of necessity decrease its basicity. So should we be surprised that pyrimidine (pK_a=1.3) is a weaker base than pyridine (pK_a=5.2)? The answer is of course no, since the conjugate acid of pyrimidine is

Introductory Heterocyclic Chemistry, First Edition. Peter A. Jacobi.
© 2019 John Wiley & Sons Ltd. Published 2019 by John Wiley & Sons Ltd.

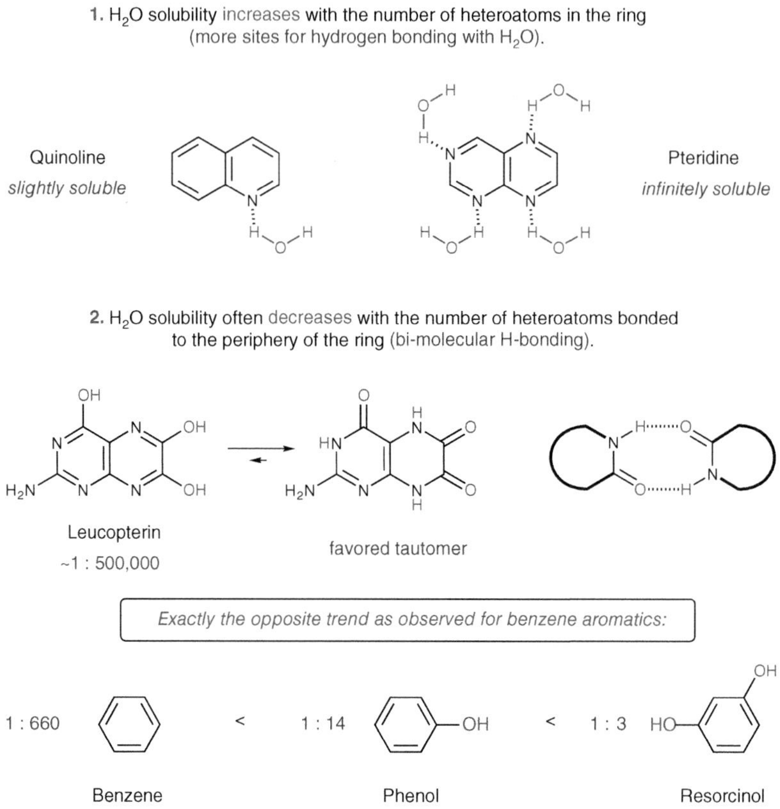

1. H_2O solubility increases with the number of heteroatoms in the ring (more sites for hydrogen bonding with H_2O).

Quinoline
slightly soluble

Pteridine
infinitely soluble

2. H_2O solubility often decreases with the number of heteroatoms bonded to the periphery of the ring (bi-molecular H-bonding).

Leucopterin
~1 : 500,000

favored tautomer

Exactly the opposite trend as observed for benzene aromatics:

1 : 660 Benzene < 1 : 14 Phenol < 1 : 3 Resorcinol

Figure 4.1 π-Deficient heterocycles: Solubility and tautomerization.

destabilized by the inductive effect of the second ring nitrogen [2]. Indeed, s-triazine is such a feeble base that it will only protonate under conditions that bring about its decomposition ($pK_a<0$).

Finally, the corollary to the arguments presented above is that any factor which stabilizes the conjugate acid of an amine will necessarily increase its basicity. This is evident in the series of pyridine bases shown at the bottom of Figure 4.2, where basicity increases significantly as we add an amino group in either the 2- or 4-position [3]. In both cases the resultant conjugate acid is stabilized by resonance delocalization involving the free electron pair on the exocyclic amino group. Can you suggest a reason why this type of resonance is particularly effective in the case of 4-aminopyridine? Recall that inductive effects drop off rapidly with distance.

Basicity decreases with the number of heteroatoms in the ring (inductive *destabilization* of conjugate acid):

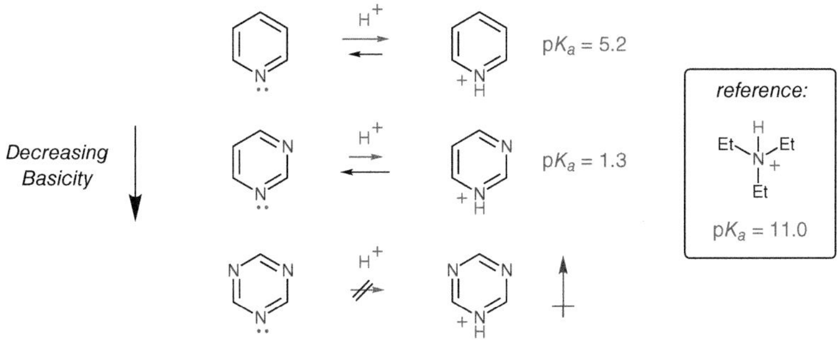

Basicity increases with periphery heteroatoms on the ring (resonance *stabilization* of conjugate acid):

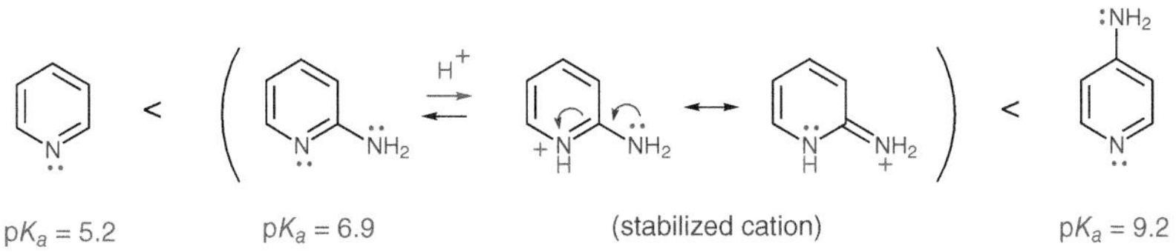

Figure 4.2 π-Deficient heterocycles: basicity.

References

1 Albert, A., *Heterocyclic Chemistry: An Introduction*, 2nd edition, Athlone Press, London, **1968**, pp. 71–72.
2 Albert, A., in *Physical Methods in Heterocyclic Chemistry*, Vol. I, Katritzky, A. R., Ed., Academic Press, New York, New York, **1963**, pp. 20–21.
3 Schofield, K., *Heteroaromatic Nitrogen Compounds: Pyrroles and Pyridines*, Butterworth, London, **1967**, pp. 146–147.

5

π-Deficient Heterocycles: De Novo Syntheses

Our discussion of the synthesis of π-deficient heterocycles will be divided into three sections, consisting of (1) preparation "de Novo" with all substituents present; (2) introduction of new substituents; and (3) manipulation of existing substituents (Figure 5.1). In the category of de Novo syntheses we shall make great use of two reactions with which you are already familiar, the aldol condensation and imine formation. The examples that follow are not meant to constitute a comprehensive survey of synthetic methodology, for which the appropriate reviews can be consulted [1]. Rather, the hope is to provide a flavor for the basic principles of π-deficient heterocyclic synthesis.

To begin, it is useful to stress that in this section we are synthesizing mainly stable aromatic species, which provides an opportunity for exercising thermodynamic control. That is, an ideal synthesis would be one in which the product is formed directly, without any need for subsequent oxidation or reduction steps (dropping into a so-called "thermodynamic well"). For aromatic species this will always be an energetically favorable process.

For example, consider the fact that glutaconic aldehyde is in the same oxidation state as pyridine, as well as having the same carbon bond connectivity (Scheme 5.1). This relationship suggests that upon reaction with ammonia there is a pathway available that will ultimately produce pyridine. This is in fact the case, for which a reasonable mechanism involves initial condensation of ammonia with one of the aldehyde groups in glutaconic aldehyde to produce an imine. Next, tautomerization to the corresponding enamine followed by intramolecular nucleophilic addition produces a hemi-aminal, which on simple dehydration affords the target aromatic ring. Note that none of these steps involves an oxidation or reduction.

For pyridine itself this synthesis is not competitive with isolation from natural sources. But for more complicated examples there are many variants of this same approach possible. If the dicarbonyl starting materials are available, they will typically undergo a high yielding condensation with ammonia to afford pyridine derivatives. Again, the purpose of this discussion is mainly to give the

Introductory Heterocyclic Chemistry, First Edition. Peter A. Jacobi.
© 2019 John Wiley & Sons Ltd. Published 2019 by John Wiley & Sons Ltd.

1. Preparation "de Novo" with all substituents present.
2. Introduction of new substituents.
3. Manipulation of existing substituents.

----

de Novo: Important bond forming reactions include aldol *condensations and* imine *formation.*

a) Aldol Condensation

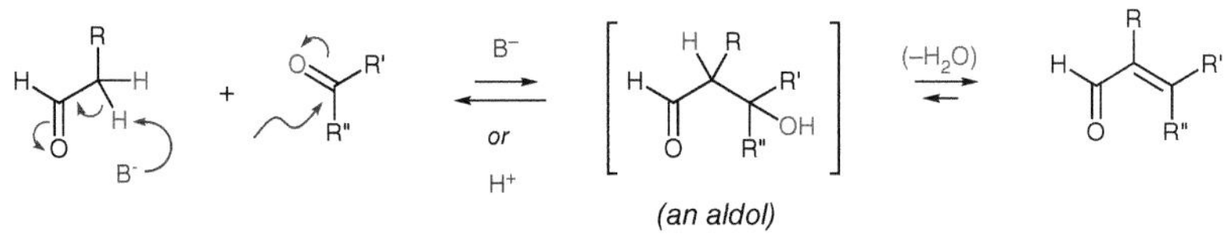

(an aldol)

b) Imine formation

(an imine)

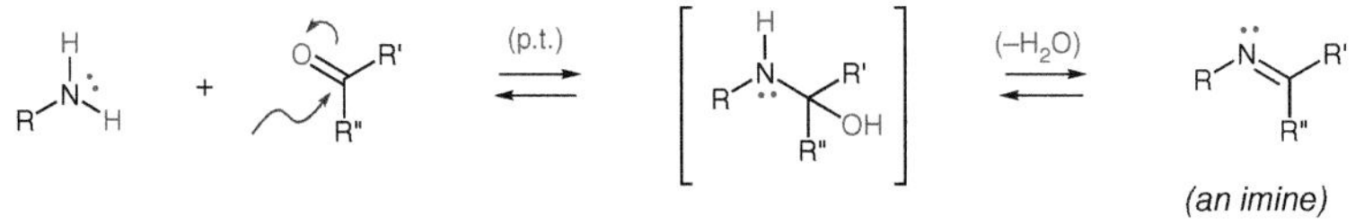

Figure 5.1 Synthesis of π-deficient heterocycles.

Glutaconic aldehyde **is in the same oxidation state as** *pyridine:*

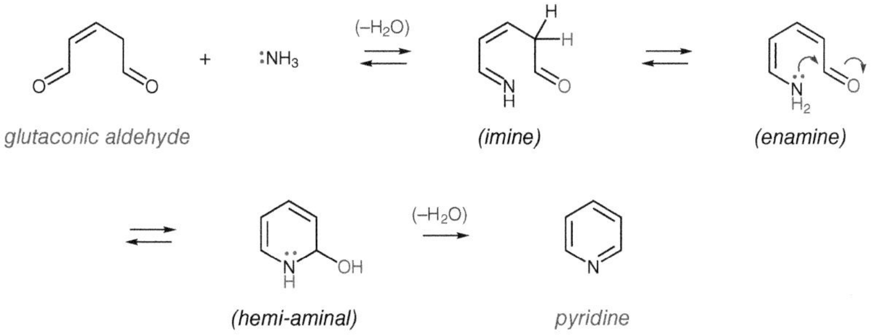

glutaconic aldehyde   *(imine)*   *(enamine)*

(hemi-aminal)   *pyridine*

Not useful for pyridine itself, but many variants of this reaction are possible:

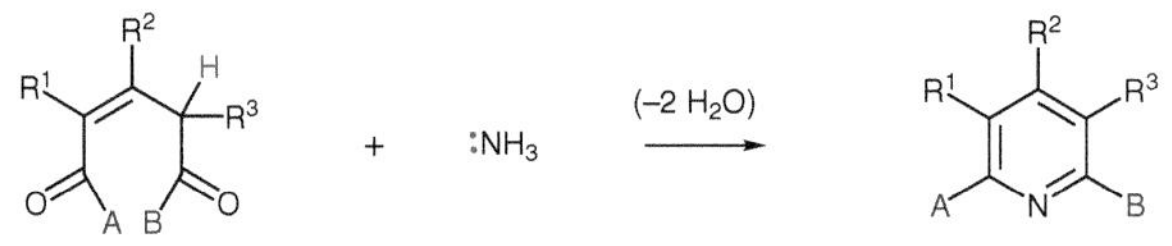

Scheme 5.1 Synthesis of π-deficient heterocycles: pyridines.

reader a feel for the kind of bond disconnections that are feasible in making these ring systems.

This principle is further illustrated by the Hantzsch pyridine synthesis, first published in 1881 and still perhaps the most general strategy of this class (Scheme 5.2) [1]. Aldol chemistry and imine formation again play a prominent role, with the starting materials consisting of a β-keto ester ($pK_a \approx 11$), an aldehyde, and an ammonia source in a ratio of 2:1:1. Simply warming these components in an appropriate solvent initiates a sequence of events involving, first, aldol-like condensation of the β-keto ester with the aldehyde to afford a highly reactive enone intermediate. Next, the second equivalent of β-keto ester comes into play, undergoing Michael addition to the enone to produce a 1,5-dicarbonyl derivative embodying the carbon skeleton of the eventual pyridine product. Condensation in situ of this adduct with ammonia then forms a cyclic hemi-aminal, which on dehydration affords the corresponding 1,4-dihydropyridine in generally good-excellent yields.

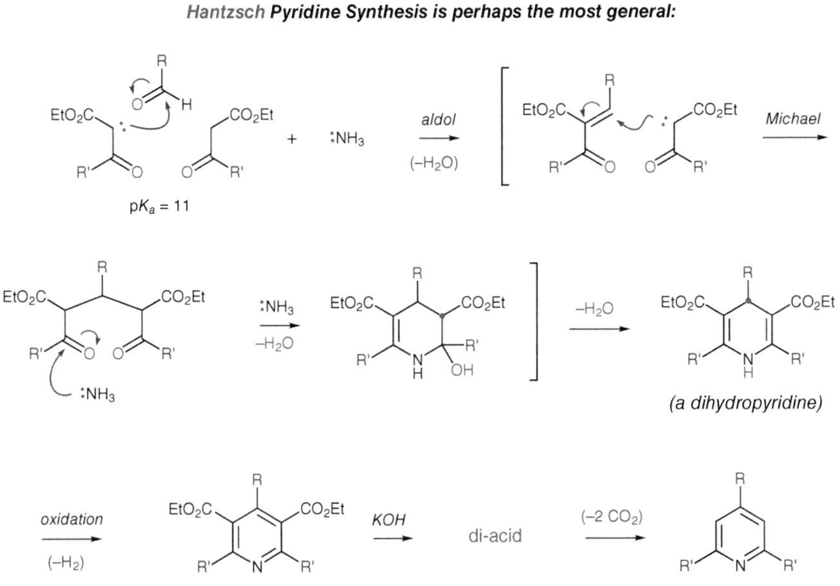

Hantzsch *Pyridine Synthesis is perhaps the most general:*

Scheme 5.2 Synthesis of π-deficient heterocycles: pyridines.

A few comments are in order. First, the mechanistic steps for this process may vary in detail depending upon the actual substrates employed, in particular with regard to the timing of initial imine formation. However, the end results are the same. Second, in this case, as in most cases with the

Hantzsch pyridine synthesis, you will recognize that we have violated one of the principles set forth in the discussion above, in that the first isolable product is not aromatic. Nevertheless, these condensations are so favorable that if carried out in the presence of an oxidant, the substrates will readily find their way to the pyridine thermodynamic well. And third, the two ester groups at positions 3 and 5 can be efficiently removed by hydrolysis and decarboxylation, producing a 2,4,6-trisubstituted pyridine. This is a reaction type that is characteristic of π-deficient heterocyclic acids, which in contrast to benzoic acids, undergo decarboxylation under relatively mild conditions.

Many variants of the Hantzsch pyridine synthesis are also available [3], some of which do produce an aromatic pyridine ring directly. For example, the combination of a 1,5-diketone with hydroxylamine is in the proper oxidation state for aromatization via dehydration [4a]:

Finally, under the category of miscellaneous, there is no lacking of alternative synthetic routes to the pyridine skeleton, some of which employ more direct bond disconnections than others. In the first group we can include the base-catalyzed condensation of cyanoacetamide with symmetrical β-diketone derivatives (equation 1 in Figure 5.2) [5]. The product of this reaction is a 2-pyridone, which is the more stable tautomer of the corresponding 2-hydroxy derivative (vide supra). Although in the lactam form, however, it is important to emphasize that this is still a 6 π-electron aromatic species, due to amide resonance of the type indicated. Regarding the mechanism for this process, it is clear that the two starting materials are in the proper oxidation state for direct conversion to the product. Leaving aside for the moment the issue of timing, the carbon-carbon double bond will be formed by abstracting an acidic proton from the methylene group of cyanoacetamide, shown in red, and subsequent aldol-like condensation. The carbon-nitrogen bond, in turn, is a consequence of imine formation involving the free electron pair on the amide nitrogen, followed by tautomerization.

Also in the "direct" category is the combination of starting materials shown in equation 2, wherein Michael addition of an enamine across the carbon-carbon triple bond of an enyone produces the vinylogous enamide shown in brackets [6]. Again, intramolecular imine formation completes the process,

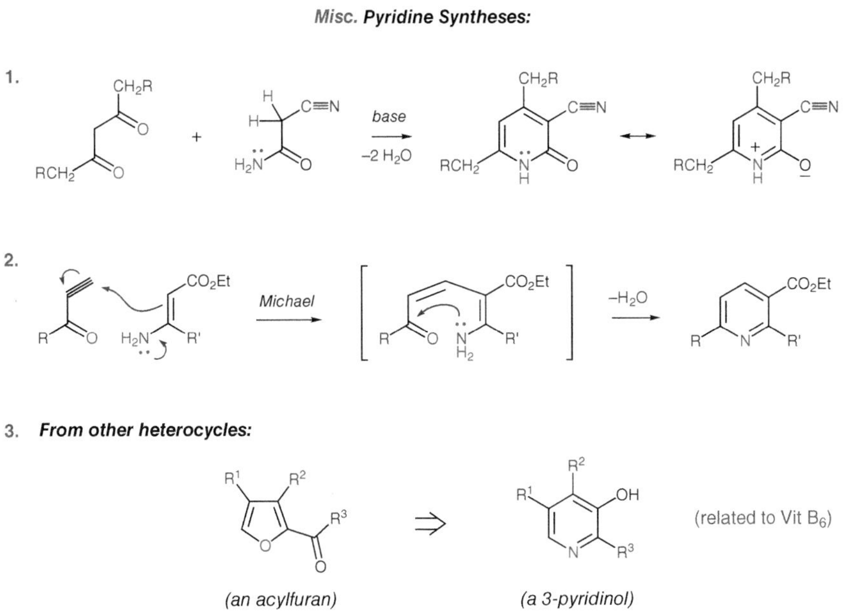

Misc. Pyridine Syntheses:

Figure 5.2 Synthesis of π-deficient heterocycles: pyridines.

producing a 2,3,6-trisubstituted pyridine derivative. As a general comment, it is useful to include as many substituents as possible in these so called de Novo syntheses, since as we will see shortly, it is not always easy to add substituents to pre-formed pyridine rings.

This brings us for the first time to the category of synthesizing heterocycles from other heterocycles, a perhaps less intuitive approach that will become increasingly important as we progress through later chapters. In equation 3 the connection is made that a 2-acylfuran is a viable precursor to a 3-pyridinol, by virtue of the fact that they have the same carbon bond connectivity and they are in the same oxidation state [7]. How do we know this? Picture the anticipated product of acid hydrolysis of a 2-acylfuran, and imagine it undergoing condensation with ammonia and cyclization (do not be discouraged—this might take some practice!). In any event, we will discuss this transformation in greater detail when exploring the chemistry of π-excessive heterocyclic systems. For the present, we might only draw attention to the fact that a hydroxyl group in the 3-position of a pyridine ring does not undergo keto-enol tautomerization, existing exclusively in the pyridinol form. In this regard it is analogous to a phenol. And, as many readers will have noticed, we have seen this substitution pattern before, in the structure of the very important B_6 group of vitamins (see Figure 1.3).

5.1   De Novo Syntheses, Pyrimidines

Moving on from pyridines, let us apply these same principles to the de Novo synthesis of pyrimidines, which are among the simplest of π-deficient ring systems to prepare. Why should this be the case? One reason is that most such syntheses involve condensation of an N-C-N component with a C-C-C unit, a so-called "3+3" approach (Figure 5.3) [8a]. For example, condensation of an amidine with a 1,3-dicarbonyl derivative affords tetrasubstituted pyrimidines directly, by a straightforward pathway involving bis-imine formation (equation 1; R=H, alkyl, aryl) [8b]. Furthermore, analogous results are obtained with ureas (R=OH), thioureas (R=SH), and guanidine (R=NH$_2$). Clearly this approach has a number of strong points. Of particular note, the final product has an axis of symmetry passing through the R and R' groups at C2 and C5. This assures that regiochemical control will not be an issue, since the A and B substituents occupy equivalent positions. Also, and not to be under-appreciated, many of the starting materials are either commercially available or easily prepared. And lastly, as a practical bonus, the desired pyrimidine derivatives generally precipitate directly from the reaction mixture.

The variants to this strategy are many. Rather than diketones, for example, we can carry out the reaction of amidines and their derivatives with

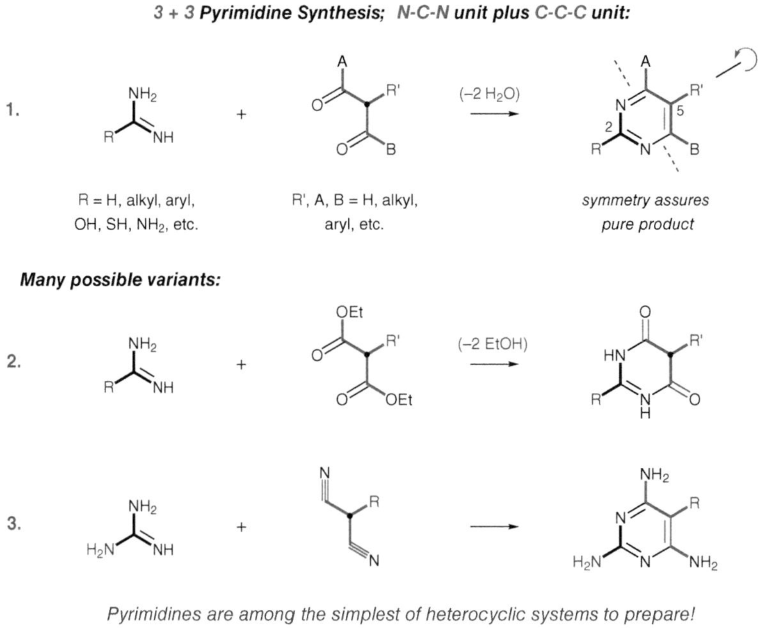

3 + 3 **Pyrimidine Synthesis;   N-C-N unit plus C-C-C unit:**

1.

R = H, alkyl, aryl,
OH, SH, NH$_2$, etc.

R', A, B = H, alkyl,
aryl, etc.

*symmetry assures
pure product*

(–2 H$_2$O)

Many possible variants:

2.

(–2 EtOH)

3.

Pyrimidines are among the simplest of heterocyclic systems to prepare!

Figure 5.3 Synthesis of π-deficient heterocycles: pyrimidines.

1,3-diesters to give the corresponding bis-pyrimidinones (equation 2) [8c]. In this case the mechanistic pathway involves nucleophilic acyl substitution and only a single regioisomer is possible (different tautomers, however, may be present). Finally, in a particularly illustrative example, condensation of guanidine with malononitrile and its derivatives affords 2,4,6-triaminopyrimidines, through a process involving nucleophilic addition and subsequent tautomerization (equation 3) [8d]. Here it is worth emphasizing that exocyclic amino groups, in contrast to hydroxyl groups, almost invariably exist in the amine tautomer. And why is this reaction so important? Mainly because the resultant substitution pattern is so common in heterocyclic systems, wherein we often see amino groups in the *ortho* position to a ring nitrogen. In all such instances, retrosynthetic analysis cries out for intramolecular addition of an amino group across a nitrile. Much of Chapter 16 will be devoted to this principle.

5.2   Fused-Ring Systems, Quinolines

Thus far we have had little to say about fused-ring heterocycles, of which quinoline and isoquinoline are perhaps the best known examples (see below). The first synthesis of the quinoline skeleton was carried out by Skraup in 1880, involving heating glycerol and aniline in concentrated H_2SO_4 in the presence of a mild oxidant [9a]:

aniline   glycerol   quinoline   isoquinoline

Glycerol in this case served as a surrogate for acrolein, to which it is slowly converted under the strongly acidic reaction conditions. Interestingly, if employed directly, acrolein undergoes polymerization faster than reaction with aniline. However, substituted enals and enones function well (Scheme 5.3) [9b]. Together with its various modifications, the Skraup synthesis is to this day probably the most widely employed route to quinoline derivatives, in part because it offers considerable flexibility.

Mechanistic studies suggest that the initial step in this conversion involves conjugate addition of the aniline amine to the terminus of the α,β-unsaturated carbonyl derivative, at least under the classical conditions [10a]. Intramolecular electrophilic substitution then produces the intermediate alcohol shown (Scheme 5.3), which on dehydration affords the corresponding 1,2-dihydroquinoline. In contrast to the Hantzsch pyridine synthesis, where 1,4-dihydropyridines can be isolated (cf. Scheme 5.2), this is

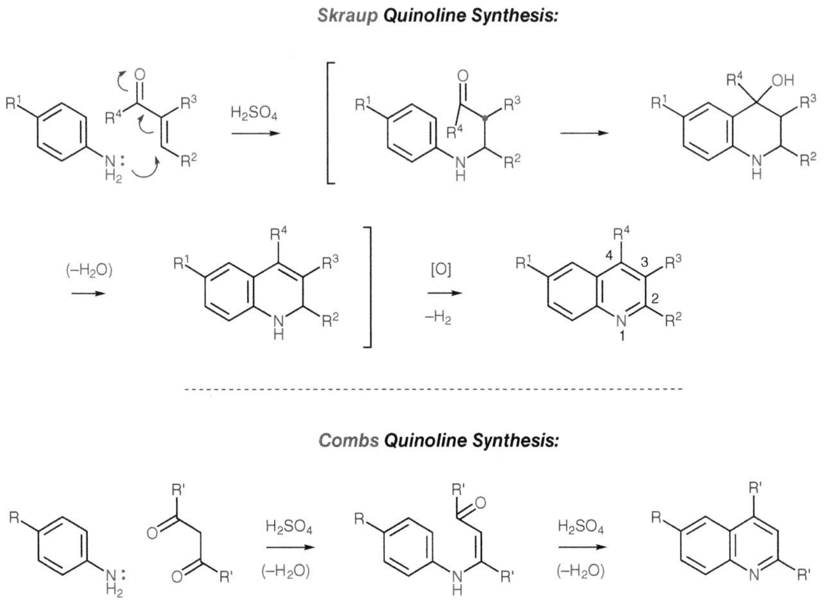

Skraup *Quinoline Synthesis:*

Combs *Quinoline Synthesis:*

Scheme 5.3 Synthesis of π-deficient heterocycles: quinolines.

hardly ever the case with the corresponding 1,2-dihydroquinolines. Rather, these species are rapidly oxidized under the reaction conditions to the aromatic quinoline derivative. Common oxidants can include air or even the co-solvent nitrobenzene. As an aside, an alternative mechanism involving initial imine formation can be ruled out on the basis of the regiochemical outcome of this reaction [10b]. For example, with R^2=alkyl, and R^3,R^4=H, the exclusive product is a 2-substituted quinoline, as opposed to the 4-substituted isomer predicted by the imine pathway. Similarly, when R^4=alkyl, the only regioisomer formed is a 4-alkyl quinoline.

Finally, while not as versatile, we should mention that there are a number of quinoline syntheses that give the desired product directly in the proper oxidation state. One of these is the Combs synthesis (Scheme 5.3), in which symmetrical β-diketones undergo condensation with aniline derivatives to give an initial enamide intermediate [11]. Frequently this can be isolated, but in general it is preferable to complete the process of cyclization, involving intramolecular electrophilic substitution followed by dehydration.

5.2.1 Isoquinolines

By this time the reader can likely propose his or her own synthetic route to the isoquinoline skeleton, building upon the same concepts as above. Again, we will want to choose our starting materials carefully, such that an advanced intermediate might cyclize directly to the aromatic product. As a start, let us analyze the two possibilities below, both of which involve dehydration as the final step:

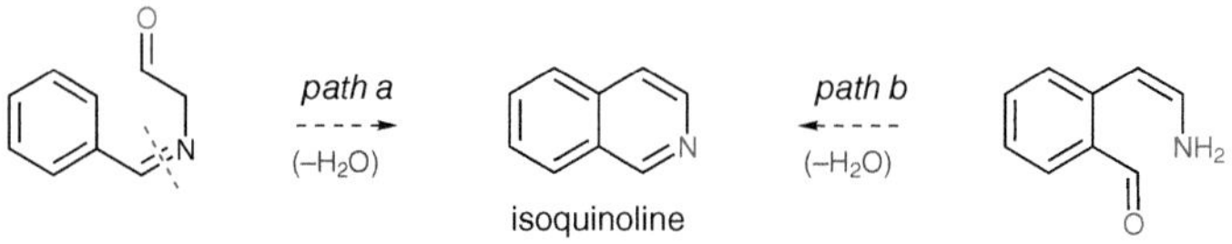

Following *path a*, the requisite starting material could presumably be synthesized by imine formation involving benzaldehyde and aminoacetaldehyde or its equivalent. Alternatively, following *path b*, the enamine starting material might be generated in situ by condensation of ammonia with the corresponding dialdehyde, itself derivable by ozonolysis of indene. Examples of both routes are in fact well known. However, *path a* has the advantage that both the aldehyde and amine precursors are readily available, and so we would probably attempt this approach first.

Perhaps this was also the thought process of Pomeranz and Fritsch, who in 1893 independently published their experimental results leading to the isoquinoline ring system (Scheme 5.4) [12a,b]. As suggested above, the first step in their synthesis involves imine formation between a benzaldehyde derivative

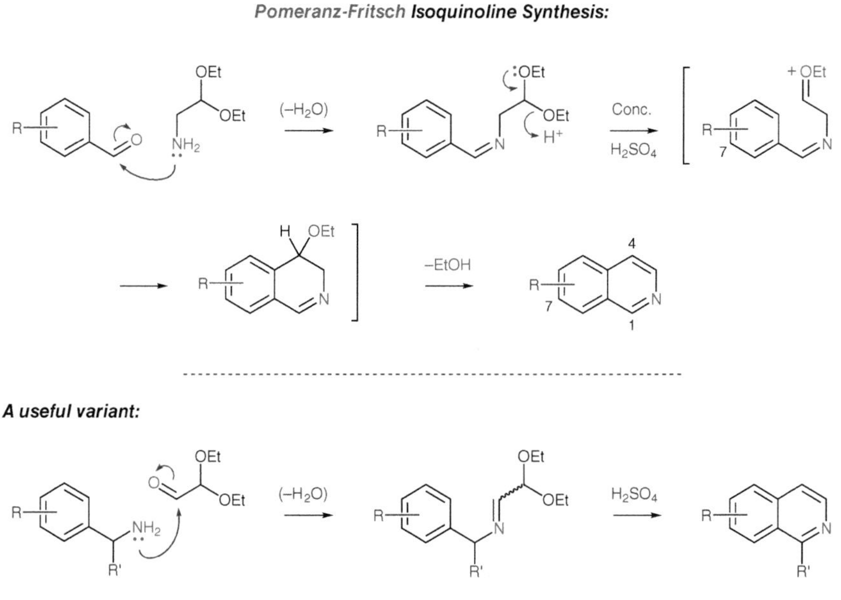

Scheme 5.4 Synthesis of π-deficient heterocycles: isoquinolines.

and aminoacetaldehyde diethyl acetal, which generally proceeds in excellent yield upon simply warming the two components together. The resultant aldimine, sometimes referred to as a Schiff base, is typically purified by distillation or crystallization, but it can also be utilized directly for the cyclization step. In the case of isoquinoline itself (R=H) this is accomplished in ~45% yield by heating at 100 degrees Celsius in concentrated H_2SO_4. Employing these "classic" conditions, yields are generally in the range of 0–80% depending upon the nature of R [12c]. More recently, improvements have been realized utilizing Lewis acid combinations, such as boron trifluoride etherate/trifluoroacetic anhydride [13].

The mechanism for the H_2SO_4-mediated cyclization deserves comment. A competing pathway in this process involves imine hydrolysis, so water must be scrupulously avoided. But how then is the acetal to be hydrolyzed prior to nucleophilic attack by the benzene ring? Or, perhaps a better question is: What is the actual species undergoing electrophilic substitution? All evidence points to the intermediacy of an oxonium cation of the type shown in brackets, which upon cyclization would produce a transient ether. Aromatization then involves elimination of a molecule of ethanol. In support of this mechanism, it has long been known that the Pomeranz-Fritsch synthesis is most effective when the benzene ring contains strongly electron donating substituents, in particular at what will become the isoquinoline 7-position. In contrast, strongly electron withdrawing groups at this position cause the reaction to fail completely. This is as would be expected for a reaction pathway involving electrophilic aromatic substitution, where the activating/deactivating effect would be most powerful at a position para to the attacking electrophile.

One weakness of the Pomeranz-Fritsch synthesis is that it usually fails with aromatic ketones, due to difficulties in forming the starting imine. Thus, we must find other means for making 1-substituted isoquinolines. Fortunately, this deficiency can be remedied by a slight modification, involving condensation of a substituted benzylamine with the highly reactive mono-acetal of glyoxal (Scheme 5.4) [14]. As in the original strategy, the resulting Schiff base is in the correct oxidation state for direct cyclization to the isoquinoline skeleton, which again is accomplished with H_2SO_4.

Finally, we close this chapter with a brief discussion of the Bischler-Napieralski reaction, an example of an isoquinoline synthesis that proceeds through a 3,4-dihydroisoquinoline intermediate (Scheme 5.5) [15]. The starting materials for this synthesis consist of a phenethylamine derivative and an acid chloride, which readily combine to produce the expected amides. These last materials are properly disposed for intramolecular cyclodehydration to afford 3,4-dihydroisoquinolines, but amide activation is required. Among other activating reagents, this can be accomplished by treatment with $POCl_3$, which produces the highly reactive imidoyl chloride shown as the first structure in brackets. In principle this species might undergo direct electrophilic

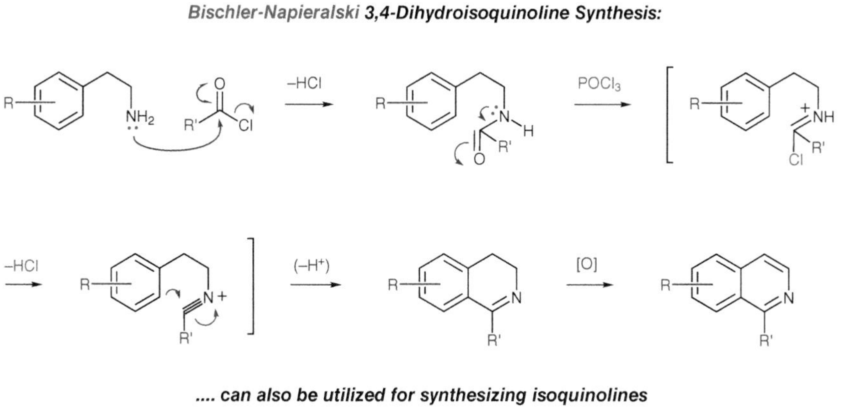

Bischler-Napieralski 3,4-Dihydroisoquinoline Synthesis:

.... can also be utilized for synthesizing isoquinolines

Scheme 5.5 Synthesis of π-deficient heterocycles: isoquinolines.

aromatic substitution. However, most evidence points to the intermediacy of the derived nitrilium salt, formed by elimination of HCl (second structure in brackets) [16]. By whichever pathway, the desired 3,4-dihydroisoquinolines are formed in good-excellent yields, especially starting with electron rich phenethylamines. A variety of mild oxidizing conditions will then generate the corresponding fully aromatic isoquinolines, including, for example, simply heating over Pd.

References

1 For extensive reviews on individual ring systems, the author recommends the highly acclaimed series of monographs *The Chemistry of Heterocyclic Compounds*, published by John Wiley & Sons, Inc. (series editors Weissberger, A., Taylor, E. C., and Wipf, P.).

2 (a) Hantzsch, A. *Chem. Ber.* **1881**, *14*, 1637–1638. (b) Hantzsch, A. *Ann. Chem.* **1882**, *215*, 1–82.

3 Li, J.-J., *Name Reactions in Heterocyclic Chemistry*, John Wiley & Sons, Inc., Hoboken, New Jersey, **2005**, pp. 304–320.

4 (a) Knoevenagel, E. *Ann. Chem.* **1898**, *303*, 223–228. See also (b) Gill, N. S.; James, K. B.; Lions, F.; Potts, K. T. *J. Am. Chem. Soc.* **1952**, *74*, 4923–4928.

5 For a representative example, see Harris, S. A.; Folkers, K. *J. Am. Chem. Soc.* **1939**, *61*, 1242–1244.

6 (a) Bohlmann, F.; Rahtz, D. *Chem. Ber.* **1957**, *90*, 2265–2272. For a useful modification, see (b) Bagley, M. C.; Dale, J. W.; Bower, J. *Synlett* **2001**, 1149–1151.

7 Gruber, W. *Can. J. Chem.* **1953**, *31*, 564–568.

8 (a) Brown, D. J., *The Pyrimidines*, John Wiley & Sons, Inc., New York, New York, **1994**. For a useful modification, see (b) Ghosh, U.; Katzenellenbogen, J. A.; *J. Heterocycl. Chem.* **2002**, *39*, 1101–1104. (c) Kenner, G. W.; Lythcoe, B.; Todd, A. R.; Topham, A. *J. Chem. Soc.* **1943**, 388–390. (d) Russell, P. B.; Hitchings, G. H. *J. Am. Chem. Soc.* **1952**, *74*, 3443–3444.

9 (a) Skraup, Z. H. Chem. *Ber.* **1880**, *13*, 2086–2087. For a review, see (b) Manske, R. H. F.; Kulka, M. *Org. React.* **1953**, *7*, 59–98.

10 (a) Manske, R. H. F. Chem. *Revs.* **1942**, *30*, 113–144. Under anhydrous conditions, however, a mechanism invoking initial imine formation has been suggested: (b) Eisch, J. J.; Dluzniewski, J. *J. Org. Chem.* **1989**, *54*, 1269–1274.

11 (a) Combes, A. *Bull. Soc. Chim. France* **1888**, *49*, 89–90. For an early review, see (b) Roberts, E.; Turner, *E. J. Chem. Soc.* **1927**, 1832–1857.

12 (a) Pomeranz, C. *Monatsch.* **1893**, *14*, 116–119. (b) Fritsch, P. *Chem. Ber.* **1893**, 26, 419–422. For a review, see (c) Gensler, W. *J. Org. React.* **1951**, *6*, 191–206.

13 Kucznierz, R.; Dickhaut, J.; Leinert, H.; von der Saal, W. *Synth. Commun.* **1999**, *29*, 1617–1625.

14 Schlittler, E.; Müller, J. *Helv. Chim. Acta* **1948**, *31*, 914–924.

15 (a) Bischler, A.; Napieralski, B. *Chem. Ber.* **1893**, *26*, 1903–1908. For a review, see (b) Whaley, W. M.; Govindachari, T. R. *Org. React.* **1951**, *6*, 74–150.

16 Fodor, G.; Nagubandi, S. *Tetrahedron* **1980**, *36*, 1279–1300, and references cited therein.

6

π-Deficient Heterocycles: Introduction of New Substituents: Nucleophilic Substitution

We are now in position to discuss the second of our synthetic approaches to π-deficient heterocycles, involving introduction of new substituents. First though, as in wine tasting, it is best to take a "cleansing rinse," for we must put aside much of what we have learned about functionalizing aromatic rings. Namely, absent strongly activating groups, electrophilic aromatic substitution is not a viable process for π-deficient heterocycles. To illustrate, consider the outcome of attempted nitration of pyridine in comparison to benzene, which affords a 95% yield of nitrobenzene at 30–40 degrees C:

Clearly the first event upon reaction of pyridine with HNO$_3$ will be protonation on nitrogen, since that is the site of highest electron density (cf. Figure 2.3). The effect of this protonation is to render an already electron deficient ring even more so, with the result that temperatures >350 degrees Celsius are required to obtain even trace amounts of 3-nitropyridine. Note that the regiochemical outcome of this reaction is the best of a bad situation, since resonance will concentrate most of the positive charge at positions 2 and 4.

What alternatives, then, do we have? Let us begin with a class of reactions that is of both theoretical and practical interest, wherein direct nucleophilic addition is followed by re-aromatization. This last step oftentimes requires addition of external oxidants (vide infra). However, with exceedingly strong nucleophiles it can also take place by ejection of a hydride anion, thereby negating one of our most ingrained notions about leaving group ability. A well-documented example of hydride ejection is found in the Chichibabin reaction, in which sodium amide adds to the 2-position of a pyridine ring (Figure 6.1) [1]. This is only possible, of course, because the developing negative charge is stabilized on nitrogen. In any event, the resultant adduct, which is formed

Introductory Heterocyclic Chemistry, First Edition. Peter A. Jacobi.
© 2019 John Wiley & Sons Ltd. Published 2019 by John Wiley & Sons Ltd.

Direct nucleophilic addition/hydride ejection is occasionally useful:

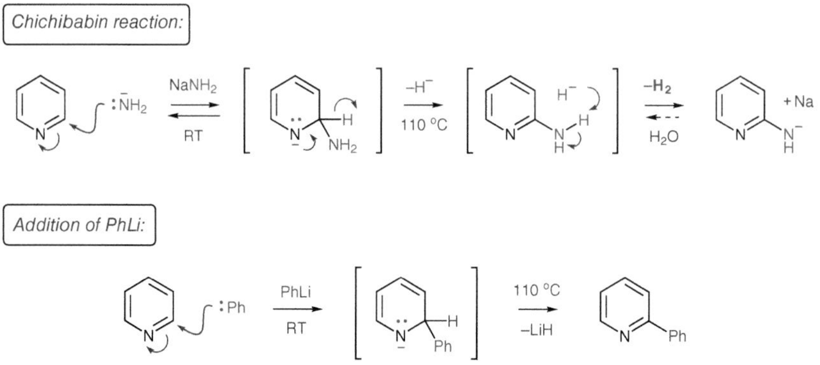

Figure 6.1 Introducing new substituents.

reversibly, undergoes aromatization via a mechanism consistent with an addition-elimination pathway. As shown, this reaction leads directly to the sodium salt of 2-aminopyridine, formed by abstraction of the relatively acidic amino proton by hydride anion. The hydrogen gas thus produced provides a convenient means of monitoring the reaction progress, and ultimately, the parent 2-aminopyridine is liberated by careful hydrolysis (66–76% yield).

In a related transformation, pyridine is converted in ~50% yield to 2-phenylpyridine upon treatment with phenyllithium (bottom Figure 6.1) [2]. Once again, the initial step in this reaction involves nucleophilic addition to the pyridine 2-position, which occurs rapidly at room temperature. The resultant adduct is not isolated, but rather upon heating undergoes aromatization by elimination of LiH. A number of organometallic reagents behave similarly, including, for example, *n*-butyllithium, where the initial 1,2-adduct has been studied by NMR [3]. Such conversions have the obvious advantage of simplicity. However, the downside of this class of reactions is that very few functional groups are compatible with such strongly nucleophilic, and basic, conditions.

Not surprisingly, the rate determining step in these transformations is elimination of hydride anion, which requires elevated temperatures. Therefore, it should be beneficial to incorporate better leaving groups at the α-position, and this is indeed the case. For example, 2-fluoropyridine undergoes smooth displacement in two hours at room temperature with a variety of lithium amides, affording the corresponding 2-aminopyridine derivatives in good yield [4]:

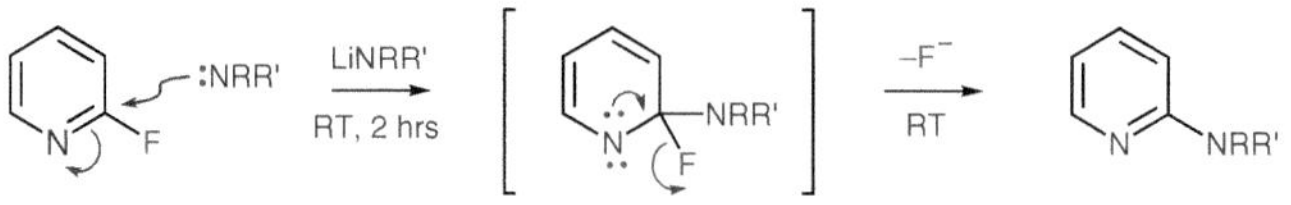

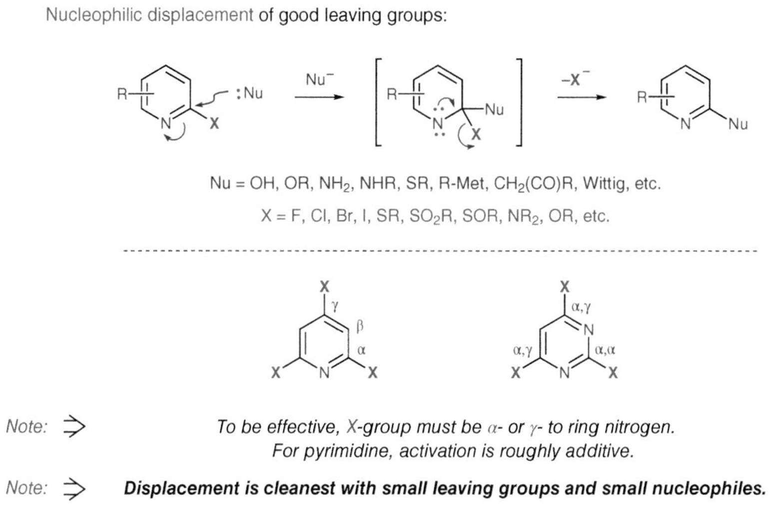

Nucleophilic displacement of good leaving groups:

Nu = OH, OR, NH$_2$, NHR, SR, R-Met, CH$_2$(CO)R, Wittig, etc.

X = F, Cl, Br, I, SR, SO$_2$R, SOR, NR$_2$, OR, etc.

Note: ⇒ *To be effective, X-group must be α- or γ- to ring nitrogen.*
For pyrimidine, activation is roughly additive.

Note: ⇒ **Displacement is cleanest with small leaving groups and small nucleophiles.**

Figure 6.2 Introducing new substituents.

This pathway is an example of a S$_N$Ar reaction, and it is adaptable to a wide range of leaving groups and nucleophiles (Figure 6.2). In addition to the α-position, leaving groups in the γ-position are also activated, since conjugate addition will again produce a stabilized anion. In this regard their reactivity pattern is analogous to that observed with *p*-fluoronitrobenzene, which undergoes nucleophilic aromatic substitution with extraordinary ease. Moreover, with pyrimidine the activating effect is roughly additive, with the α,γ-derivatives being particularly reactive. However, we must be cautious in interpreting these last results, since more than one mechanistic pathway may be operative (vide infra).

Neutral nucleophiles will also participate in S$_N$Ar reactions with suitably functionalized heterocycles, although at a slower rate. For example, 2-chloropyridine undergoes clean displacement with N-methylethanolamine at 150 degrees Celsius, affording a 95% yield of the aminopyridine derivative shown in Scheme 6.1 [5]. This last material was then converted in four steps to rosiglitazone, a once widely employed type-2 diabetes drug that functions by increasing the body's sensitivity to insulin. Unfortunately, though, long-term studies showed that rosiglitazone also increases the risk of heart attacks, and it is now only prescribed with great caution.

It remains to comment on the fact that displacement is cleanest with small leaving groups and nucleophiles (bold, bottom Figure 6.2). This makes intuitive sense, since we are going through a relatively crowded transition state in creating a tetrahedral carbon. However, even forewarned, the consequences of such crowding can be surprising.

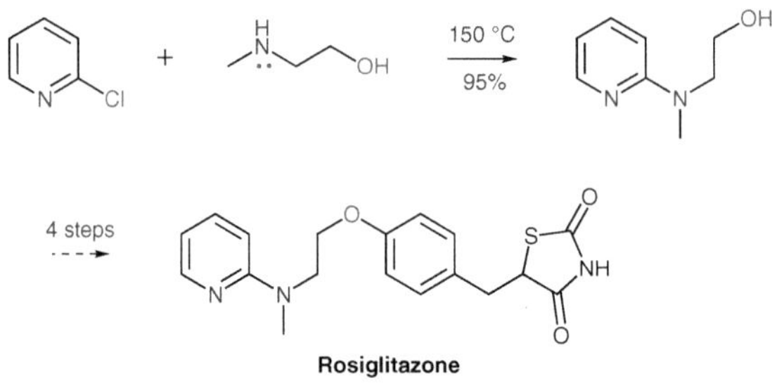

Scheme 6.1

For example, consider the S_NAr reaction of 2,6-difluropyridine with excess potassium amide (top, Scheme 6.2). In this case both the leaving group F^- and the entering nucleophile $^-NH_2$ are relatively small, so there is little steric hindrance to initial nucleophilic addition at C2(6). As anticipated, then, the predominant reaction pathway involves bis-displacement to give 2,6-diaminopyridine [6].

However, the results of the same nucleophilic displacement beginning with 2,6-dibromopyridine are less straightforward (bottom, Scheme 6.2). In this case the only identifiable product is 4-amino-2-methylpyrimidine (24%), which clearly does not arise by initial displacement of bromide [6]. Also, no trace of

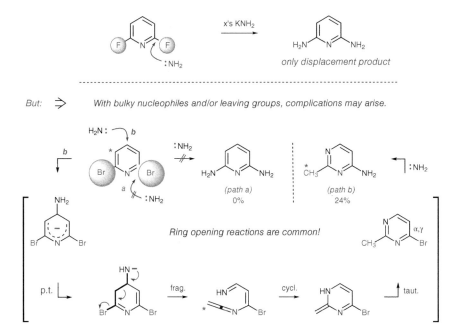

Scheme 6.2 Nucleophilic displacement: potential complications.

2,6-diaminopyridine could be detected by careful gas chromatographic (gc) analysis (*path a*). How do we rationalize this divergence in reaction pathway? The key resides in the fact that the α-positions in 2,6-dibromopyridine are significantly shielded by the steric bulk of the bromine atoms, and nucleophilic addition is more favorable at the electron deficient 4-position (*path b*). Although we do not have a leaving group at C4, the resultant anion is still stabilized by delocalization onto nitrogen. Next, by the simple process of proton transfer (p.t.), we can generate an intermediate that is ideally constituted for a concerted fragmentation reaction with loss of bromide. Note that this is an irreversible reaction, but the resultant conjugated ketimine can still find a pathway to an aromatic ring. Thus, electrocyclization, followed by tautomerization, affords 4-bromo-2-methylpyrimidine, which is now doubly activated toward final S_NAr displacement.

Ring opening-ring closing processes of this type are so common in heterocyclic chemistry that they are often described as following an ANRORC mechanism (Addition of Nucleophile, Ring Opening, and Ring Closure) [7]. Indeed, careful labeling studies have shown this mechanism to be operative even when there is no apparent skeletal rearrangement. By way of illustration, 4-bromo-6-phenylpyrimidine undergoes clean reaction with ⁻NH_2 to produce 4-amino-6-phenylpyrimidine:

83% of product   17% of product

A clear-cut example of an S_NAr pathway, right? Actually not. When this transformation was reinvestigated utilizing ^{15}N-labeled pyrimidine (*), it was found that ~83% of the product was labeled on the exocyclic 4-amino group [7]. Can we propose a mechanism to account for this fact? Yes, but before doing so, let us examine the fragmentation step in Scheme 6.2 in more detail.

Recall that E2-elimination reactions have a preferable geometry, in which the hydrogen undergoing abstraction and the leaving group X adopt an anti-periplanar conformation (Figure 6.3). This allows for a concerted process involving synchronous breaking of the C-H and C-X bonds, concomitant with re-hybridization to form the π-bond. Fragmentation reactions, which generally involve breaking of a C-C bond, have their own stereoelectronic requirements. For example, in the Grob fragmentation illustrated in Figure 6.3, the free electron pair on nitrogen, the C-C bond being cleaved, and the C-X bond all must be antiperiplanar for proper orbital overlap (sometimes referred to as a "zig-zag" conformation) [8]. Thus, it is no coincidence that we see the same arrangement in the structure shown with bold bonds in Scheme 6.2.

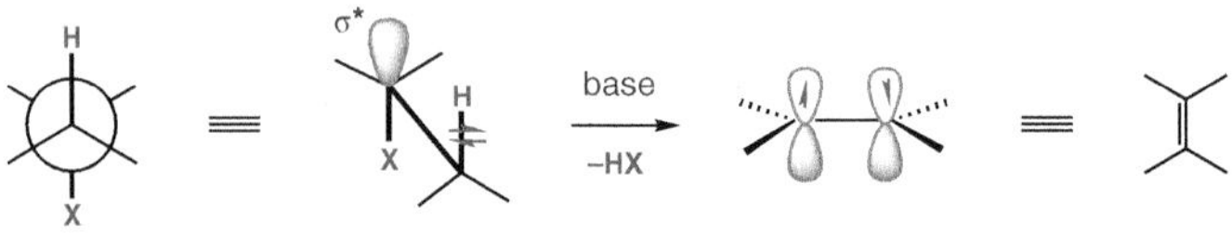

1,2-Elimination: The antiperiplanar conformation allows for overlap of the C-H σ-bond with the C-X σ*-orbital necessary for a developing π-bond.

Fragmentation: Involves breaking of a C-C bond, which must be antiperiplanar to the leaving group. The example shown is a Grob fragmentation.

Figure 6.3 Orbital requirements for fragmentation.

Now, returning to the case of 4-bromo-6-phenylpyrimidine, where is $^-NH_2$ most likely to add? Of course, addition at C4 would correspond to the normal S_NAr mechanism, with direct displacement of bromide. However, the labeling studies show that only 17% of product was formed via this pathway, presumably due to steric hindrance (vide supra). The alternative is nucleophilic addition at C2, which also generates a highly stabilized anion (Scheme 6.3). Moreover, proton transfer now reveals an intermediate with the proper orbital arrangement for fragmentation (bold bonds). The resultant amidine nitrile is then ideally situated for intramolecular cyclization, following the familiar pathway of nucleophilic addition and subsequent tautomerization. As required, the labeled ring nitrogen formerly at position 3 now occupies the C4-position as an exocyclic amino group.

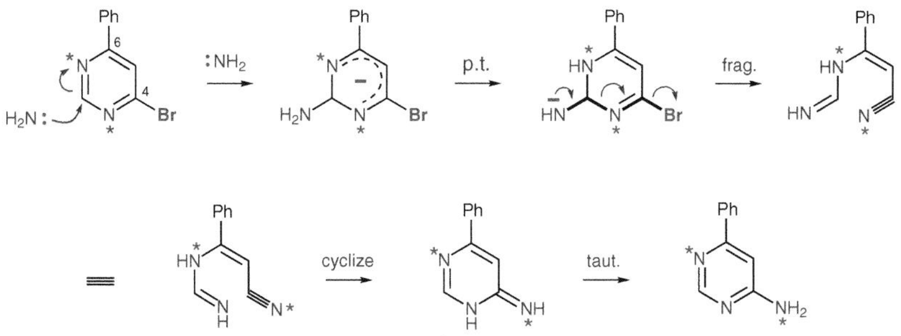

Scheme 6.3

Moving on from ANRORC transformations, let us examine a complication of a different sort, but also having its roots in steric hindrance. The reaction in question involves nucleophilic attack of sodio diethylmalonate on

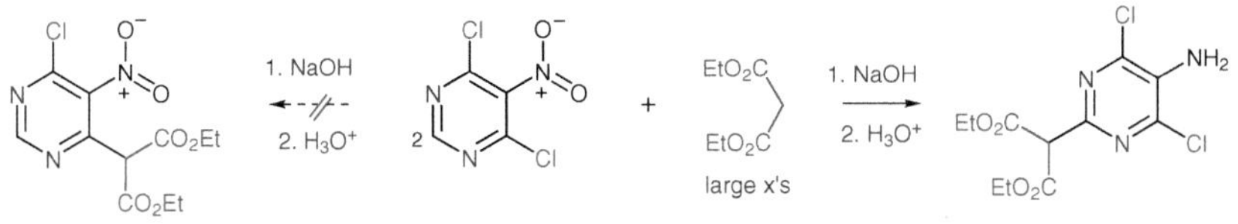

Figure 6.4 A complication of a different sort.

4,6-dichloro-5-nitropyrimidine, which was initially reported as affording the substitution product shown to the left in Figure 6.4 [9]. In fact, this was not an unrealistic expectation, given that the chlorides at C4 and C6 are "triply" activated. That is, they not only occupy α- and γ-positions to the ring nitrogens, but they are also *ortho* to the nitro group. In addition, there was precedent for the desired displacement when the C2-position was blocked. Both of these considerations might have influenced the original structural assignment. Upon re-examination, though, the actual product was proven to be the fully substituted pyrimidine shown to the right in the figure [10]. Not only had no displacement taken place, but the nitro group had been reduced to an amino group!

With the benefit of hindsight, it is easy to see why addition at C2 should be so favorable, since this position is also "triply" activated and has virtually no steric crowding (Scheme 6.4). In a sense the only thing lacking is a group to displace. Nonetheless, the drive to aromatic stability is not to be denied, and in this instance that is accomplished by internal redox chemistry. Thus, loss of a proton from C2 provides an electron pair for donation into the ring, in the process effecting reduction of the 5-nitro group to a nitroso group. But having reached this point, by what means do we accomplish the final reduction of the nitroso group to an amine? As the authors suggest [10], a reasonable pathway

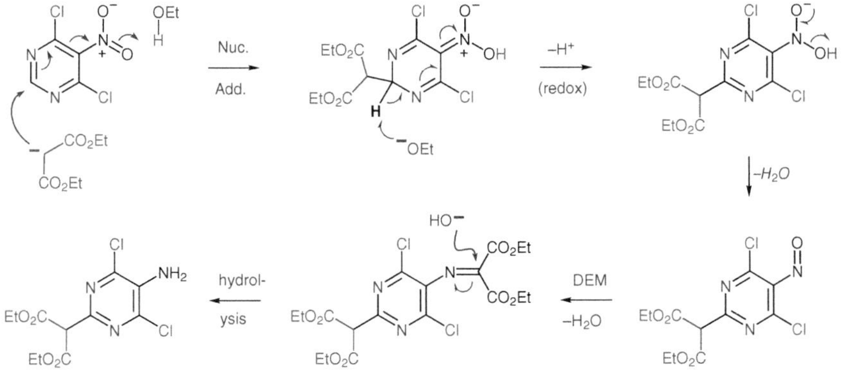

Scheme 6.4 Nucleophilic displacement: a second complication.

presents itself in the form of excess diethylmalonate, which under the reaction conditions undergoes rapid addition to the nitroso group to form a highly reactive imine. Hydrolysis of this last material then produces the observed pyrimidine amine, forming diethyl ketomalonate as the by-product.

Are there other pathways available for nucleophilic substitution? Thus far we have emphasized addition-elimination reactions, wherein the initial negative charge is stabilized by delocalization onto nitrogen. This is the most common mechanism of displacement for α- and γ-substituted pyridines and, in the absence of steric crowding, for other π-deficient heterocycles as well. See, for example, the clean conversion of 2-chloropyridine to 2-aminopyridine with KNH$_2$ (Figure 6.5) [11]. Clearly, though, this mechanism is not feasible for β-leaving groups on pyridine. No matter how many resonance forms we draw, we can never accommodate the negative charge on nitrogen. And yet, the facts are that even 3-chloropyridine will undergo substitution with sufficiently basic nucleophiles, albeit not selectively. How does this occur? As indicated, the intermediate in this case is 3,4-pyridyne, formed by initial 1,2-elimination of HCl. Many readers will recognize the close structural resemblance between 3,4-pyridyne and benzyne, and their chemistry is very similar. Both will react in such a manner as to regain aromaticity. In the present case this involves addition of ⁻NH$_2$ across the formal triple bond, to give an ~2:1 mixture of 4- and 3-aminopyridine [11].

Is 2,3-pyridyne also formed in the 1,2-elimination of 3-chloropyridine? Computational studies, as well as the complete absence of 2-aminopyridine as a product, effectively rule this possibility out. In any event, several other procedures are available for generating 3,4-pyridyne in a regioselective fashion. A particularly attractive method begins with the commercially available pyridine

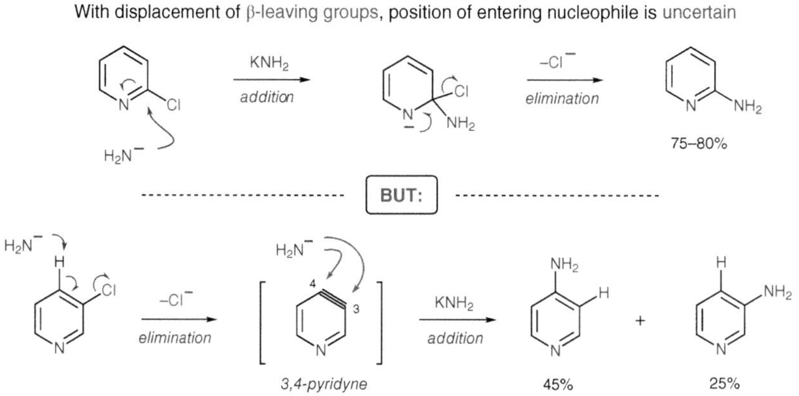

Figure 6.5 Elimination-addition: a competing mechanism.

derivative shown in Figure 6.6, which upon treatment with CsF undergoes clean 1,2-elimination [12a,b]. In the example illustrated, trapping with N-methylaniline provides a 77% yield of the corresponding addition products, again in ~2:1 ratio:

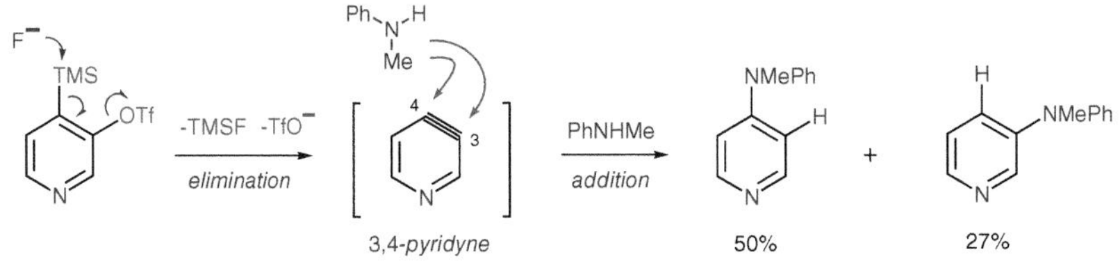

Figure 6.6 Unequivocal generation of 3,4-pyridyne.

Finally, we close this chapter with some problems for practice. Problem solving is an integral part of learning heterocyclic chemistry. Try working the mechanisms below, which are based upon material presented in this chapter. Provide as much detail as possible, including such features as direction of electron flow and proper orbital overlap.

Problems for Practice [13]

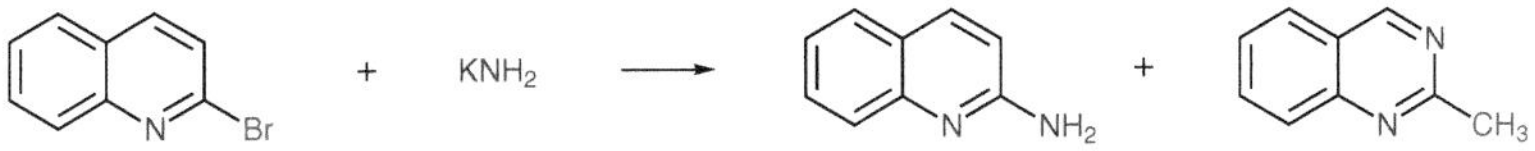

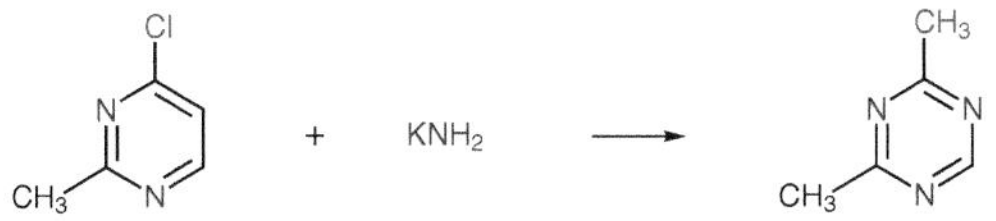

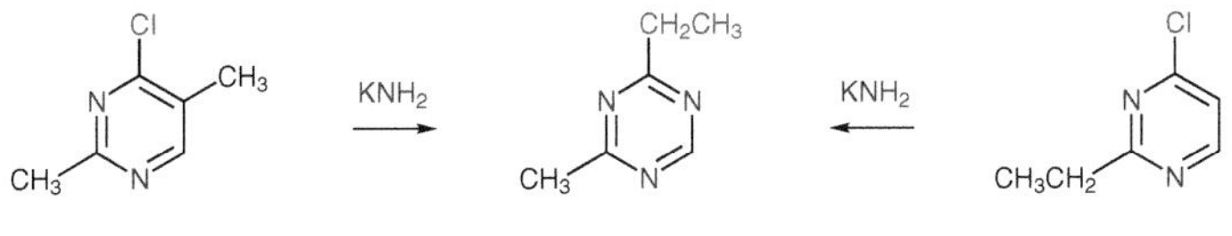

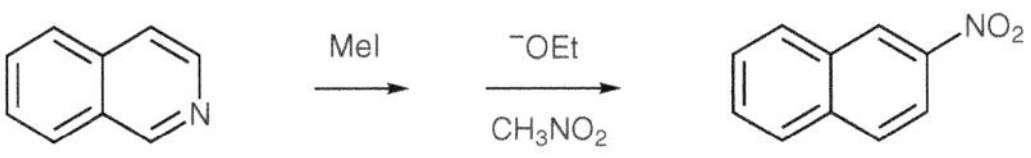

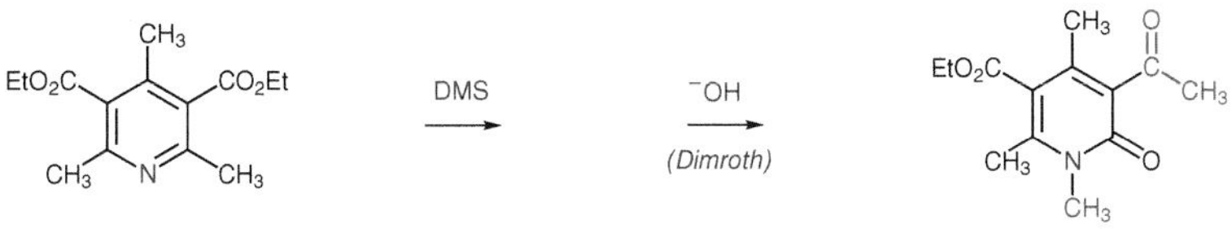

References

1 Leffler, M. T. *Org. Reactions* **1942**, *1*, 91.
2 Evans, J. C. W.; Allen, C. F. H. *Org. Synth., Coll. Vol. 2* **1943**, 517–518.
3 Fraenkel, G.; Cooper, J. C. *Tetrahedron Lett.* **1968**, *15*, 1825–1830.
4 Pasumansky, L.; Hernández, A. R.; Gamsey, S.; Goralski, C. T.; Singaram, B. *Tetrahedron Lett.* **2004**, *45*, 6417–6420.
5 Cantello, B. C. C.; Cawthorne, M. A.; Haigh, D.; Hindley, R. M.; Smith, S. A.; Thurlby, P. L. *Bioorg. Med. Chem. Lett.* **1994**, *4*, 1181–1184.
6 Den Hertog, H. J.; van der Plas, H. C.; Pieterse, M. J.; Streef, J. W. *Rec. Trav. Chim.* **1965**, *84*, 1569–1576.
7 Van der Plas, H. C. *Acc. Chem. Res.* **1978**, *11*, 462–468.
8 Grob, C. A. *Angew. Chem. Int. Ed.* **1969**, *8*, 535–546.
9 Rose, F. L. *J. Chem. Soc.* **1954**, 4116–4126.
10 Rose, F. L.; Brown, D. *J. J. Chem. Soc.* **1956**, 1953–1956.
11 Pieterse, M. J.; den Hertog, H. J. *Rec. Trav. Chim.* **1961**, *80*, 1376–1386.
12 (a) Goetz, A. E.; Garg, N. K. *J. Org. Chem.* **2014**, *79*, 846–851. See also, (b) Goetz, A. E.; Shah, T. K.; Garg, N. K. *Chem. Commun,* **2015**, *51*, 34–45.
13 Problems for practice 1: (a) den Hertog, H. J.; Buurman, D. J. *Rec. Trav. Chim.* **1967**, *86*, 187–192. (b,c) Van der Plas, H. C.; Haase, B.; Zuurdeeg, B.; Vollering, M. C. *Rec. Trav. Chim.* **1966**, *85*, 1101–1113. (d) Leonard, N. J.; Leubner, G. W. *J. Am. Chem. Soc.* **1949**, *71*, 3405–3408. (e) Mumm, O.; *Ann. Chem.* **1925**, *443*, 272–309.

7

π-Deficient Heterocycles: Introduction of New Substituents: Heterocyclic N-Oxides

We turn now to one of the most fascinating topics in heteroaromatic chemistry, that of heterocyclic N-oxides [1]. By the late 1930s these species were already a well-known class of compounds, but their chemistry had been little explored. World War II was on the horizon and scientific communication between the opposing nations was breaking down. Thus it developed that great advances in this area of chemistry would remain little-known until well after peace was achieved.

Pyridine N-oxide itself was first prepared in 1926, by the action of perbenzoic acid on pyridine (Figure 7.1) [2]. Given the structure of this compound, with its positive charge on nitrogen, how might its properties differ from those of pyridine? As we already know, pyridine is extremely reluctant to undergo electrophilic aromatic substitution, affording, for example, an ~5% yield of 3-nitropyridine at temperatures above 350 degrees Celsius. A priori, then, we might expect even lower reactivity for pyridine N-oxide. But what about the effect of the negative charge on oxygen? Could this have an impact on reactivity? In 1942, Ochiai and his collaborators in Japan provided the answer [3], demonstrating that pyridine N-oxide gave high yields of 4-nitropyridine N-oxide upon heating at ~100 degrees Celsius with H_2SO_4/HNO_3. Unfortunately, though, this landmark paper, and many follow-up studies published in Japanese, went unrecognized in the west for nearly a decade. It remained until 1950 for den Hertog and his Dutch colleagues to independently publish their nearly identical findings on the electrophilic nitration of pyridine N-oxide [4]. Other heteroaromatic N-oxides react similarly.

Is it possible the N-oxide functionality could also facilitate direct nucleophilic addition? The answer is yes, but the initial adducts frequently undergo ring opening reactions [5]. More often, such additions are preceded by "tying up" the extra electron pair on oxygen, at the same time providing a better leaving group for re-aromatization. As one example, pyridine N-oxide is rapidly acylated with *i*-butyl chloroformate (IBCF) at 20 degrees Celsius, producing an extremely electron deficient pyridinium ring (middle Figure 7.1). Without isolation, this species is captured with PhMgBr at -50 degrees Celsius, affording a 1,2-adduct which undergoes elimination and aromatization upon warming to 20 degrees

Introductory Heterocyclic Chemistry, First Edition. Peter A. Jacobi.

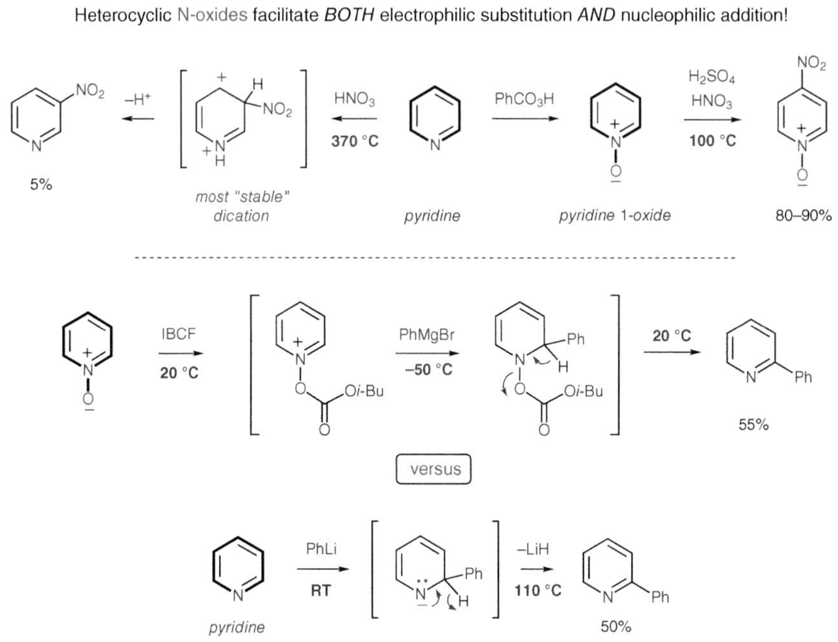

Heterocyclic N-oxides facilitate *BOTH* electrophilic substitution *AND* nucleophilic addition!

Figure 7.1 Introducing new substituents.

Celsius [6]. Compare these conditions with those required for effecting nucleophilic addition of PhLi to pyridine itself (bottom Figure 7.1).

We shall see numerous other examples of this reactivity pattern in the future. But first, how can we rationalize the fact that pyridine N-oxide can function as both a nucleophile and an electrophile? An important clue is found in the dipole moment of this species, which at 4.24 D is about twice that of pyridine (Figure 7.2). On first impression this value seems low, especially when compared to aliphatic model systems such as trimethylamine and its corresponding oxide (top right in Figure 7.2). For this pair the dipole moment of the oxide is nearly eight times that of the free amine, reflecting the very polar nature of the N-O bond. Note in particular that with aliphatic amine oxides the negative charge is localized on oxygen, since there is no orbital on nitrogen available for overlap. Such is not the case, however, with aromatic N-oxides, where resonance involving the π-cloud is very favorable. This was first recognized by Linton in 1940, who postulated that the low dipole moment of pyridine N-oxide is a result of back-donation of an electron pair on oxygen into the pyridine ring [7]. As shown in the center of Figure 7.2, electron density is particularly increased at the α- and γ-positions, which are thus activated toward electrophilic addition.

In a seeming paradox, though, we can also draw resonance structures that bear a formal positive charge at the α- and γ-positions, thereby activating toward nucleophilic addition (bottom Figure 7.2). So which is it to be? The

Dipole moments **provide a clue to** reactivity:

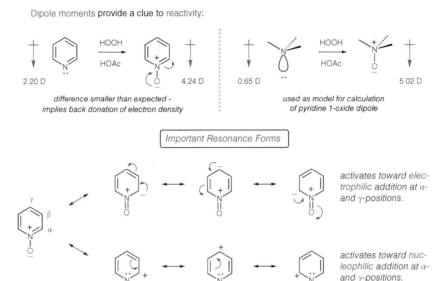

Figure 7.2 Pyridine N-oxides.

answer in large part depends upon the reagent, and the electronic demands that the reagent imposes. In the case of electrophilic addition, we can consider the oxide group to be a strong electron donor, much as, say, the hydroxyl group in phenol. Now, picture an electrophile approaching the α- or γ-position in pyridine N-oxide. As bonding begins, the developing cation will be stabilized by back donation of electron density from oxygen (top right Figure 7.3). This is exactly analogous to the *o,p*-directing effect observed with phenol. On the other hand, nucleophiles will be attracted to the α- and γ-positions because they are in direct conjugation with the positively charged nitrogen (top left Figure 7.3). This is particularly the case when back donation is inhibited, rendering the pyridine ring even more electron deficient (cf. Figure 7.1).

The N-oxide group in these examples can be viewed as an activating auxiliary, whether the ultimate goal is electrophilic aromatic substitution or nucleophilic addition with subsequent re-aromatization. Suppose, for example, we wished to convert pyridine to 4-nitropyridine (bottom Figure 7.3). We know we cannot accomplish this by direct nitration, since the pyridine ring is far too unreactive in electrophilic aromatic substitution. However, a slight detour solves the problem. This consists of initial oxidation to pyridine N-oxide, which affords high yields of 4-nitropyridine N-oxide with H_2SO_4/HNO_3. Next comes the step where we must remove the activating group, and here is where a little caution is in order. Formally this is simply a reductive cleavage, and there are many reagents that could qualify for "[H]" in Figure 7.3. Once you have removed

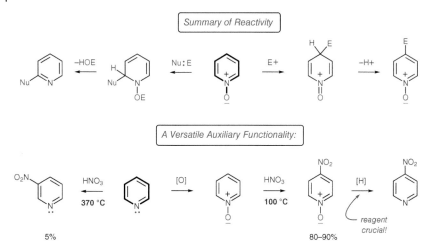

Figure 7.3 **Pyridine N-oxides.**

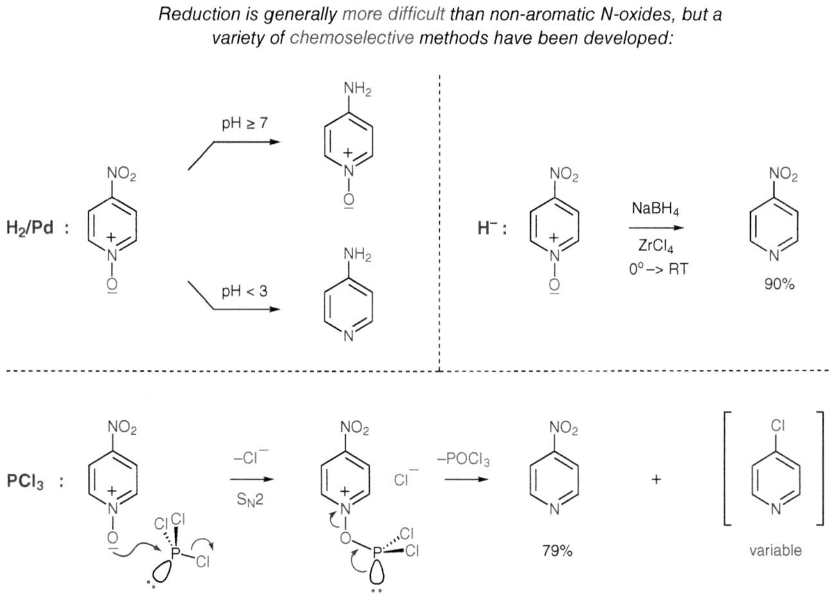

Reduction is generally more difficult than non-aromatic N-oxides, but a variety of chemoselective methods have been developed:

Figure 7.4 **Pyridine N-oxides, reduction.**

the N-oxide functionality you are returned to the pyridine oxidation state and you have achieved the transformation that you set out to accomplish. But the choice of reagent is crucial, since the nitro group is also very easily reduced.

To illustrate this point, let us start at the top left of Figure 7.4 with a procedure that is ill suited to providing the desired selectivity, catalytic hydrogenation. Employing palladium as catalyst the course of hydrogenolysis varies with pH, but under no circumstances are we able to effect selective cleavage of the N-oxide functionality. Thus, under neutral or alkaline conditions, the nitro group is actually reduced preferentially, affording 4-aminopyridine N-oxide. On the other hand, at pH<3 both the nitro group and the N-oxide are reduced, providing an efficient route to 4-aminopyridine [3b]. These results serve to highlight the fact that aromatic N-oxides are more difficult to reduce than their aliphatic counterparts. However, because of their versatility, a variety of chemoselective methods has been developed.

One of these makes use of the reagent combination $NaBH_4$ with $ZrCl_4$, which provides a 90% yield of 4-nitropyridine under very mild conditions (top right Figure 7.4) [8]. Interestingly, this reduction may involve intramolecular delivery of hydride through a zirconium complexed intermediate, followed by 1,2-elimination:

A second method takes advantage of the strong nucleophilicity of the N-oxide oxygen atom, which will readily displace Cl⁻ from oxophilic reagents such as PCl_3 (bottom Figure 7.4). This is an energetically favorable step since we have substituted a relatively weak P-Cl bond with a strong P-O bond. Moreover, as a consequence of surrendering its free electron pair, the N-O bond is now much more susceptible to reductive cleavage. But from where does the reducing agent come? Actually, it has been there all along, in the form of the free electron pair on phosphorus. Note from the drawing that this electron pair is ideally situated for cleaving the N-O bond, by a process strongly reminiscent of an E1cB mechanism. The difference here is that no strong base is required, and the leaving group is a neutral molecule of 4-nitropyridine.

So if the N-oxide group has been reduced, what has been oxidized? The answer of course is phosphorus, which has gone from PCl_3 to $POCl_3$. $POCl_3$ is an important reagent in its own right, with a reactivity pattern quite different from PCl_3 (vide infra). At any rate, Ochiai cited a yield of 79% for this reduction [3b], but depending upon the reaction conditions, variable amounts of 4-chloropyridine are also formed. In this case not only has the N-oxide functionality been reduced, but the NO_2 group has been replaced by Cl. It turns out that nucleophilic displacements of this type are quite general:

In fact, nitro groups are comparable to halogens in their leaving group ability when bonded to π-deficient heterocycles. This explains why 4-nitropyridine is relatively unstable, undergoing bimolecular self-displacement on standing at room temperature. Hydrolysis then yields N-(4'-pyridyl)-4-pyridone [9]:

Electrophilic aromatic nitration thus serves an important role in introducing a good leaving group into the 4-position of pyridine N-oxide, which we shall see has considerable utility in natural product synthesis. For now, though, let us continue our discussion of nucleophilic addition to N-oxides, wherein aromatization is regained by formal elimination of the elements of water. In Figure 7.1 we saw one example of this process involving the very strong nucleophile PhMgBr, in which the N-oxide group was activated by acylation with *i*-butyl chloroformate. With suitable activation, even very weak nucleophiles will undergo addition to heteroaromatic N-oxides.

Such is the case with the phosphonium bromide derivative PyBroP, which undergoes rapid S_N2 displacement with a variety of pyridine and isoquinoline N-oxides (Figure 7.5) [10]. The resultant salts are now strongly polarized toward in situ nucleophilic addition, and the adducts thus formed have an energetically favorable pathway to re-aromatization. This involves E2 elimination of the excellent leaving group phosphoryltripyrrolidine ($Py_3P=O$), under the action of N,N-diisopropylethylamine. It is worth noting the remarkably simple experimental conditions for these transformations, which involves simply mixing all of the reagents together at room temperature in an inert solvent. Reaction is complete within fifteen hours, producing exclusively the 2-substituted derivatives. The authors ascribe this selectivity to charge

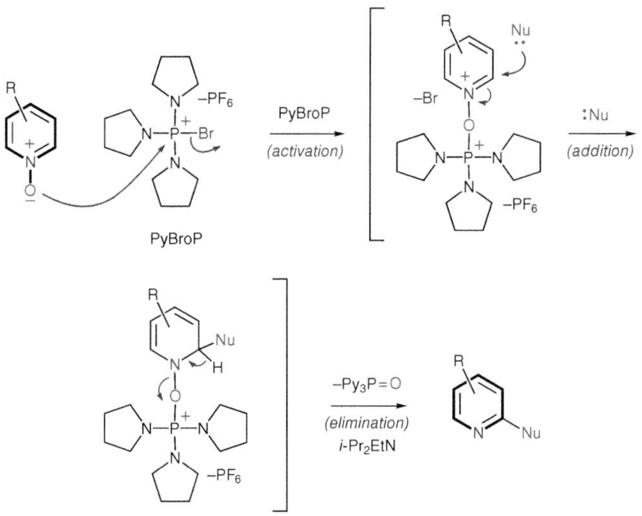

Nu = amine, phenol, sulfonamide, malonate, pyridone, thiol, etc. (up to 96% yield)

Figure 7.5 A novel activating agent.

association of the incoming nucleophile with the activated complex. In addition, the PyBroP methodology is compatible with an impressive array of nucleophiles, generally in the pK_a range of ~10–20. Other examples of "external" activating agents include tosyl chloride and tosyl anhydride, but with these reagents pre-activation is necessary.

Importantly, though, activation need not require an external reagent. Oftentimes the activating species and the nucleophilic component will be part of the same molecule, as for example in the case of trimethylsilyl cyanide (TMSCN) reacting with pyridine N-oxide (top Figure 7.6) [11]. The first step in this sequence involves S_N2 displacement of ¯CN by the oxide oxygen, a reaction that is driven by the greater strength of the Si-O bond. With the oxygen electron pair now taken out of play, ¯CN is sufficiently nucleophilic to undergo addition to the electron deficient 2-position, producing a transient covalent adduct. This last intermediate then undergoes rapid 1,2-elimination in the presence of NEt$_3$ to afford 2-cyanopyridine in 80% yield.

In analogous fashion, POCl$_3$ is "self-activating" in its reaction with pyridine N-oxide, undergoing initial displacement of Cl¯ through an acyl substitution-like process (bottom Figure 7.6) [12]. Once again we form a strong P-O bond at the expense of a weak P-Cl bond, much as we saw with PCl$_3$ in Figure 7.4. However, from this point all similarities between the two reagents end. In particular, the POCl$_3$ adduct lacks a free electron pair on phosphorus that might be employed in reductive cleavage of the N-O bond. Rather, the now strongly

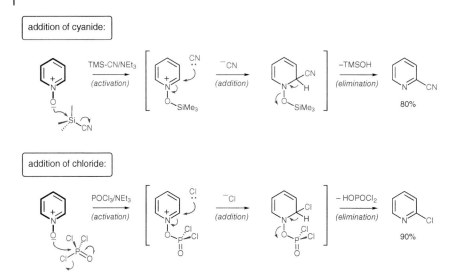

Figure 7.6 For weaker nucleophiles, activation is crucial.

electron deficient pyridine ring undergoes addition at the α-position by the only nucleophile present, the Cl⁻ anion produced on initial activation. Subsequent NEt_3-induced elimination of $HOPOCl_2$ then gives a 90% yield of 2-chloropyridine, which as we have seen is a valuable precursor to other 2-substituted pyridine derivatives (vide supra).

To gain some experience, let us apply what we have learned so far about heteroaromatic N-oxides to analyzing a short natural product synthesis. Our target is the highly substituted pyridine derivative ricinine, a toxic alkaloid isolated from the seeds of the castor oil plant (Figure 7.7) [13]. We will start with 3-methylpyridine (β-picoline), which is commercially available on large scale.

To begin, a well-designed synthesis takes into account both strategy and tactics, much like a game of chess. Strategy can be roughly defined as an overall

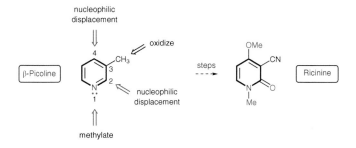

Figure 7.7 Planning a synthesis of ricinine.

plan of action for achieving a major objective. Tactics, on the other hand is the step-by-step means by which a strategy is carried out. In our synthesis of ricinine, the overall objective is the regioselective introduction of three new substituents into the skeleton of β-picoline, along with modification of the C-3 methyl group to a nitrile. In what order should these steps be implemented for maximum efficiency? This is a matter of strategy. On the tactical side, however, it is essential that we execute each step based on well-founded precedent. That is, we must know our reagents and how to apply them—just as in a board game you have to take best advantage of the capabilities of each piece. In Figure 7.7, arrows are drawn to those sites where we need to plan our moves, along with some indication as to what those moves might entail. For the moment, let us consider each of these transformations individually, starting with the C-3 methyl group. We can think of these as model studies.

The nitrile group at C-3 in ricinine is in the same oxidation state as a carboxylic acid, suggesting that nicotinic acid might be a viable precursor to our ricinine model compound (equation 1, Figure 7.8). To be sure, it will require a powerful oxidizing agent to carry out the transformation of β-picoline to nicotinic acid, but fortunately the pyridine ring is up to the challenge. Thus, with a resonance energy of ~28 kcal/mol, pyridine is comparable to benzene in its ability to withstand reagents that will rapidly degrade attached alkyl groups. Possible oxidizing agents include $KMnO_4$ or Cr(VI). The pathway forward now seems clear, involving conversion of nicotinic acid to the corresponding amide nicotinamide, followed by dehydration. The reagent $POCl_3$ is highlighted in bold, since in our tactical planning this may serve more than one purpose [14].

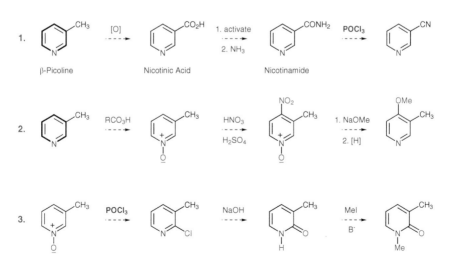

Figure 7.8 Synthesis of ricinine: tactical considerations.

Next to consider is the methoxy substituent at C-4 in ricinine, which in principle can be derived by nucleophilic displacement (cf. Figure 7.7). So far we know of only one option for introducing a leaving group into the 4-position of pyridine, and that involves N-oxide chemistry (equation 2, Figure 7.8). Our first goal, therefore, is oxidation of β-picoline to the corresponding N-oxide, a transformation that precedent tells us can be accomplished with a peracid (RCO_3H). Nitration should now proceed in a regioselective fashion, introducing a good leaving group into the 4-position. Once again, there is ample precedent for displacement of a 4-nitro substituent with NaOMe (vide supra), leaving us only with the task of N-oxide reduction. This should be readily achieved utilizing either PCl_3 or $NaBH_4/ZrCl_4$ (cf. Figure 7.4) [3b,8].

Finally, by what means can we introduce the N-methyl group and the C-2 carbonyl? Ideally, of course, we would like to find a way of combining these steps, but for now let us carry them out sequentially (equation 3, Figure 7.8). Beginning with the same picoline N-oxide prepared in equation 2, treatment with $POCl_3$ will afford the 2-chloro derivative with concomitant cleavage of the N-oxide. Note again that $POCl_3$ is highlighted in bold, reminding us that this reagent might also serve for amide dehydration [14]. In any case, hydrolysis of the 2-chloro group would then lead directly to the corresponding lactam, leaving N-methylation for the final step. Typical conditions for methylation involve treatment with methyl iodide (MeI) in the presence of base, but we must be aware of the possibility that O-methylation could be a competing process.

We thus have a good idea of how to implement each of the individual steps in a synthesis of ricinine, and it is time to develop a strategy. Give some thought to this, in particular the order in which you might introduce various substituents. Then we will compare our approach with the published route by Edward C. Taylor, of Princeton University, one of the masters of heterocycle synthesis [15]. Bear in mind that an important goal is to devise the shortest synthesis possible, but we must also take into account functional group compatibility.

From Figure 7.8, we know that our first step will involve oxidation of β-picoline, either to nicotinic acid (equation 1) or to the corresponding N-oxide (equations 2 and 3). Does it matter which operation we carry out first? To answer, we need to look ahead to the steps that follow, especially with regard to introducing the 4-methoxy substituent. Ultimately this will involve displacement of a nitro group, which must be introduced by electrophilic aromatic substitution on an N-oxide.

One means forward would involve the intermediacy of the 4-nitropyridine derivative shown in the box in Figure 7.9, which should be especially susceptible to nucleophilic displacement by methoxide. Which pathway in Figure 7.9 do you think would be most suitable for preparing this intermediate?

Following *path a*, we begin with peracid oxidation, which should readily afford β-picoline N-oxide. Nitration, followed by vigorous oxidation with

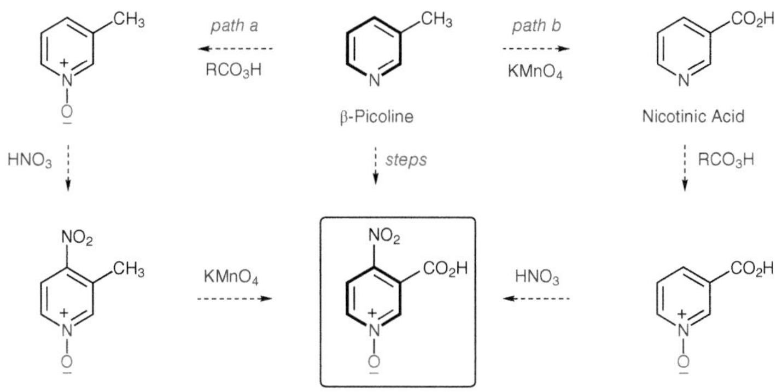

Figure 7.9 Strategic considerations in the synthesis of ricinine.

$KMnO_4$, would then lead to the desired carboxylic acid derivative. Alternatively, following *path b*, we start with $KMnO_4$ oxidation of β-picoline to give nicotinic acid. Treatment of this material with RCO_3H would next give the corresponding N-oxide, which on nitration *might* afford the same ricinine precursor as produced following *path a*. But probably not, at least not in reasonable yield. Where does the difference lie? Simply in the fact that electrophilic nitration of pyridine N-oxides is very sensitive to the nature of existing groups, just as we see in benzene chemistry. In *path a*, nitration will be facilitated by the presence of the methyl group, which is an electron donor. In contrast, the 3-carboxyl group in *path b* will be a strong deactivator.

We have some flexibility in the timing of subsequent steps, but before continuing, it is worth noting that we are already in the proper overall oxidation state for synthesizing ricinine. At any point we could carry out the conversion of the 3-carboxyl group to a nitrile, following the general protocol outlined in equation 1 of Figure 7.8. Or, we could elect to displace the nitro group first, according to equation 2. For the present, we will put off the transformation of the N-oxide to the 2-chloro derivative (cf. equation 3), since it is best to introduce the most sensitive functionality last. Also, we are hopeful that $POCl_3$ might bring about more than one transformation.

To better realize this last possibility let us introduce the methoxy group first, which should be easily accomplished since the 4-position is doubly activated by the N-oxide and 3-carboxy substituent (Figure 7.10). Subsequent amidation would then produce an intermediate that with $POCl_3$ might be converted to 2-chloro-3-cyano-4-methoxypyridine in a single step. Mixtures of regioisomers here are also possible, but there is good precedent for the 2-chloro isomer being favored [14]. In any event, we are now ready to address the "end-game," involving hydrolysis of the 2-chloro substituent and selective N-methylation. Here we must take great care, since both the nitrile and the methoxy groups are also subject to hydrolysis. Nonetheless, it seems reasonable that aqueous

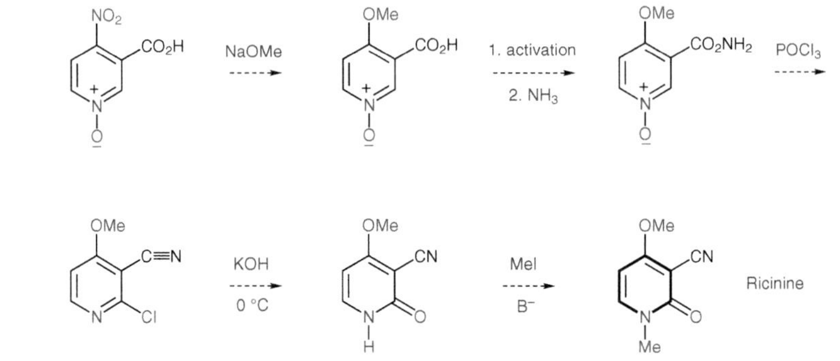

Figure 7.10 Ricinine synthesis: a potential end game.

KOH at 0 degrees Celsius might provide the desired chemoselectivity. By the same token, the final N-methylation step will undoubtedly require experimentation, but as a start we can try MeI in the presence of a strong non-nucleophilic base.

How did we fare in our synthetic planning as compared to Professor Taylor's solution [15]? It turns out, not bad, at least in the early stages. As in our analysis, the literature synthesis began with the known β-picoline N-oxide, which was readily prepared from β-picoline by oxidation with peracetic acid (Scheme 7.1). Nitration of the N-oxide under standard conditions then

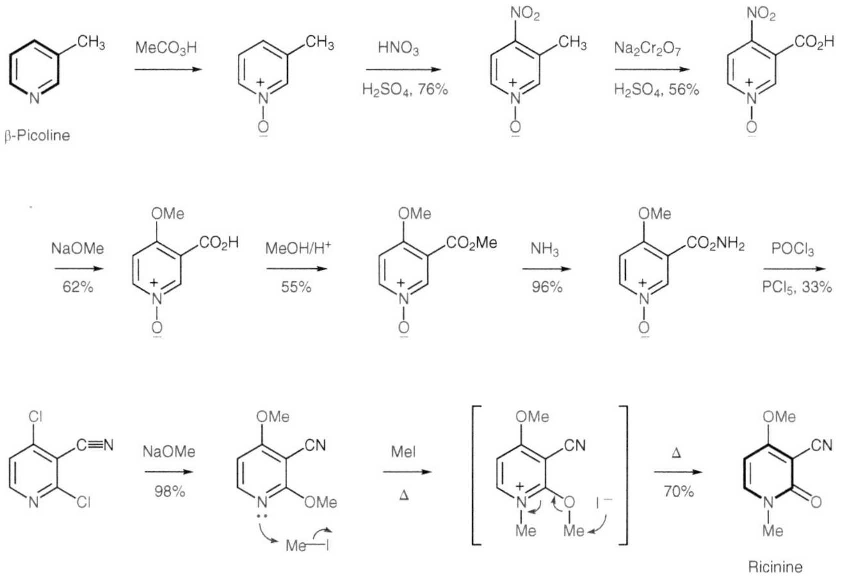

Scheme 7.1 Synthesis of ricinine: putting it all together.

gave a 76% yield of the 4-nitro derivative on multigram scales. An interesting challenge now presented itself in the oxidation of this last material to the corresponding carboxylic acid, in that alkaline conditions had to be scrupulously avoided. This was due to the susceptibility of the nitro group to displacement, which of course would be put to good purpose in a following step. Ultimately a solution was found in the use of $Na_2Cr_2O_7$ dissolved in concentrated H_2SO_4, which afforded the desired carboxylic acid in 56% yield. Now came the time to displace the nitro group, and this was smoothly accomplished simply on warming with NaOMe in MeOH. As noted above, this step was undoubtedly facilitated by the activating influence of both the N-oxide and carboxy groups.

By this juncture all substituents were in place for completing the synthesis of ricinine, and what remained was mostly functional group manipulation. This was initiated by conversion of the carboxy group to the corresponding amide, which was brought about in straightforward fashion via esterification followed by aminolysis. The stage was thereby set for reaction with $POCl_3$, which effected both amide dehydration and α-chlorination. Also, as an added measure, the 4-methoxy group was substituted by chloride. The outcome of this sequence was of considerable significance for two reasons. First, only a single regioisomer was obtained. And second, the resultant 2,4-dichloro derivative was a known precursor to ricinine. Thus, it is here that the Taylor synthesis diverges markedly from our proposed route, which required both selective hydrolysis and N-methylation (cf. Figure 7.10). Rather, both chloride groups were next displaced by methoxide to afford a 98% yield of 2,4-dimethoxy-3-cyanopyridine. In this compound, methylation can only occur at the free electron pair on nitrogen, producing the pyridinium salt shown in brackets. Finally, upon warming, this last material undergoes in situ S_N2 displacement by iodide at the 2-methoxy substituent, in which the leaving group is ricinine. From both a strategic and a tactical perspective, this synthesis gets very high marks.

7.1 Further Reactions of N-Oxides

We have thus seen that $POCl_3$ is an excellent reagent for introducing a chloride substituent into the α-position of a pyridine ring, via the corresponding N-oxide. A related transformation is known as the Katada rearrangement, in which the N-oxide is simply heated in acetic anhydride (Figure 7.11) [3b]. The product of this reaction is 2-acetoxypyridine, formed in essentially quantitative yield via a mechanism exactly analogous to that observed with $POCl_3$ (cf. Figure 7.6). In fact, by now we recognize a familiar pattern in all such deoxygenation-substitution reactions of heteroaromatic N-oxides: activation, followed by nucleophilic addition, and finally elimination. In the case of the

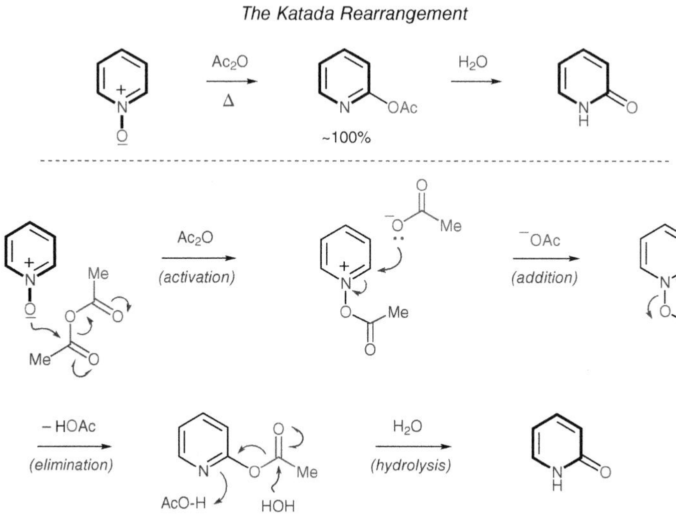

The Katada Rearrangement

Figure 7.11 Action of Ac$_2$O on N-oxides.

Katada rearrangement, the activating electrophile is acetic anhydride, which functions by acylating the strongly nucleophilic oxide anion. With the electron pair thus tied up, the pyridine ring is even more susceptible to nucleophilic addition at the α-position. And what nucleophiles are present? Chiefly the acetate anion liberated in the acylation step. Thus, acetate addition, followed by 1,2-elimination of acetic acid generates the observed 2-acetoxypyridine. While species of this type can be isolated, more often than not the reaction is quenched under aqueous conditions, which effects rapid hydrolysis to the corresponding 2-pyridone.

Interestingly, the closely related compound 2-methylpyridine N-oxide reacts with acetic anhydride in an entirely different fashion, affording high yields of 2-acetoxymethylpyridine (Figure 7.12). None of the 6-acetoxypyridine derivative corresponding to a Katada rearrangement is observed. This is an example of a Boekelheide rearrangement [16], and it is quite general for heteroaromatic N-oxides substituted in the α- or γ-position with alkyl groups.

The mechanism for this transformation was a matter of considerable debate in the early years following its discovery (1953), but there was little doubt about the first step. This involves acylation of the N-oxide, which again has the effect of rendering the pyridine ring highly electron deficient. In principle, acetate anion might now undergo nucleophilic addition to the 6-position, followed by re-aromatization with loss of acetic acid. Instead, though, acetate functions as a base in abstracting a very acidic proton from the C-2 methyl group. The product of proton abstraction is a so-called anhydro-base, shown to the right in the middle row of Figure 7.12. From this

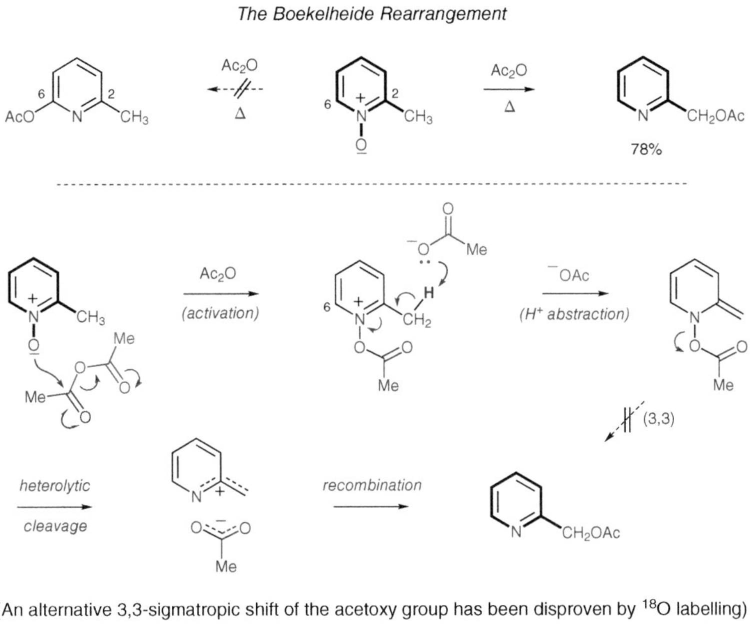

The Boekelheide Rearrangement

78%

(An alternative 3,3-sigmatropic shift of the acetoxy group has been disproven by ^{18}O labelling)

Figure 7.12 Action of Ac$_2$O on N-oxides.

point a number of pathways are feasible leading to product, including a concerted 3,3-sigmatropic shift. However, this mechanism was excluded on the basis of detailed labeling studies, in which all three oxygen atoms in acetic anhydride were enriched to the same degree with ^{18}O [17]. The intermediate anhydro-base was thereby labeled with ^{18}O at the carbonyl oxygen, while the original N-oxide oxygen had natural isotopic distribution:

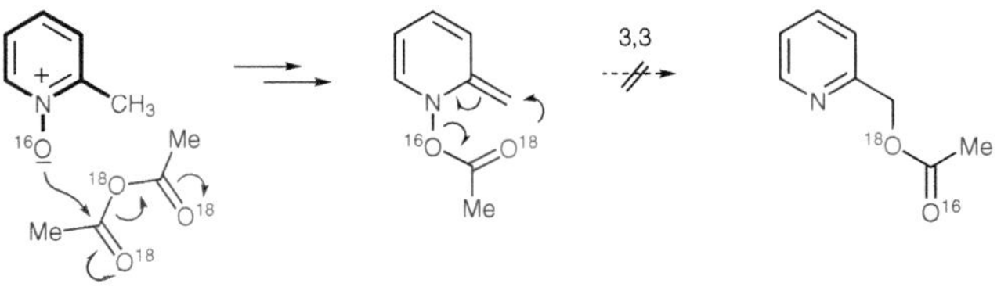

Were a 3,3-sigmatropic shift operative, all of the ^{18}O label in the product should end up attached to the terminus of the C-2 methylene group, as shown above. However, when the experiment was run, it was found that only half of the label occupied this position, with the other half residing in the carbonyl group. We can thus rule out a concerted intramolecular mechanism leading to 2-acetoxymethylpyridine, and by similar reasoning, we can also exclude an intermolecular S_N2' route involving doubly labeled acetate anion:

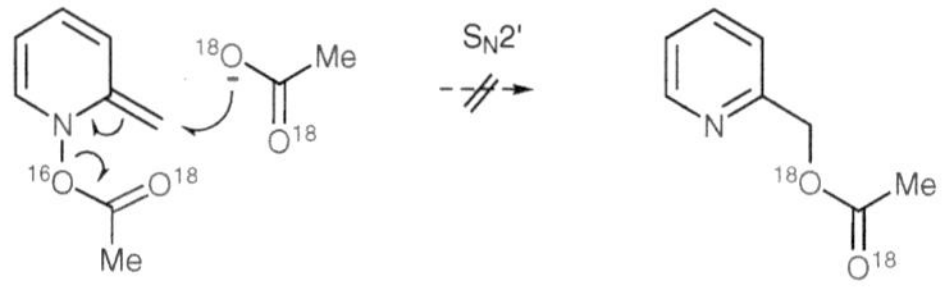

So where does this leave us? As shown in Figure 7.12, the preponderance of evidence points to an intramolecular pathway involving initial heterolytic cleavage of the N-O bond to give an ion pair. In such a species the two acetate oxygens are equally available for bonding, which explains the ^{18}O distribution in the labeling studies. In this case, an alternative radical pair mechanism is deemed less likely on three counts. First is the fleeting existence of acetoxy radicals, which undergo rapid decarboxylation [18]; second is the observation that electron donating groups on the ring accelerate the reaction; and third, in related systems carbocation derived products have been isolated [19].

Rearrangements of this type are not limited to acetic anhydride or pyridine N-oxides. For example, simply warming an ethereal solution of 2-methylquinoline N-oxide and benzoyl chloride gives a 49% yield of 2-benzoyloxymethylquinoline (Figure 7.13) [20]. Similar results were obtained in benzene and other non-polar solvents. As indicated, the initial step in this transformation involves benzoylation of the N-oxide, with subsequent

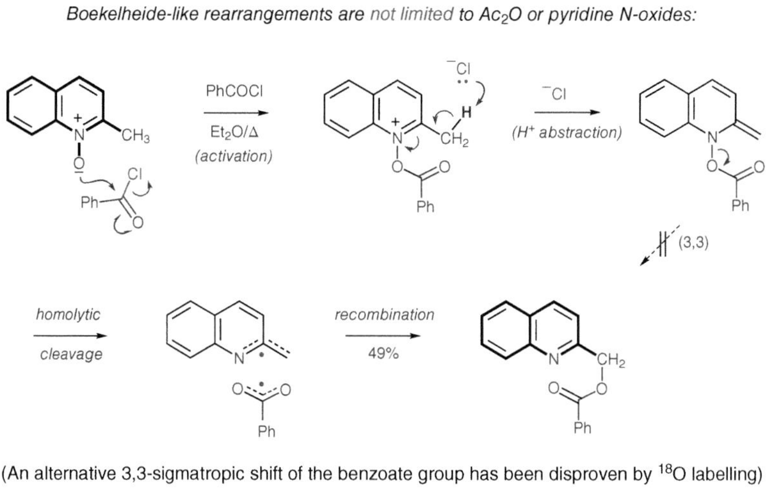

Boekelheide-like rearrangements are not limited to Ac₂O or pyridine N-oxides:

(An alternative 3,3-sigmatropic shift of the benzoate group has been disproven by ^{18}O labelling)

Figure 7.13 Related transformations.

proton abstraction to generate the anhydro-base shown in the top right of the figure. Thus far, we are following a mechanistic pathway analogous to the Boekelheide rearrangement diagrammed in Figure 7.12. And once again, detailed labeling studies rule out the possibility of a concerted 3,3-sigmatropic shift leading to product. Thus, when carried out with ^{18}O-enriched benzoyl chloride, approximately 50% of the label appeared in each of the oxygen atoms of the rearranged material [20].

It is clear, then, that N-O bond cleavage must precede benzoyloxy group migration, but will this step be heterolytic or homolytic in nature? The fact that rearrangement is so favorable in apolar solvents makes a strong case for homolytic cleavage, as opposed to a mechanism involving ion pairs. In addition, the radical cage pathway is supported by the isolation of small quantities of benzoic acid and 2-methylquinoline, both derived by hydrogen atom abstraction. Finally, it is worth noting the far greater stability of a benzoyloxy radical toward decarboxylation as compared to an acetoxy radical [18]. Stability issues were one of the factors arguing against a radical cage mechanism in the acetic anhydride induced rearrangement shown in Figure 7.12. However, in the case of benzoyloxy group migration, the data is fully consistent with the homolytic pathway outlined in Figure 7.13.

We might also wish to carry out the selective chlorination of an alkyl group at the α- or γ-position to an N-oxide, with concomitant deoxygenation. Reagents such as POCl$_3$ can effect this transformation, but ring chlorination is almost always a competing process when α-free positions are available (cf. Figure 7.6). Also, as we have just seen, benzoyl chloride follows a markedly different course in its reaction with aromatic N-oxides (Figure 7.13). Interestingly, though, aryl and alkyl *sulfonyl* chlorides provide the desired chemoselectivity. For example, 2-methylpyridine N-oxide is cleanly converted to 2-chloromethylpyridine upon heating in benzene with benzenesulfonyl chloride (Figure 7.14) [21]. The mechanism for this transformation has a familiar feel, at least in the first two steps leading to anhydro-base formation. These involve initial sulfonylation of the oxide group followed by proton abstraction. From here, there is a straightforward pathway to product involving S$_N$2' displacement of the excellent leaving group benzenesulfonic acid by chloride anion.

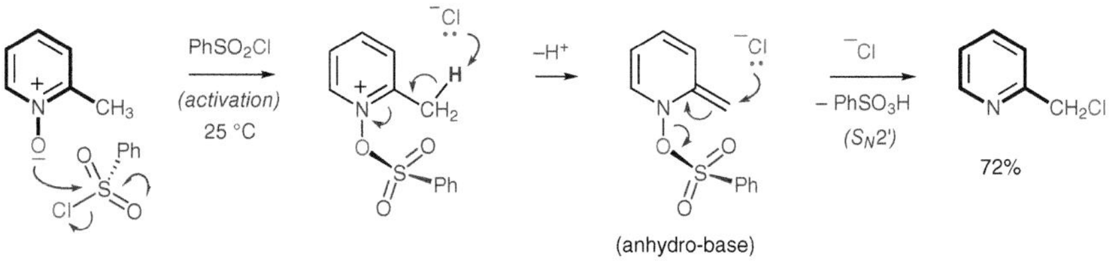

Figure 7.14 Synthesis of 2-chloromethylpyridine.

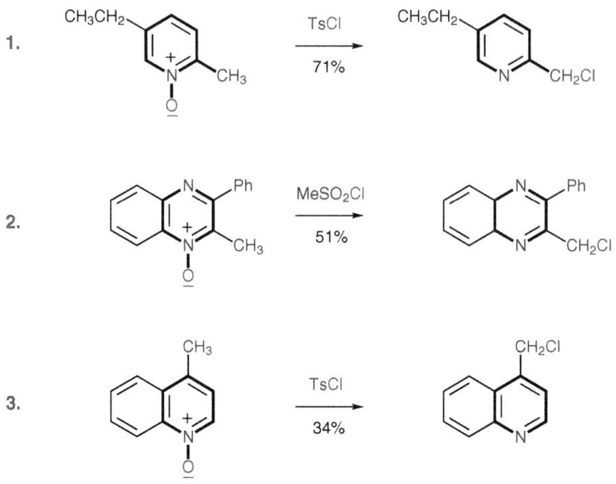

Figure 7.15 Further reactions with sulfonyl chlorides.

Similarly, 2-methyl-5-ethylpyridine N-oxide reacts with *p*-toluenesulfonyl chloride (TsCl) to afford a 71% yield of the corresponding 2-chloromethyl derivative (equation 1 in Figure 7.15) [22]. It is noteworthy that no chlorination is observed on the β-ethyl group. Also, the quinoxaline N-oxide in equation 2 is converted by methanesulfonyl chloride to 2-chloromethyl-5-phenylquinoxaline [23]. And finally, in an interesting example of γ-selectivity, 4-methylquinoline N-oxide is converted in 34% yield at room temperature to 4-chloromethylquinoline (equation 3) [24]. The relatively modest yield in this last reaction may have more to do with product stability than chemoselectivity.

Are there ever circumstances where it might be beneficial to carry out consecutive N-oxide rearrangements? The answer not surprisingly is yes. As one example, consider the case where we have a 2-methylpyridine derivative that we wish to oxidize to the corresponding aldehyde (Figure 7.16). The classic means of accomplishing this would involve SeO$_2$, or perhaps ceric ammonium nitrate. But what if other functional groups G present on the ring were not compatible with such powerful oxidizing agents. Do we have any recourse? As a start we could employ a Boekelheide rearrangement to prepare the corresponding 2-acetoxymethyl derivative, which is one oxidation state higher than the starting material. The first step in this sequence would involve oxidation of the pyridine ring to the corresponding N-oxide, employing the very mild oxidizing agent RCO$_3$H. Subsequent treatment with acetic anhydride would then bring about acetoxy group migration, in what we predict would be a completely chemoselective manner (cf. also Figure 7.12). Now comes the interesting part, though, where we re-oxidize the pyridine ring and carry out a second reaction with Ac$_2$O. To where does the second acetoxy group migrate? It turns out that rearrangement takes place to the same carbon, generating a

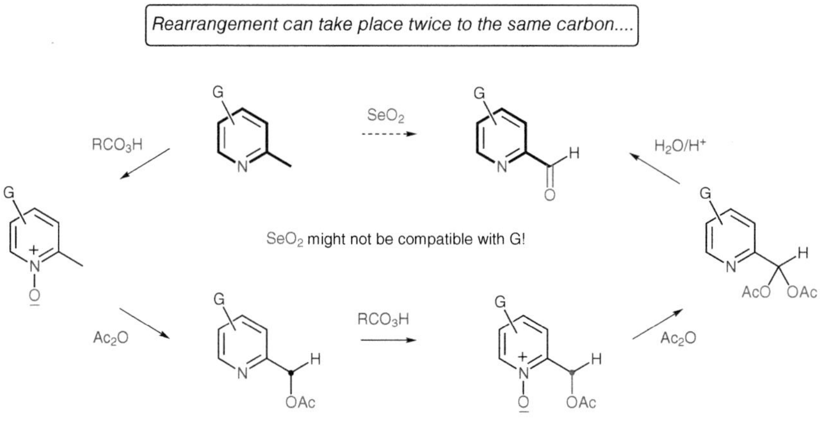

Figure 7.16 Multiple Boekelheide rearrangements.

diacetoxymethyl substituent [16]. Simple aqueous hydrolysis then completes the desired transformation. The net result is that we have gone from the oxidation state of a methyl group to the oxidation state of an aldehyde under very mild conditions.

A logical question, then, is what happens if alkyl groups occupy both α-positions? What pattern will we observe upon consecutive Boekelheide rearrangement? The experimental outcome is that an acetoxy group is introduced sequentially to two different carbons [16]:

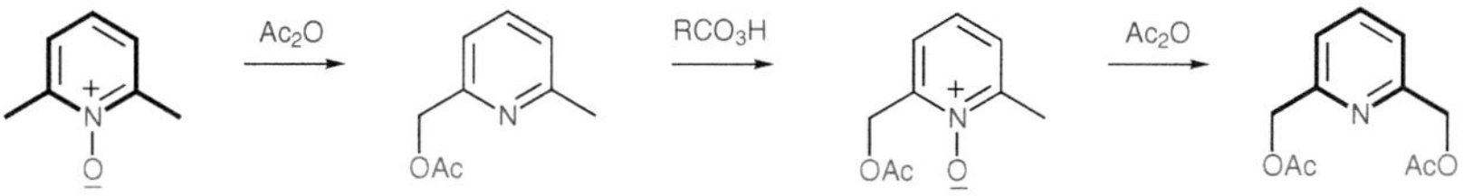

For reference, the "rules" for Ac_2O rearrangements are summarized in Figure 7.17. In the Katada rearrangement there are no α- or γ-alkyl groups, so the product is a 2-acetoxypyridine (equation 1). Alkyl substituents in the β-position are not affected, and hydrolysis yields the corresponding 2-pyridone. In the Boekelheide rearrangement, there must be at least one alkyl group in either the α- or γ-position. If only one α-alkyl group is present, then the first rearrangement will give a 2-acetoxyalkylpyridine (equation 2). For the case where R=H, a second migration will also take place to the same carbon, affording a 2-diacetoxymethylpyridine. Hydrolysis would then give the corresponding aldehyde. Finally, as just discussed, if there are two α-alkyl groups, rearrangement will take place sequentially to each α-position (equation 3).

Not long after its discovery, the Boekelheide rearrangement was put to good use by Büchi et al. in a novel synthesis of muscopyridine [25], an unusual

Katada Rearrangement

1. *If there are no α- or γ-alkyl groups, the entering nucleophile will prefer the α-ring position:*

Boekelheide Rearrangements

2. *If there is one α-alkyl group, the 1st rearrangement will involve the methylene group.*
For R = H, the 2nd rearrangement will also involve the methylene group:

3. *If there are two α-alkyl groups, rearrangement will take place sequentially at each position:*

Figure 7.17 Rules for Ac₂O arrangements.

metapyridinophane isolated in small quantities from glandular secretions of the Asian musk deer:

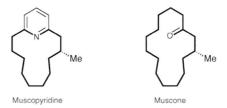

Muscopyridine   Muscone

The major component of musk oil is the macrocyclic ketone muscone, and mixtures of these compounds have been used in perfumery and medicine for thousands of years. However, since obtaining natural musk often involves sacrificing an endangered species, nearly all of the odoriferous principles used in perfumery today are synthetic.

The Büchi synthesis proceeded through the intermediacy of nor-muscopyridine, and required introduction of a methyl group at a position two bonds removed from the pyridine ring juncture (Scheme 7.2) [25]. A timely solution to this challenge was now available, involving initial oxidation of the pyridine ring to the corresponding N-oxide utilizing peracetic acid (~90% yield

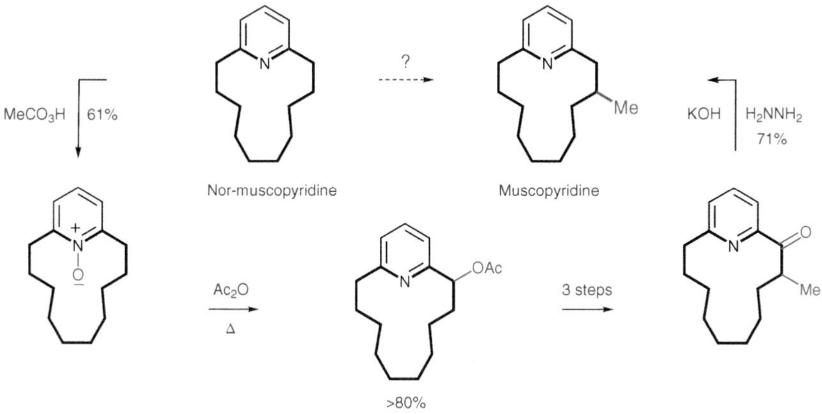

Scheme 7.2 Application of the Beokelheide rearrangement to the synthesis of muscopyridine.

based upon recovered starting material). This last material then underwent a smooth Boekelheide rearrangement upon brief heating in acetic anhydride, providing the anticipated alkyl ring acetoxy derivative in >80% yield. Hydrolysis, followed by alcohol oxidation and methylation then set the stage for a final Wolff-Kishner reduction, which afforded *d,l*-muscopyridine in 71% yield.

We have now covered quite a bit of ground relating to the chemical reactivity of heterocyclic N-oxides. However, before leaving this topic, let us briefly explore how these species might be generated under non-oxidative conditions. Typically this involves cyclization methodology, and again, the drive toward aromaticity is never to be underestimated.

For example, we generally do not think of nitro groups as being very good electrophiles. However, in the presence of sufficiently strong nucleophiles, intramolecular condensations can occur. Such is the case with the biphenyl derivative shown at the top of Figure 7.18, in which an ester enolate anion undergoes reversible nucleophilic addition to a proximal nitro group. The equilibrium for this process likely lies to the left, but that is not so important as the step that follows. Thus, we now have an energetically favorable pathway to an aromatic N-oxide, which is formed in 53% yield by dehydration [26].

Nitro groups also play a prominent role in a second cyclization strategy leading to N-oxides and related species, but this time as precursors to hydroxylamines. One example is shown at the bottom of Figure 7.18, in which the illustrated nitro ester is subjected to reduction employing the reagent combination Pd/C/NaBH$_4$. Under normal circumstances, one might expect this reaction to proceed rapidly to the amine oxidation level. However, intramolecular acylation intercedes, and the hydroxylamine group is trapped in situ to produce a species known as a hydroxamic acid (59% yield) [27]. The structure shown is the most stable tautomer of a 2-hydroxy N-oxide.

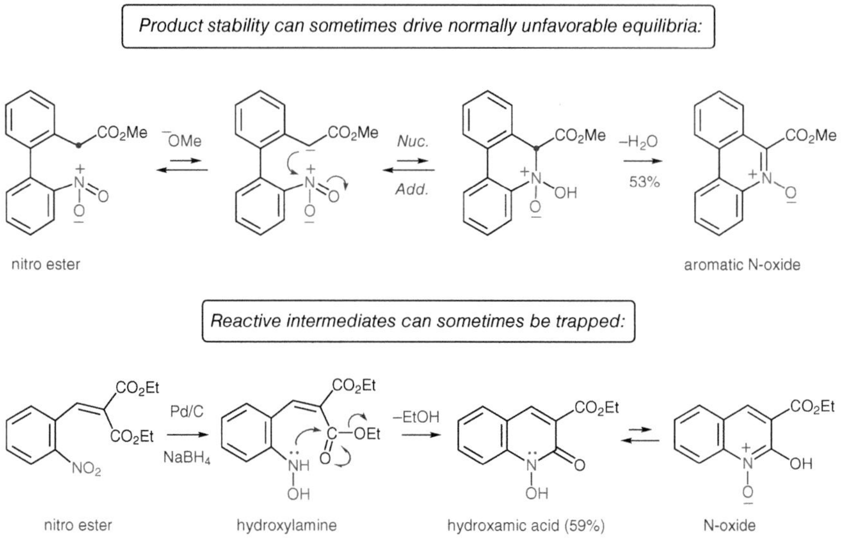

Figure 7.18 Cycloaddition routes to N-oxides.

Finally, we close this chapter with an elegant solution to what had been a long-standing problem in pteridine chemistry, that of preparing biologically important 6-substituted derivatives in isomerically pure form. Historically, the classic route to these ring systems has been the Isay synthesis, in which a pyrazine ring is fused onto an existing pyrimidine by bis-imine formation [28]:

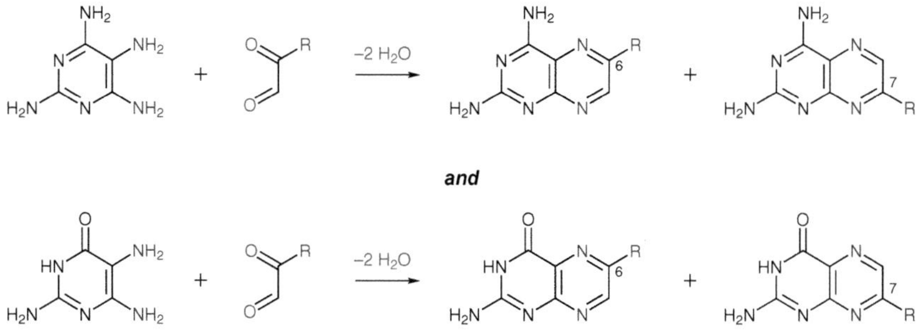

Isay Synthesis

Yields for such reactions are generally high, but this approach has the inherent flaw of producing nearly 50:50 mixtures of 6- and 7-substituted derivatives. Moreover, such mixtures are difficultly separable.

Taylor et al. took a fundamentally different approach, developing unequivocal syntheses of 5-substituted pyrazine 1-oxides, which were suitably functionalized for annulation of a pyrimidine ring. One variant involved condensation of aminomalononitrile with an α-oximinoketone, which we can safely assume

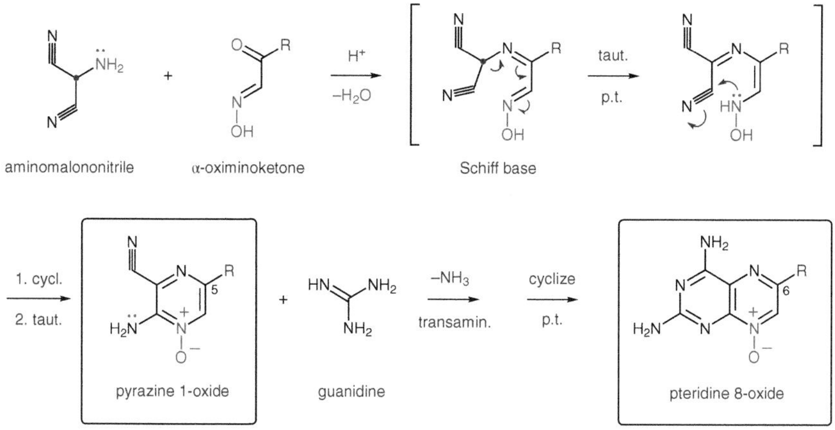

Figure 7.19 Cycloaddition routes to N-oxides.

forms the Schiff base shown in brackets (Figure 7.19) [29]. But where from here? Can this molecule find a route forward to produce an aromatic ring? Clearly yes, since the illustrated pyrazine 1-oxides are formed in uniformly high yields. A reasonable mechanistic pathway involves initial tautomerization of the oxime to the corresponding hydroxylamine, which is ideally situated for intramolecular cyclization across the nitrile group. In so doing, aromatization is achieved, and we also produce a 2-amino group in an *ortho*-relationship to a ring nitrile. We know from our adenine synthesis in Scheme 1.3 that this is a convenient handle for appending a pyrimidine ring, in this case employing guanidine. Thus, transamination, followed by cyclization and tautomerization affords the corresponding pteridine 8-oxides in excellent yield, and as isomerically pure materials [29].

From this point two options are available. The 8-oxide group is readily reduced to afford the parent pteridine ring system, providing access to a wide range of analogs of the potent anti-cancer drug methotrexate (cf. also Figure 1.4). Alternatively, this functionality can be exploited for introducing additional substituents into the 7-position (vide supra):

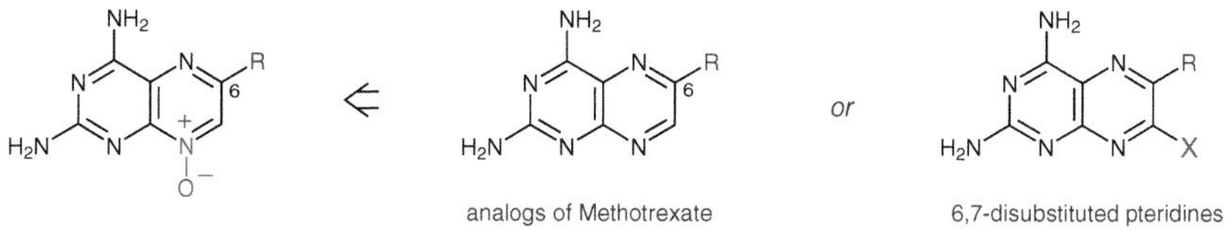

analogs of Methotrexate   or   6,7-disubstituted pteridines

In analogous fashion, but using ethyl aminocyanoacetate, condensation with α-oximinoketones affords the corresponding 2-amino-3-carboethoxypyrazine 1-oxides, which on cyclization with guanidine provide unequivocal access to 6-substituted pterin 8-oxides (Figure 7.20) [30]. Once again, reduction

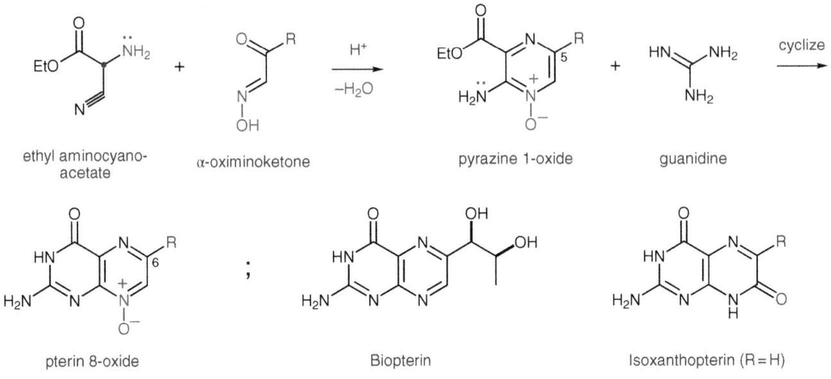

Figure 7.20 Cycloaddition routes to N-oxides.

provides the parent pterin ring system, which is widespread in Nature. This methodology has been employed in synthesizing numerous natural products, including the hydroxylase enzyme cofactor biopterin [31], and, by intermediate Katada rearrangement, the ubiquitous insect pigment isoxanthopterin [32].

Problems for Practice [33]

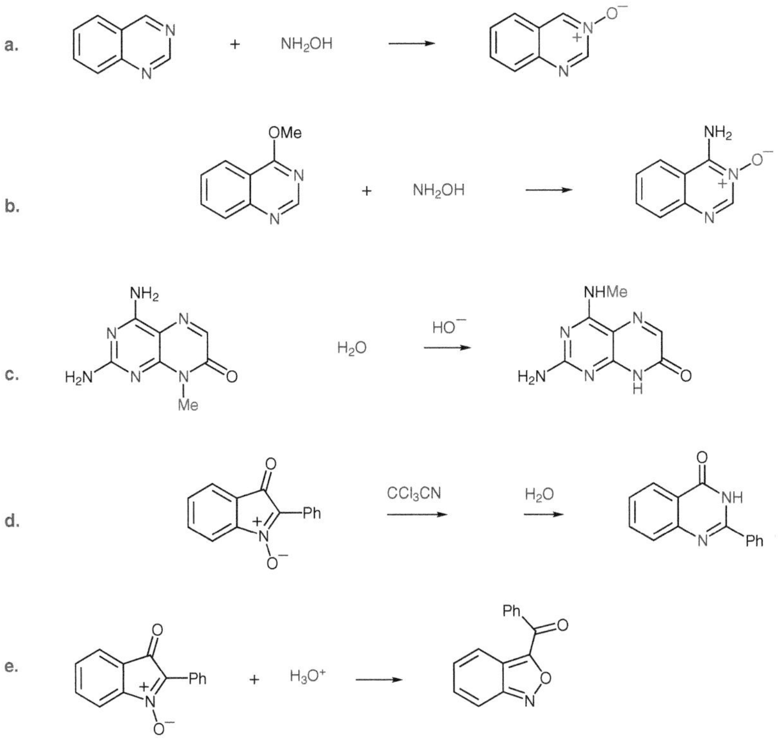

References

1 For excellent reviews of this area, see Katritzky, A. R.; Lagowski, J. M., *Chemistry of the Heterocyclic N-Oxides*, Academic Press, London, **1971**; Ochiai. E., *Aromatic Amine Oxides*, Elsevier, New York, **1967**.
2 Meisenheimer, J. *Chem. Ber.* **1926**, *59*, 1848–1853.
3 (a)Ochiai, E.; Ishikawa, M. O. *Proc. Japan. Acad.* **1942**, *18*, 561. For an English summary, see(b) Ochiai, E. *J. Org. Chem.* **1953**, *18*, 534–551.
4 Den Hertog, H. J.; Overhoff, J. *Rec. Trav. Chim.* **1950**, *69*, 468–473.
5 Van Bergen, T. J.; Kellogg, R. M. *J. Org. Chem.* **1971**, *36*, 1705–1708.
6 Webb, T. R. *Tetrahedron Lett.* **1985**, *26*, 3191–3194.
7 Linton, E. P. *J. Am. Chem. Soc.* **1940**, *62*, 1945–1948.
8 Chary, K. P.; Mohan, G. H.; Iyengar, D. S. *Chem. Lett.* **1999**, *28*, 1339–1340.
9 Den Hertog, H. J.; Broekman, F. W.; Combé, W. P. *Rec. Trav. Chim.* **1951**, *70*, 105–111.
10 Londregan, A. T.; Jennings, S.; Wei, L. *Org. Lett.* **2011**, *13*, 1840–1843.
11 Vorbrüggen, H.; Krolikiewicz, K. *Synthesis*, **1983**, 316–319.
12 Jung, J.-C.; Jung, Y.-J.; Park, O.-S. *Syn. Commun.* **2001**, *31*, 2507–2511.
13 Evans, T. J. *Am. Chem. Soc.* **1900**, *22*, 39–46.
14 Taylor, E. C.; Crovetti, A. J. *J. Org. Chem.* **1954**, *19*, 1633–1640.
15 Taylor, E. C.; Crovetti, A. J. *J. Am. Chem. Soc.* **1956**, *78*, 214–217.
16 Boekelheide, V.; Linn, W. J. *J. Am. Chem. Soc.* **1954**, *76*, 1286–1291.
17 Oae, S.; Kitao, T. Kitaoka, Y. *J. Am. Chem. Soc.* **1962**, *84*, 3359–3362.
18 Pacansku, J.; Brown, D.W. *J. Phys. Chem.* **1983**, *87*, 1553–1559.
19 Bodalski, R.; Katritzky, A. R. *Tetrahedron Lett.* **1968**, *9*, 257–260.
20 Oae, S.; Kozuka, S. *Tetrahedron* **1964**, *20*, 2671–2676.
21 Vozza, J. F. *J. Org. Chem.* **1962**, *27*, 3856–3860.
22 Matsumura, E.; Hirooka, T.; Imagawa, K. *Nippon Kagaku Zasshi* **1961**, *82*, 616–619.
23 Taylor, E. C.; Cheeseman, G. W. H. *J. Am. Chem. Soc.* **1964**, *86*, 1830–1835.
24 Tanida, H. *J. Pharm. Soc. Japan* **1958**, *78*, 611–613.
25 Biemann, K.; Büchi, G.; Walker, B. H. *J. Am. Chem. Soc.* **1957**, *79*, 5558–5564.
26 Muth, C. W.; Ellers, J. C.; Folmer, O. F. *J. Am. Chem. Soc.* **1957**, *79*, 6500–6504.
27 Coutts, R. T.; Wibberley, *D. G. J. Chem. Soc.* **1963**, 4610–4612.
28 Isay, O. *Chem. Ber.* **1906**, *39*, 250–265
29 Taylor, E. C.; Perlman, K. L.; Kim, Y. H.; Sword, I. P.; Jacobi, P. A. *J. Am. Chem. Soc.* **1973**, *95*, 6413–6418.
30 Taylor, E. C.; Perlman, K. L.; Sword, I. P.; Sequin-Frey, M.; Jacobi, P. A. *J. Am. Chem. Soc.* **1973**, *95*, 6407–6412.
31 Taylor, E. C.; Jacobi, P. A. *J. Am. Chem. Soc.* **1976**, *98*, 2301–2307.
32 Taylor, E. C.; Abdulla, R. F.; Tanaka, K.; Jacobi, P. A. *J. Org. Chem.* **1975**, *40*, 2341–2347.

33 Problems for practice 2: (a) Adachi, K. *Yakugaku Zasshi* **1957**, *77*, 507–510. (b) Adachi, K. *ibid.* **1957**, *77*, 510–513 (c) Taylor, E. C.; Abdulla, R. F.; Jacobi, P. A. *J. Org. Chem.* **1975**, *40*, 2336–2340. (d) Noland, W. E.; Modler, R. F. *J. Am. Chem. Soc.* **1964**, *86*, 2086–2087. (e) Pinkus, J. L.; Woodyard, G. G.; Cohen, T. *J. Org. Chem.* **1965**, *30*, 1104–1107.

8

π-Deficient Heterocycles: Introduction of New Substituents: Quinolines and Isoquinolines

Quinoline and isoquinoline undergo all of the functionalization reactions discussed in Chapters 6 and 7, but they also have special properties that merit closer attention. First, though, a little background information is in order.

In 1898, Claisen described a new synthesis of benzoyl cyanide, in which ethereal solutions of benzoyl chloride were treated with anhydrous HCN and pyridine (Figure 8.1) [1]. No mention is made of the role of pyridine, other than that pyridine hydrochloride is formed as a byproduct. However, we now know that pyridine is not an idle spectator in such transformations. Rather, it undergoes rapid reaction with benzoyl chloride to form the pyridinium salt shown in the brackets, which is relatively stable in the absence of nucleophiles. On the other hand, when called upon this species is a powerful benzoylating agent, since the leaving group is a neutral molecule of pyridine. Thus, in the presence of cyanide, it undergoes rapid acyl substitution to afford benzoyl cyanide.

Based on this observation, one might expect quinoline and isoquinoline to behave in the same manner, but we would be only partly correct. Yes, quinoline also undergoes rapid reaction with benzoyl chloride, generating benzoyl quinolinium chloride (Figure 8.2). However, the fate of this salt with cyanide anion is entirely different. Little or no benzoyl cyanide is observed by direct attack at the benzoyl carbonyl group. Instead, as first shown by Reissert [2], cyanide undergoes nucleophilic addition to C-2 of the quinolinium nucleus to generate a stable, crystalline adduct (a so-called Reissert compound). It is noteworthy that the same product is formed in 96% yield utilizing anhydrous HCN as the cyanide source [3]. These are essentially the conditions employed by Claisen in his synthesis of benzoyl cyanide (vide supra), but with quinoline substituting for pyridine.

As we shall see, Reissert compounds undergo many interesting transformations [4], but for the moment let us focus on the C-2 hydrogen depicted in bold in Figure 8.2. In the parent quinoline ring this hydrogen has only very modest acidity, and direct lithiation is not practical. However, H-2 in the corresponding Reissert derivative is readily abstractable by alkyl- and aryllithiums.

Introductory Heterocyclic Chemistry, First Edition. Peter A. Jacobi.
© 2019 John Wiley & Sons Ltd. Published 2019 by John Wiley & Sons Ltd.

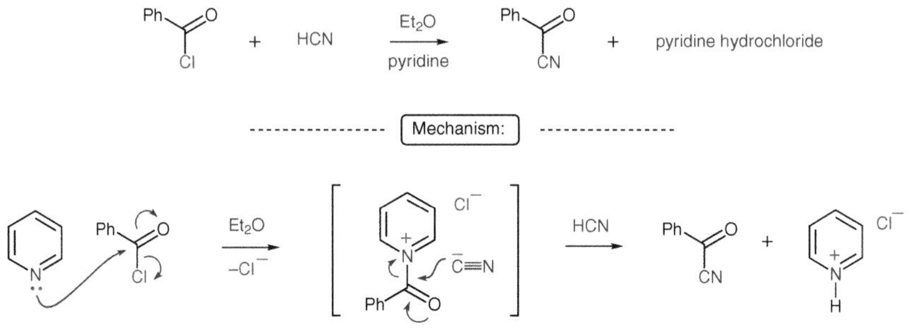

Figure 8.1 Claisen's synthesis of benzoyl cyanide.

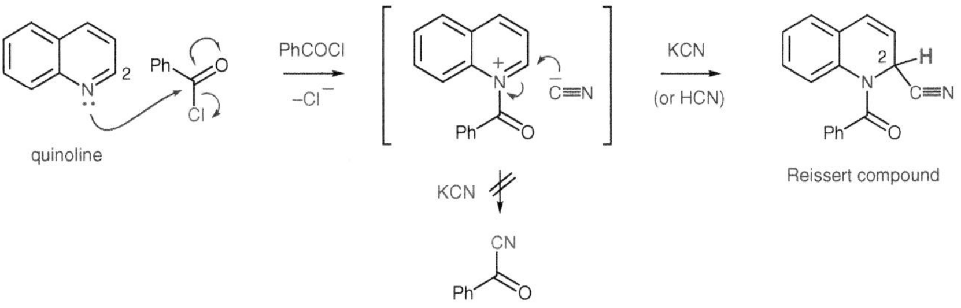

Figure 8.2 Formation of a Reissert compound from quinoline.

Why should this be the case? There are several effects operable here, but probably the most important is the fact that the newly introduced nitrile group stabilizes a developing anion by resonance. What is more, the conjugate base is flanked by both a double bond and the inductively electron withdrawing ring nitrogen.

Isoquinoline also affords a stable, relatively acidic Reissert derivative, this time by nucleophilic addition of cyanide to C-1 (Figure 8.3; the acidic proton is highlighted in bold). However, it is worth emphasizing that pyridine does not react in this fashion [4]. The energy cost in lost aromatic stabilization is simply too high, and the more favorable path is benzoylation (cf. Figure 8.1).

With this as introduction, let us explore some of the chemistry of these species, starting with perhaps the simplest transformation of all. This involves reversal of their formation, which is readily brought about with ethanolic KOH:

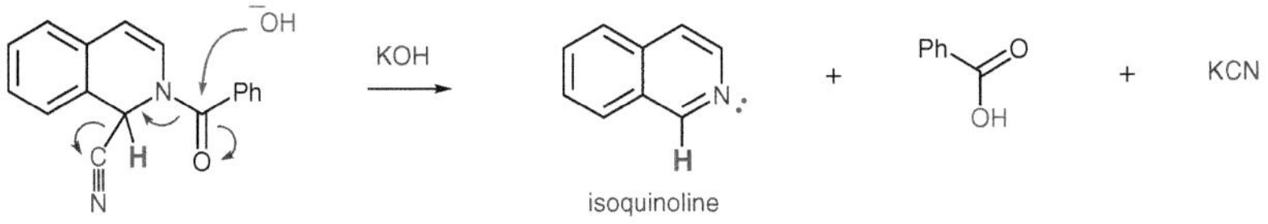

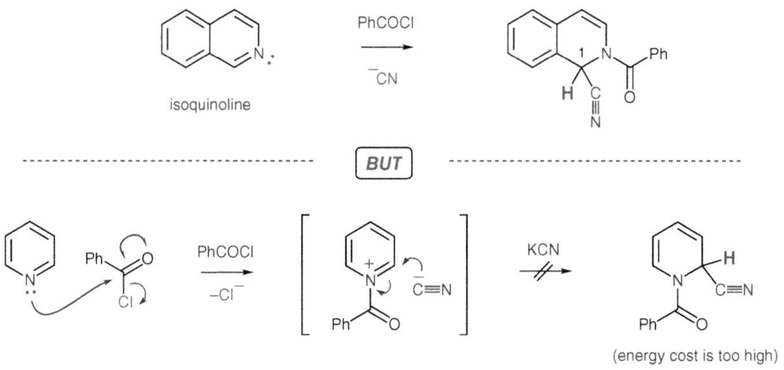

Figure 8.3 Pyridine does not form a Reissert compound.

The mechanism for this process is straightforward, involving initial attack by hydroxide at the benzoyl carbonyl group. Cleavage of the amide then provides an electron pair that can be used for ejecting cyanide anion. But why should we wish to carry out such a transformation, in which the net result has been to hydrolyze benzoyl chloride to benzoic acid, with return of our starting isoquinoline? Clearly there are more efficient (and less expensive!) ways to accomplish this.

The answer to the above question has all to do with timing. Suppose, for example, we wished to introduce an alkyl group at C-1 in isoquinoline. In principle this might be achieved by direct α-lithiation, followed by trapping with a suitable alkyl halide. However, this route is complicated by the fact that alkyl- and aryllithiums tend to add to the ring of π-deficient heterocycles, rather than abstract a proton (cf. Figure 6.1). Also, alternative bases such as lithium diisopropylamide (LDA) produce rapid dimerization. In no case was it possible to trap the presumed 2-lithio derivative [5]:

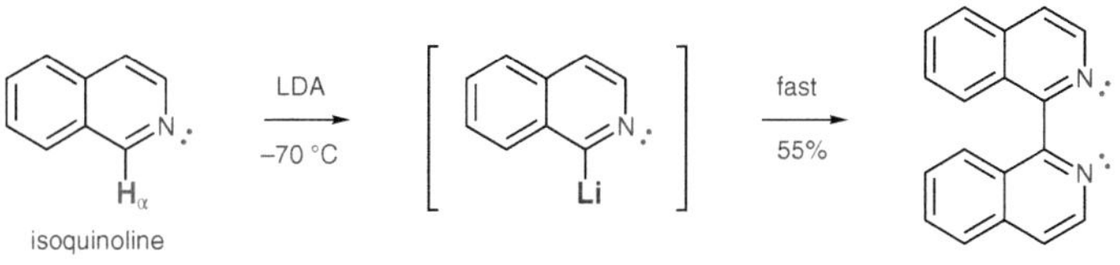

So what options are left? Definitely not Friedel-Crafts chemistry (top Figure 8.4). But by now, the obvious solution is to convert isoquinoline to the corresponding Reissert derivative, which is readily de-protonated to afford the stabilized anion shown in the brackets (bottom Figure 8.4). Alkylation with a variety of primary alkyl halides then yields exclusively the 1-substituted derivatives. Lastly comes the step involving regeneration of the aromatic

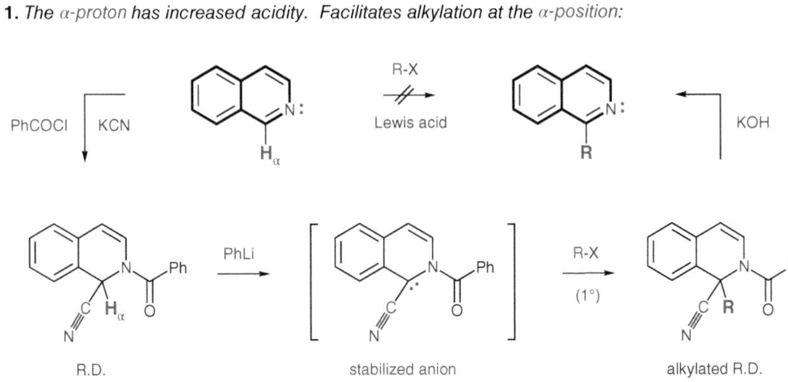
1. The α-proton has increased acidity. Facilitates alkylation at the α-position:
R-X
Lewis acid
PhCOCl
KCN
KOH
PhLi
R-X
(1°)
R.D.
stabilized anion
alkylated R.D.

Figure 8.4 Chemistry of Reissert derivatives.

isoquinoline ring, which is efficiently effected with ethanolic KOH (vide supra) [4].

Among numerous other examples, this strategy was prominently featured in a novel synthesis of papaverine [6], an isoquinoline alkaloid having potent antispasmodic properties:

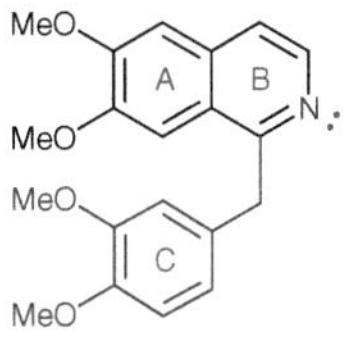
MeO
MeO
MeO
MeO
A
B
C

Papaverine

Papaverine was initially isolated in 1848 from the opium poppy [7], the same species that produces the morphine class of analgesic alkaloids. However, the structure and pharmacological properties of this material are very different, and it is currently used as a smooth muscle relaxant in vascular surgery.

There were two stages to the Reissert-based synthesis of papaverine by Popp and McEwen [6], the first being preparation of the requisite A,B-ring starting material, 6,7-dimethoxyisoquinoline (Scheme 8.1). This was accomplished in excellent overall yield beginning with commercially available homoveratrylamine, which underwent clean formylation, and Bischler-Napieralski cyclization, to afford a 97% overall yield of 3,4-dihydro-6,7-dimethoxyisoquinoline. This last material then gave an 82% yield of 6,7-dimethoxyisoquinoline upon catalytic dehydrogenation with Pd/C.

Preparation of 6,7-dimethoxyisoquinoline

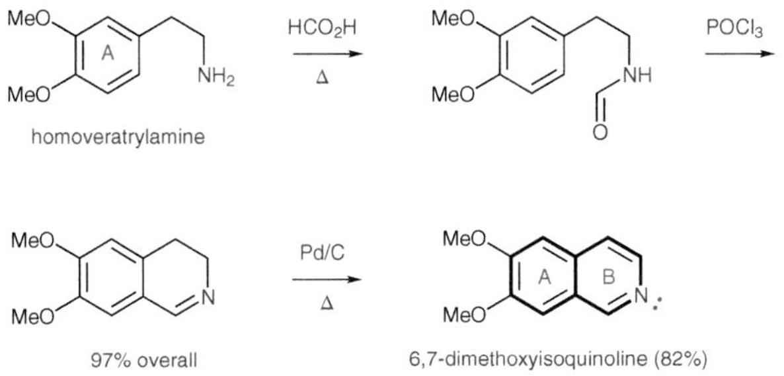

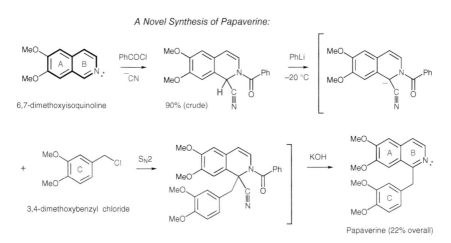

Scheme 8.1

With the A,B-ring portion complete, the corresponding Reissert derivative was formed in ~90% yield utilizing benzoyl chloride and potassium cyanide, although this product was somewhat labile (Scheme 8.2). For example, crystallization from ethanol caused significant reversion to starting material, presumably accelerated by the electron donating properties of the two methoxy groups. Nevertheless, by working at low temperatures, there was obtained moderate overall yields of papaverine from the purified Reissert compound. This involved lithiation, followed by alkylation with 3,4-dimethoxybenzyl chloride, and finally, cleavage of the benzoyl and cyanide groups with ethanolic KOH.

A Novel Synthesis of Papaverine:

Scheme 8.2 Aklylation of Reissert derivatives.

Building upon this methodology, Neumeyer et al. devised a concise synthesis of aporphine [8], the parent skeleton of a large family of tetrahydroisoquinoline alkaloids:

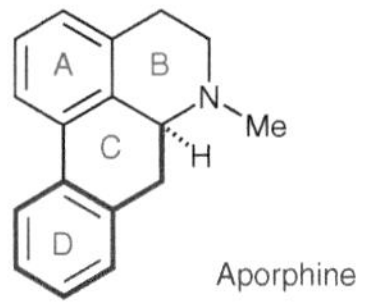

Aporphine

Once again the literature synthesis was divided into roughly two stages, beginning with the well-known Reissert derivative of isoquinoline (Scheme 8.3). In this case, it proved advantageous to deprotonate the Reissert compound utilizing NaH in DMF, generating the sodium salt shown in brackets. This species turned out to be far more reactive than the corresponding lithium derivative, although solvent effects may play an important role here as well. In any event, quenching of the sodium salt with *o*-chloromethylnitrobenzene afforded an 80% yield of the product of S_N2 alkylation, which with ethanolic KOH gave a 73% yield of 1-(2-nitrobenzyl)isoquinoline. This three step sequence thereby provided an ~60% overall yield of an intermediate possessing the A, B, and D-rings of aporphine. However, there was still work to be done.

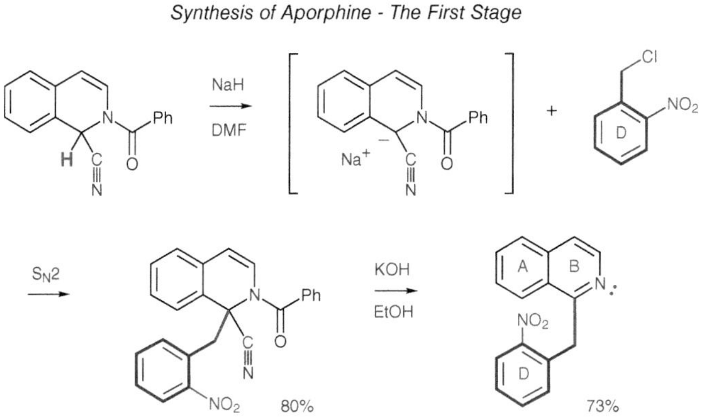

Synthesis of Aporphine - The First Stage

Scheme 8.3 Aklylation of Reissert derivatives.

In the second stage it was necessary to address three objectives, consisting of N-methylation, reduction of ring B to the tetrahydroisoquinoline oxidation state, and finally, construction of ring C. The only question was in what order to carry out these steps. The authors chose to introduce the N-methyl group first, which was accomplished in 87% yield by alkylation with methyl iodide (Scheme 8.4). This was a good strategic decision on two counts. First,

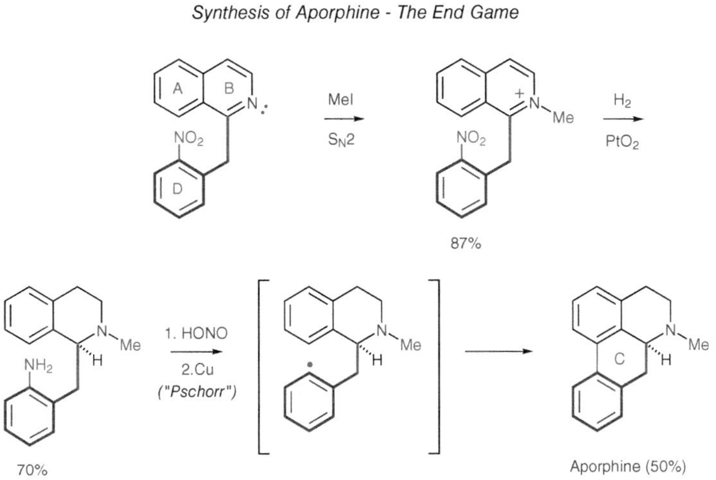

Synthesis of Aporphine - The End Game

Scheme 8.4 Aporphine - the final steps.

the free electron pair on the isoquinoline ring is the only nucleophilic site at this point in the synthesis; and second, formation of the methiodide salt renders ring B more susceptible to catalytic hydrogenation. This was effected in 70% yield utilizing H_2/PtO_2, with concomitant reduction of the nitro group to the corresponding amine. Finally, the crucial juncture between rings A and D leading to aporphine was established utilizing a Pschorr cyclization (50% yield). While the mechanism for this last transformation is complex, it is generally believed to proceed via a phenyl radical intermediate, formed by Cu-catalyzed decomposition of the corresponding diazonium salt [9].

In many ways the chemistry of Reissert anions parallels that of other active methylene compounds, and is not limited to S_N2 alkylations. Thus, Reissert compounds derived from both quinoline and isoquinoline undergo base-mediated condensation with a wide range of aldehydes and ketones, although the initial adducts are generally not isolable (Figure 8.5) [10]. Rather, as in the example shown, the alkoxy anion initiates a benzoyl migration from nitrogen to oxygen, providing a ready pathway to re-aromatization. The resultant benzoate esters are then typically hydrolyzed with KOH to provide the corresponding alcohols.

This chemistry was showcased in an efficient synthesis of papaverinol [6], beginning with the Reissert compound derived from 6,7-dimethoxyisoquinoline (Scheme 8.5). Lithiation of this compound with PhLi at -20 degrees Celsius led to smooth formation of the corresponding lithio derivative, which was trapped in situ with veratraldehyde. Upon warming to room temperature, the aldehyde adduct underwent clean rearrangement to form papaverinol

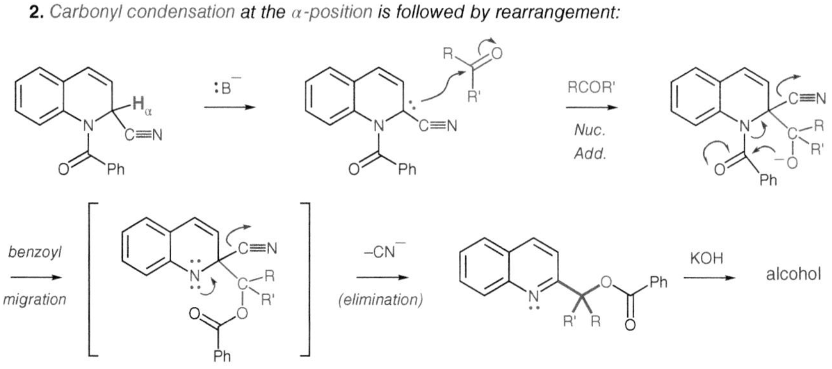

2. *Carbonyl condensation at the α-position is followed by rearrangement:*

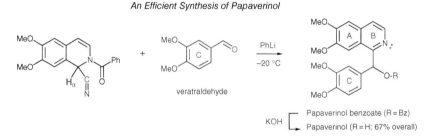

Figure 8.5 Chemistry of Reissert derivatives.

An Efficient Synthesis of Papaverinol

veratraldehyde

Papaverinol benzoate (R = Bz)

KOH

Papaverinol (R = H; 67% overall)

Scheme 8.5 Chemistry of Reissert derivatives.

benzoate, via the mechanism outlined in Figure 8.5. Hydrolysis with KOH then afforded a 67% overall yield of the parent alcohol papaverinol.

Do there exist other pathways by which Reissert compounds might regain aromaticity? So far we have seen two, both of which involve cleavage of the amide functionality and ejection of cyanide (cf. Figures 8.4 and 8.5). A third means follows an even more direct path, in which the Reissert anion is simply allowed to warm to ambient temperature in the absence of added electrophiles (Figure 8.6) [11]. Under these conditions, the benzoyl group undergoes intramolecular transfer from N-1 to C-2, via the intermediacy of the 3-membered ring shown in brackets. Fragmentation, with loss of cyanide, completes the aromatization process.

This rearrangement was put to good use in preparing a key intermediate for a total synthesis of O-methyldauricine, an unusual bisbenzylisoquinoline alkaloid isolated from the bark of *Colubrina asiatica* (Scheme 8.6) [12]. The target compound was prepared in two steps beginning with 6,7-dimethoxyisoquinoline,

3. *With no added electrophiles, R.D. anions re-aromatize by benzoyl migration:*

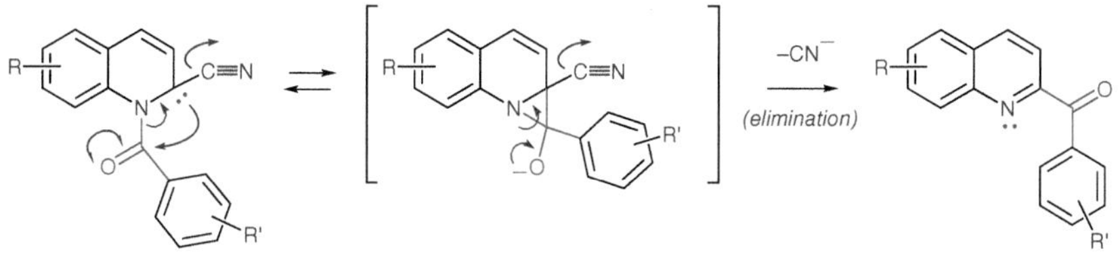

Figure 8.6 Chemistry of Reissert derivatives.

A Key Intermediate for O-Methyldauricine

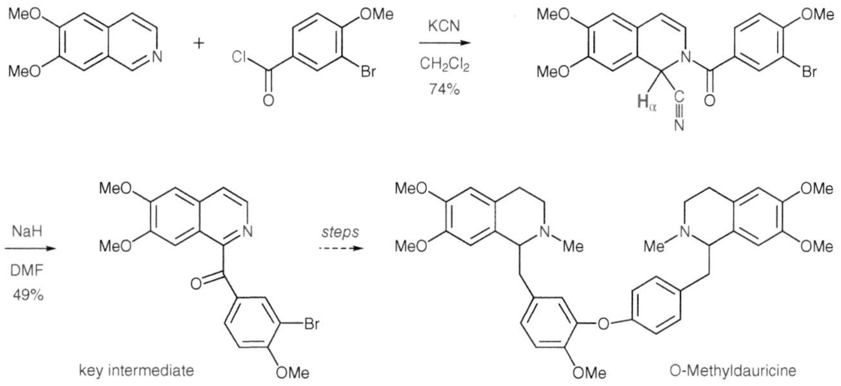

Scheme 8.6 Chemistry of Reissert derivatives.

via the Reissert derivative shown at the top right of the scheme. Rearrangement with NaH in DMF then afforded an ~50% yield of the desired highly substituted isoquinoline.

Finally, a remarkable transformation ensues upon acid hydrolysis of Reissert compounds, as first demonstrated for the adduct derived from quinoline and benzoyl chloride (Figure 8.7) [2]. One product from this reaction is quinaldic acid, in which the quinoline ring has found a pathway to aromaticity. Thus, from the heterocyclic chemist's point of view, we have at hand a convenient means of functionalizing quinoline at the 2-position. However, a second product is benzaldehyde, which is formed in essentially quantitative yield. From a different perspective, then, this reaction constitutes a selective means for reducing benzoyl chloride to benzaldehyde.

And how do we rationalize this reaction, which turns out to be quite general for a wide range of Reissert derivatives? In each case an acid chloride is reduced

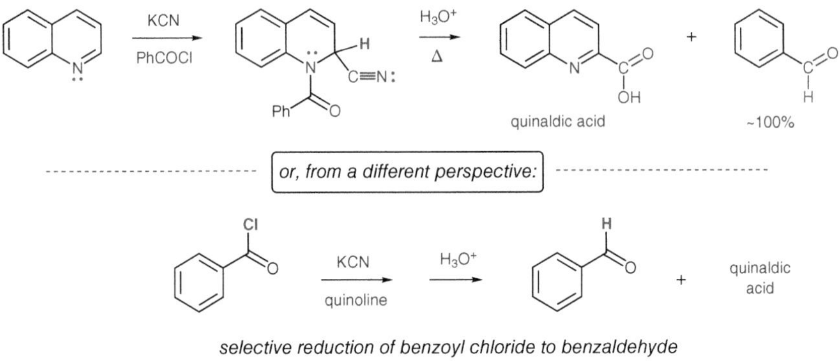
KCN
PhCOCl
H_3O^+
Δ
quinaldic acid
~100%
or, from a different perspective:
KCN
quinoline
H_3O^+
quinaldic acid
selective reduction of benzoyl chloride to benzaldehyde

Figure 8.7 A very selective reduction.

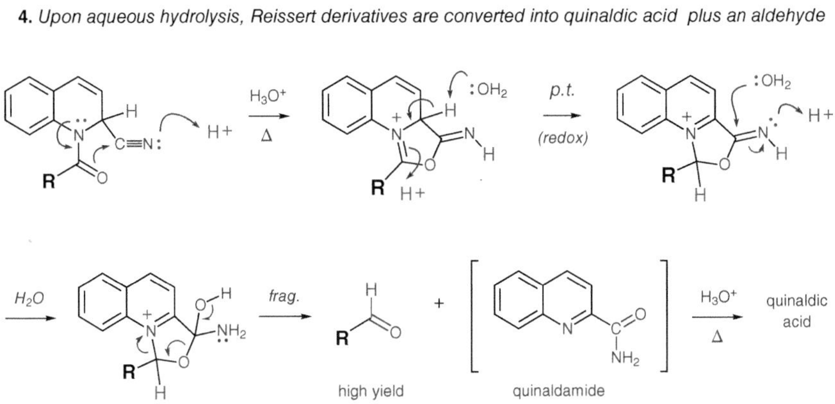
4. Upon aqueous hydrolysis, Reissert derivatives are converted into quinaldic acid plus an aldehyde
H_3O^+
Δ
p.t.
(redox)
H_2O
frag.
high yield
quinaldamide
H_3O^+
Δ
quinaldic acid

Figure 8.8 Chemistry of Reissert derivatives.

to an aldehyde, while a dihydroquinoline (or dihydroisoquinoline) undergoes oxidation (Figure 8.8). But yet, no reducing or oxidizing agents are present! Detailed mechanistic studies have shown that this transformation involves internal redox chemistry, initiated by protonation of the free electron pair on the nitrile group [13]. This provides the driving force for concomitant nucleophilic attack by the adjacent amide carbonyl group, generating the oxazoline ring shown in the top center of the figure. So far, there has been no change in oxidation state at any of the positions corresponding to our original substrate, but this is about to change. Thus, consider the consequences of a simple proton transfer (p.t.) between C-2 of the dihydroquinoline ring and the iminium carbon bearing the **R** group. We expect this transfer to be energetically favorable, since it re-establishes aromaticity in the quinoline ring. In the process, of course, C-2 has undergone oxidation, but if that is the case, what has been reduced? Look carefully at the carbon bonded to the **R** group, which is no longer sp^2-hybridized. By the simple act of adding a proton this carbon has gone from the oxidation state of a carboxylic acid to that of an aldehyde (top right of Figure 8.8). It remains now only to effect hydrolytic cleavage to generate the final aldehyde product, and this is accomplished via the orthoamide intermediate shown in the bottom left of the figure. As indicated, fragmentation of this species liberates the aldehyde in generally high yield, and at the same time produces quinaldamide. This last material can be isolated if reaction times are limited, but generally hydrolysis is continued to afford quinaldic acid [13].

Are there instances where this methodology might be advantageous over traditional approaches to reducing acid chlorides to aldehydes, including hydride reagents and catalytic hydrogenation with Pd/BaSO$_4$? At least three come to mind, not the least of which is that over-reduction is not possible. Also, reduction is completely chemoselective for acid chlorides. That is, other easily reducible groups such as nitro, alkenes or ketones are not affected (equations 1 and 2, Figure 8.9) [3,4]. Lastly, simply substituting D$_3$O$^+$ for H$_3$O$^+$ provides a convenient means for synthesizing specifically deuterium-labeled aldehydes (equation 3) [14].

Reissert derivatives thus have many potential applications, only a few of which have been highlighted in this chapter. For reference these are summarized in Figure 8.10 using isoquinoline as the parent ring. However, the analogous quinoline derivatives behave in identical fashion. Leading the way is the fact that the α-proton in Reissert derivatives is relatively acidic, allowing for facile abstraction with PhLi or NaH (Equation 1). The resultant anion can then be captured with a variety of electrophiles, including primary alkyl halides. In the case of alkylation, the benzoyl and cyanide

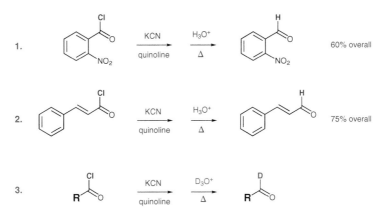

Figure 8.9 Some selective reductions employing Reissert derivatives.

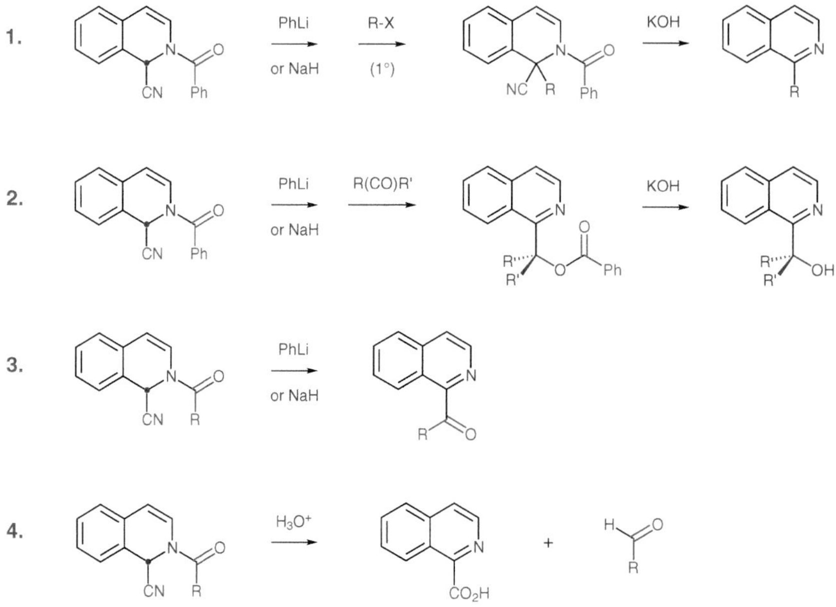

Figure 8.10 Chemistry of Reissert derivatives.

groups are subsequently cleaved with ethanolic KOH, regenerating the aromatic isoquinoline or quinoline ring (cf. Figure 8.4). In equation 2, the Reissert anion is condensed with aldehydes or ketones, producing an intermediate alkoxy anion. Recall that these anions undergo spontaneous benzoyl group migration with concomitant ejection of cyanide to generate benzoyl esters. Hydrolysis with KOH then liberates the corresponding alcohols (cf. Figure 8.5). And what transpires in the absence of external electrophiles (Equation 3)? In that case we have intramolecular transfer of the N-acyl group to the α-position, coupled with re-aromatization (cf. Figure 8.6). Finally, as just described, acid hydrolysis affords isoquinaldic acid, plus the aldehyde corresponding to reduction of the starting acid chloride (Equation 4; cf. also Figure 8.8). With D_3O^+ the corresponding deuterated aldehydes are obtained.

References

1 Claisen, L. *Chem. Ber.* **1898**, *31*, 1023–1024.
2 Reissert, A. *Chem. Ber.* **1905**, *38*, 1603–1614.
3 Grosheintz, J. M.; Fischer, H. O. L. *J. Am. Chem. Soc.* **1941**, *63*, 2021–2022.
4 MeEwen, W. E.; Cobb, R. L. *Chem. Rev.* **1955**, *55*, 511–549.

5 Clarke, A. J.; McNamara, S.; Meth-Cohn, O. *Tetrahedron Lett.* **1974**, *27*, 2373–2376.

6 Popp, F. D.; McEwen, W. E. *J. Am. Chem. Soc.* **1957**, *79*, 3773–3777.

7 Merck, G. *Annalen der Chemie und Pharmacie* **1848**, *66*, 125–128.

8 Neumeyer, J. L.; Oh, K. H.; Weinhardt, K. K.; Neustadt, B. R. *J. Org. Chem.* **1969**, *34*, 3786–3788.

9 Lewin, A. H.; Cohen, T. *J. Org. Chem.* **1967**, *32*, 3844–3850.

10 Walters, L. R.; Iver, N. T.; McEwen, W. E. *J. Am. Chem. Soc.* **1958**, *80*, 1177–1181.

11 Popp, F. D.; Wefer, J. M. *J. Het. Chem.* **1967**, *4*, 183–187.

12 Popp, F. D.; Gibson, H. W.; Noble, A. C. *J. Org. Chem.* **1966**, *31*, 2296–2299.

13 Cobb, R. L.; McEwen, W. E. *J. Am. Chem. Soc.* **1955**, *77*, 5142–1548.

14 (a) Wahren, M. *Abh. Dtsch. Akad. Wiss. Berlin, Kl. Chem., Geol. Biol.* **1965**, *7*, 687–692. See also (b) Wahren, M. *Chem. Abstr.* **1967**, *66*, 37577.

9

π-Deficient Heterocycles: Manipulation of Existing Substituents

We are now well along in our discussion of π-deficient heterocycles, having learned much about their properties, and having made significant progress in the category of synthesis. Indeed, we have completed two of the three sections we set out as our initial goal. These consisted of (1) "Preparation 'de Novo' with all substituents present" (Chapter 5), and (2) "Introduction of new substituents" (Chapters 6–7):

1. Preparation "de Novo" with all substituents present.
2. Introduction of new substituents.
3. **Manipulation of existing substituents.**

That brings us to the final section on synthesis, involving manipulation of existing substituents. Certain of these transformations have their counterparts in benzene chemistry, but most are unique to π-deficient heterocycles. Why should this be? For one thing, ring systems of this class are strongly electron withdrawing, meaning that substituents are subject to inductive polarizing effects. Also, resonance delocalization can come into play, particularly with groups in the α- or γ-ring positions.

As an example, consider the case of 2,5-dimethylpyridine shown in Figure 9.1. Both of these methyl groups are relatively electron deficient, a result of inductive electron withdraw. Because of this, we can expect enhanced acidity as compared to the methyl group in toluene. But the α-methyl group in position 2 is particularly acidic, undergoing rapid proton extraction with LDA at −60 degrees Celsius. This is due to favorable resonance stabilization of the conjugate base shown in brackets, which places a negative charge directly onto nitrogen. Note the similarity of this species to a ketone enolate. The regiochemical outcome of this deprotonation was proven by trapping with t-butyldimethylsilyl chloride (TBSCl), which afforded the mono-silylated C2-derivative as the only observed product [1]. In contrast, deprotonation at the C5 methyl group is very slow, since resonance stabilization will be relatively unimportant. These results are representative of a general phenomenon that alkyl groups attached to the α- or γ-ring positions have enhanced acidity.

Introductory Heterocyclic Chemistry, First Edition. Peter A. Jacobi.
© 2019 John Wiley & Sons Ltd. Published 2019 by John Wiley & Sons Ltd.

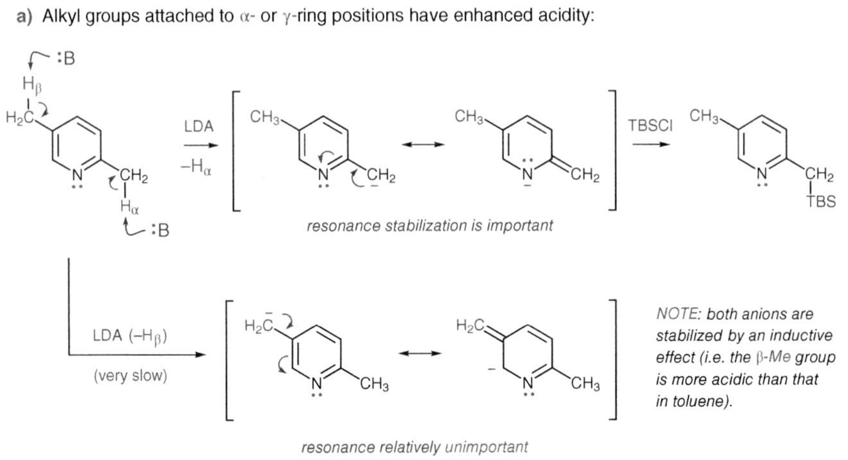
a) Alkyl groups attached to α- or γ-ring positions have enhanced acidity:
LDA
−Hα
resonance stabilization is important
TBSCl
LDA (−Hβ)
(very slow)
resonance relatively unimportant
NOTE: both anions are stabilized by an inductive effect (i.e. the β-Me group is more acidic than that in toluene).

Figure 9.1 Manipulation of existing substituents: acidic alkyl groups.

This principle is further illustrated with the anion derived from 2-methylpyridine (α-picoline), which undergoes capture with a wide range of electrophiles (Figure 9.2). These include ketones and aldehydes [2a,b], where the products of normal aldol-like condensations are observed, as well as typical

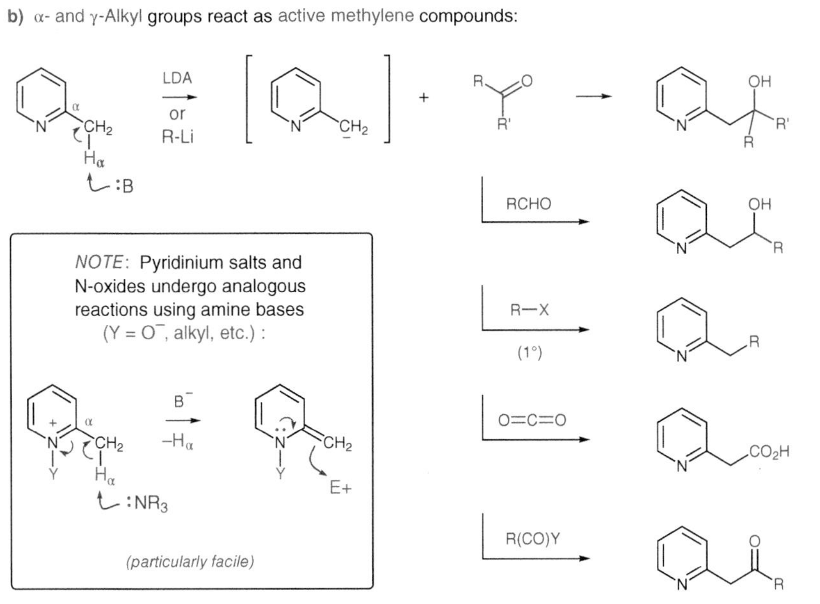
b) α- and γ-Alkyl groups react as active methylene compounds:
LDA
or
R-Li
RCHO
R—X
(1°)
O=C=O
R(CO)Y
NOTE: Pyridinium salts and N-oxides undergo analogous reactions using amine bases (Y = O⁻, alkyl, etc.) :
−Hα
E+
(particularly facile)

Figure 9.2 Manipulation of existing substituents: active methylene chemistry.

S_N2 alkylations with primary alkyl halides [2c]. Also, quenching with carbon dioxide affords 2-pyridylacetic acid [2d], and treatment with acylating agents gives the corresponding ketone derivatives [2e,f]. These last compounds are valuable synthetic intermediates in their own right.

Finally, as shown in the box, pyridinium salts and N-oxides undergo analogous reactions utilizing KOH or even amine bases, a consequence of the greatly increased acidity of the α-methyl substituent. In these examples, abstraction of a proton serves to neutralize the positive charge on the ring nitrogen, rendering the conjugate base particularly stable [3].

Included in "active methylene chemistry" are transformations that take place under neutral (or nearly so) conditions, such as the oxidative cleavage of alkyl substituents to afford carboxylic acids. This is illustrated at the top of Figure 9.3 for the case of 3,4-dimethylpyridine (3,4-lutidine), which affords 3,4-pyridinedicarboxylic acid upon reaction with the powerful oxidant $KMnO_4$ [4]. Comparable results are obtained with long chain alkyl groups. It goes without saying that "hard ball" chemistry of this type is only possible because of the great stability of the pyridine ring, and it is completely non-selective. That is, alkyl groups occupying any position on the ring are oxidized with equal facility. On the other hand, SeO_2 is selective for oxidizing methyl groups in the α- and γ-positions to the corresponding aldehydes [5a], or with excess reagent, carboxylic acids [5b].

What other types of transformations might be effected under roughly neutral conditions? Let us examine the reaction shown at the bottom of Figure 9.3, where we simply mix together 2-methylquinoline, aqueous

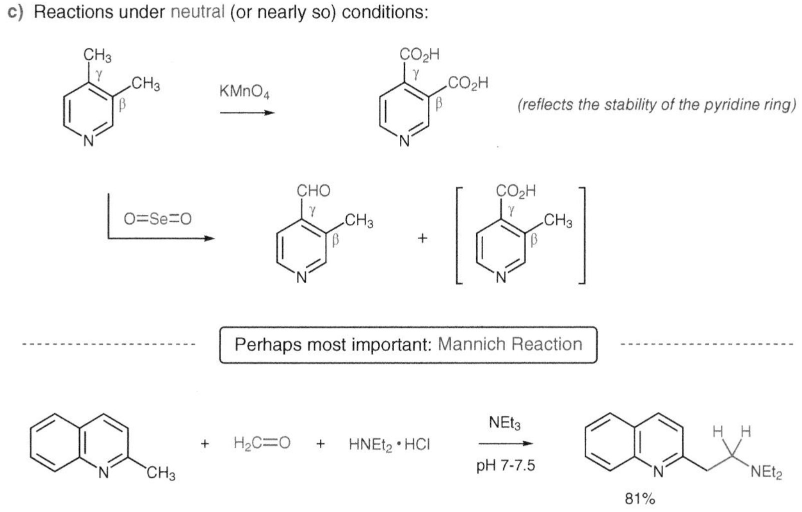

c) Reactions under neutral (or nearly so) conditions:

Figure 9.3 Active methylene chemistry.

formaldehyde and diethylamine hydrochloride, and adjust the pH to 7-7.5 with triethylamine. After two hours at 60 degrees Celsius one obtains a high yield of 2-(2-diethylaminoethyl)quinoline [6]. This is an example of a Mannich reaction, which in its many variants is one of the most important bond forming processes in alkaloid chemistry. Again, for π-deficient heterocycles, the Mannich reaction is specific to methyl groups that are either α or γ to a ring nitrogen.

Mechanistically, the key feature here is that we are generating both a moderately strong nucleophile and a potent electrophile under the same conditions (Figure 9.4). The nucleophilic component has the enamine-like structure shown at the top right of the figure, derived by isomerization of 2-methylquinoline. This interconversion finds direct analogy in keto-enol tautomerization. The equilibrium for this step undoubtedly lies far to the left, but that is not a problem, since we only require low concentrations of the enamine tautomer. The reason for this is because in the same vessel we have produced an exceedingly reactive iminium derivative, by condensation of formaldehyde with diethylamine. Nucleophilic capture of this species then affords the so-called Mannich base (bottom Figure 9.4).

Mannich bases of this type are of importance not only because we have extended a carbon chain, but also because they are excellent precursors to the corresponding vinyl substituents. For example, the closely related quinoline shown in Figure 9.5 undergoes facile quaternization, and Hofmann elimination, to afford an 81% yield of the 2-vinyl derivative [7]. This last material was then converted in several steps, including a Pd-catalyzed Heck coupling, to the potent LTD_4 receptor antagonist L-699,392.

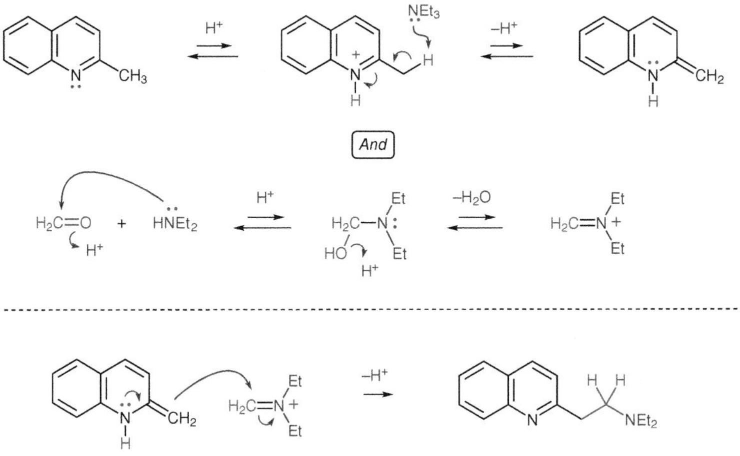

Figure 9.4 Mechanism of the Mannich reaction.

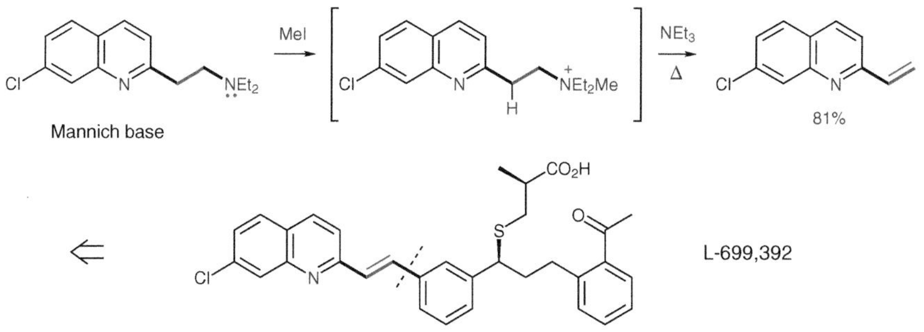

Figure 9.5 Synthesis of L-699,392.

Aside from their ready participation in Heck coupling chemistry, what other properties can we expect for vinyl groups attached to the α- and γ-positions of π-deficient heterocycles? As illustrated for 2-vinylpyridine, electrophilic addition is relatively unfavorable, since one is faced with the prospect of producing either a primary carbocation (path a), or one conjugated to a very electron withdrawing carbon-nitrogen double bond (path b):

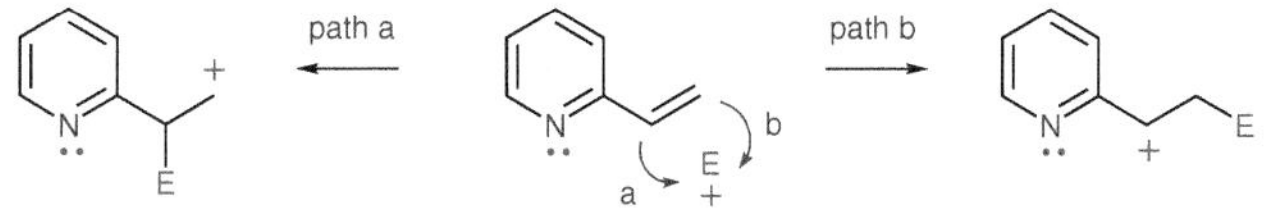

Also, of course, with many electrophiles, the initial site of attack will be the free electron pair on nitrogen.

In fact, such double bonds are very electron deficient, much akin to those found in α,β-unsaturated carbonyl compounds (Figure 9.6). Not surprisingly, then, 2-vinylpyridine, and related materials, undergo conjugate addition with a host of nucleophilic species, wherein the developing negative charge is delocalized directly onto the ring nitrogen [8]. Even secondary amines will add, to generate the corresponding Mannich base.

Be aware, though, that conjugate addition is limited to those vinyl groups that occupy α- or γ-positions. β-Vinyl groups, while still electron deficient, do not participate in this kind of chemistry. To convince yourself of this, try drawing a favorable intermediate for such a process (bottom Figure 9.6). No matter how many resonance structures you draw, the negative charge will never end up on nitrogen.

At this juncture, let us take a brief foray into an area of heterocyclic chemistry we have not yet explored, involving the concept of "latent functionality." Look up the adjective "latent" and you will likely find something along the lines of "potentially existing but not presently evident." As applied to functional groups, the term generally implies a masked equivalency. In any event, the timing for this discussion is appropriate, since pyridine derivatives have found

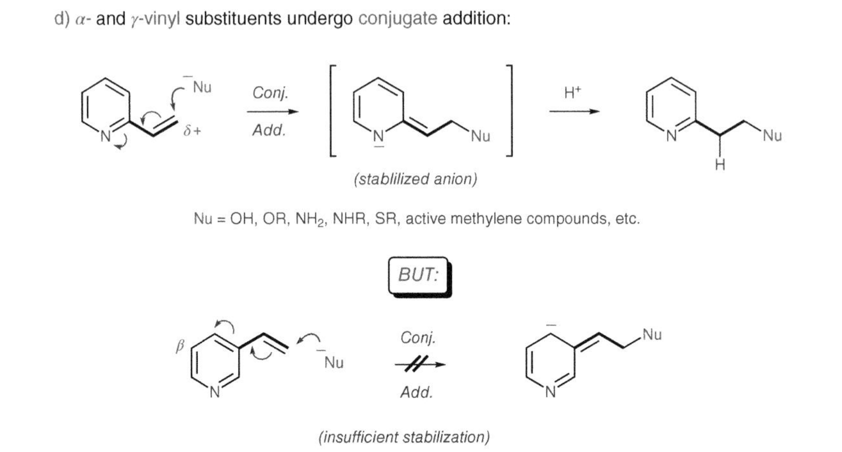

d) α- and γ-vinyl substituents undergo conjugate addition:

(stablilized anion)

Nu = OH, OR, NH$_2$, NHR, SR, active methylene compounds, etc.

BUT:

(insufficient stabilization)

Figure 9.6 Manipulation of existing substituents: vinylpyridines.

much use in synthesizing non-heterocyclic compounds. Their utility derives from the fact that 1,4-dihydropyridines can be considered the synthetic equivalent of 1,5-dicarbonyl compounds. For example, pyridine itself is readily converted into glutaraldehyde dioxime by a simple two-step process consisting of dissolving metal reduction, followed by treatment with hydroxylamine (Figure 9.7). The relatively unstable dialdehyde is then freed with nitrous acid (equation 1) [9]. In analogous fashion, 1,5-diketones are derivable from the corresponding 2,6-disubstituted pyridine derivatives, by 1,4-reduction followed by hydrolysis (equation 2). Subsequent aldol condensation then provides entry to the cyclohexenone skeleton [10].

And how does this relate to vinyl pyridines? Danishefsky et al. took this analysis one step further with the postulate that 2-methyl-6-vinylpyridine is a stable surrogate for 3-vinyl-2-cyclohexenone (Figure 9.8) [11]. Suppose, for

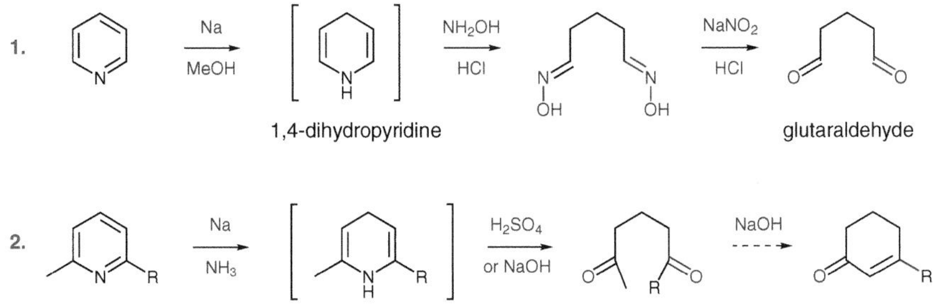

Figure 9.7 1,4-Dihydropyridines as latent 1,5-dicarbonyl compounds.

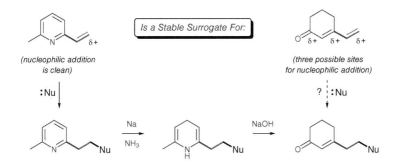

Figure 9.8 Synthetic utility of vinyl pyridines.

example, one wished to selectively add a nucleophile to the terminus of 3-vinyl-2-cyclohexenone. This could present a challenge, since the extended enone undergoes facile polymerization, and also since there are three sites available for nucleophilic addition. In contrast, one would expect clean conjugate addition to the readily prepared, and stable, 2-methyl-6-vinylpyridine. Moreover, we are now only two steps removed from our target cyclohexenone, involving dissolving metal reduction, followed by in situ hydrolysis and aldol condensation.

This methodology was first applied to a synthesis of the tricyclic dienone shown in the bottom right of Scheme 9.1, prepared as part of a model study for steroid synthesis [11]. Here the nucleophilic component was the pyrrolidine enamine derived from cyclohexanone, which underwent ready conjugate addition to 2-methyl-6-vinylpyridine. A noteworthy feature of this reaction is that it was carried out on 50 gram scales under essentially neutral conditions. With the A- and C-rings thus in place, the carbonyl group was protected as the ethylene ketal, setting the stage for dissolving metal reduction. In preliminary studies, each of the steps that followed was examined individually, and it was

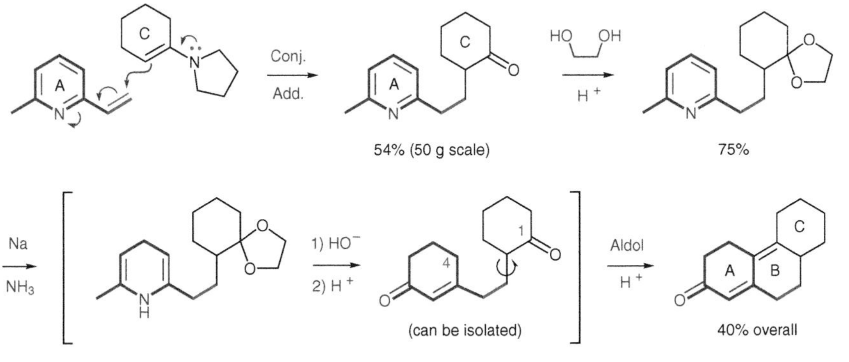

Scheme 9.1 Vinyl pyridines as bis annulation reagents.

possible to isolate and characterize the second intermediate shown in the brackets. However, in practice it proved advantageous to carry out all of these transformations without isolation of intermediates, whereby the desired tricyclic dienone was obtained in 40% overall yield.

This strategy was readily extended to an efficient synthesis of D-homoestrone, again utilizing 2-methyl-6-vinylpyridine as the electrophilic A-ring component [12]. Rings C and D were fashioned from the well-known monoketal of the Wieland-Miescher ketone:

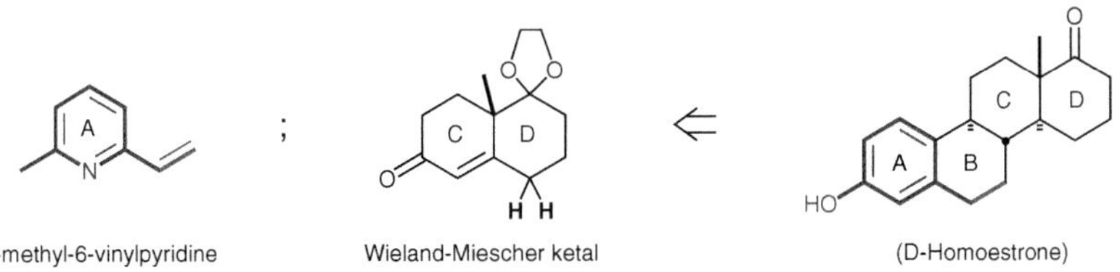

| 2-methyl-6-vinylpyridine | Wieland-Miescher ketal | (D-Homoestrone) |

While commercially available, it is worth visiting the synthesis of this last material, since it provides an opportunity for introducing (or reviewing) a reaction known as a Robinson annulation (Scheme 9.2) [13a]. This is one of the most versatile means available for synthesizing fused cyclohexenone ring systems. For the case at hand, the initial step involves Michael addition of the anion derived from 2-methyldihydroresorcinol to methyl vinyl ketone, followed by intramolecular aldol condensation to complete the formation of ring C [13b]. Selective ketalization with ethylene glycol then affords the Wieland-Miescher ketal [13c], wherein the most acidic protons are highlighted in bold. Note then, that under basic conditions, the starred (*) position is potentially nucleophilic.

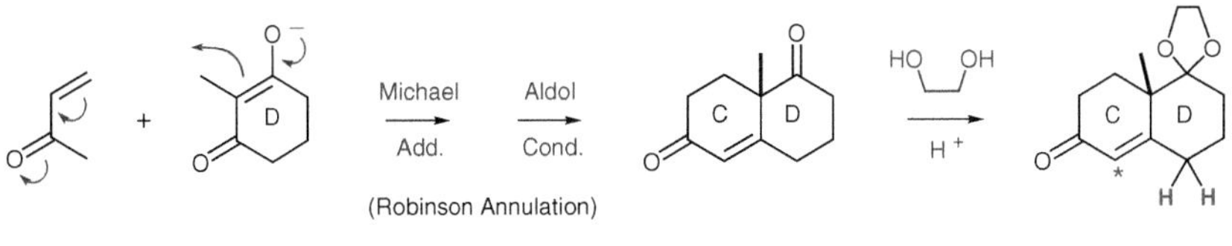

(Robinson Annulation)

Scheme 9.2

The authors took full advantage of this fact in the first step of assembling the D-homoestrone skeleton, which involved base-mediated conjugate addition of the Wieland-Miescher ketal to 2-methyl-6-vinylpyridine (curly arrows in Scheme 9.3). Acid workup then afforded an 80% yield of the A,C,D-adduct shown at the top center of the scheme, which incorporated all 19 carbon atoms destined to be found in the final product [12]. The path was now readied for unmasking of the pyridine ring, which involved selective reduction of the D-ring carbonyl, catalytic hydrogenation, and ketalization of the remaining ketone in ring C. Remarkably, we are now only six steps removed from D-homoestrone,

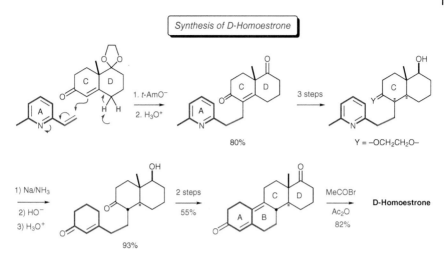

Scheme 9.3 Vinyl pyridines as bis annulation reagents.

the first three of which were accomplished in 93% overall yield. These consisted of Na/NH$_3$ reduction, base catalyzed hydrolysis/aldol condensation, and acid catalyzed ketal hydrolysis. For reference, these steps follow the general pathway outlined in equation 2 of Figure 9.7. Finally, Jones oxidation, followed by aldol condensation, gave the dienone shown in the bottom right of the scheme, which was smoothly isomerized to D-homoestrone under the influence of acetyl bromide/acetic anhydride. As testimony to the efficiency of this synthesis, the overall yield from the Wieland-Miescher ketal was 15% [12].

We have thus seen that alkyl and vinyl groups have special properties when attached to π-deficient heterocycles, and many other examples exist. For instance, suppose you were tasked with carrying out the decarboxylation of benzoic acid, which is exceedingly slow at temperatures below 450 degrees Celsius. Typically, recourse is taken to cuprous ion catalysis, which still requires heating to reflux in quinoline [14]:

Not so with picolinic acid (2-pyridinecarboxylic acid), which loses carbon dioxide at or near its melting point of 137 degrees Celsius (top Figure 9.9) [15]. Why the dramatic difference? Mainly this is due to the highly polarized nature of this compound, which exists predominantly in the zwitterionic form in the solid state, as well as in most solvents (much like aliphatic amino acids). As a consequence, the protonated pyridine ring has the means to stabilize a developing negative charge in the transition state leading to decarboxylation. It is important to emphasize that there exists no resonance effect, as for example, in the

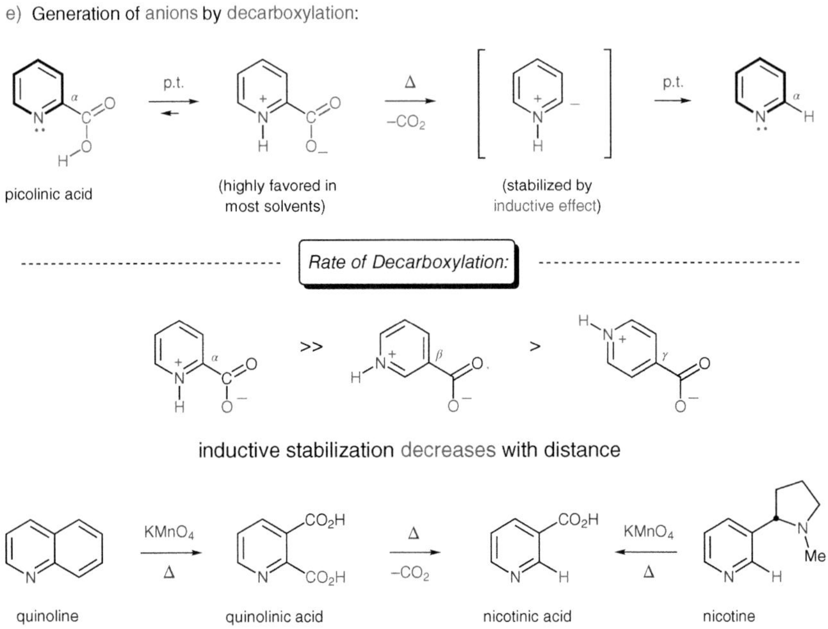

e) Generation of anions by decarboxylation:

picolinic acid   (highly favored in most solvents)   (stabilized by inductive effect)

Rate of Decarboxylation:

inductive stabilization decreases with distance

quinoline   quinolinic acid   nicotinic acid   nicotine

Figure 9.9 Manipulation of existing substituents.

decarboxylation of β-ketoacids. Rather, the positively charged iminium bond provides a powerful inductive stabilization, leading to the initial formation of the ylide intermediate shown in brackets. Proton transfer then affords pyridine.

And what of the isomeric pyridine carboxylic acids? Since inductive effects drop off rapidly with distance, it comes as no surprise that decarboxylation is much slower with nicotinic acid (3-pyridinecarboxylic acid), and slower still with isonicotinic acid (4-pyridinecarboxyic acid) (cf. middle Figure 9.9) [16]. A telling example of this phenomenon is provided by quinolinic acid, the pyridine 2,3-dicarboxylic acid derived by $KMnO_4$ cleavage of quinoline (bottom Figure 9.9). Upon mild thermolysis (~100 degrees Celsius), this material undergoes selective decarboxylation of the 2-carboxyl group to give nicotinic acid, which is identical to the compound obtained by $KMnO_4$ oxidation of nicotine [17].

To this point, an obvious question is whether reactions of this type might be put to further work, and the answer is yes. Thus, the ylide derived from decarboxylation of picolinic acid is sufficiently long-lived to be trapped with aldehydes and ketones (top Figure 9.10). This is an example of a Hammick reaction, which provides a useful means for substituting an α-carboxyl group with a secondary or tertiary alcohol [18]. Formally, this reaction corresponds to nucleophilic addition of a 2-metalated pyridine across a carbon-oxygen double bond. However, it is carried out under neutral conditions, simply upon heating the two components together neat, or in an inert solvent. Analogous results are

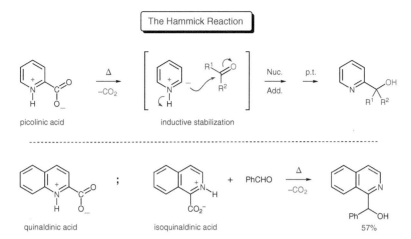

Figure 9.10 Manipulation of existing substituents.

obtained with quinaldinic acid and isoquinaldinic acid, with the latter, for instance, affording a 57% yield of the adduct shown on thermolysis with benzaldehyde (bottom Figure 9.10) [18b].

Intramolecular variants of this reaction are also known, but this is an area that seems ripe for further development. One example involves O-acylation of 3-hydroxypicolinic acid with the isopropylidine derivative of acetoacetic acid, which generates the β-ketoester illustrated in brackets (equation 1, Figure 9.11) [18a]. This intermediate was not isolated, but rather undergoes decarboxylative condensation across the ketone carbonyl group to afford a Hammick adduct in relatively modest yield. The less than favorable outcome in this case is likely due to the acidic nature of the hydrogens highlighted in bold, which are ideally situated to quench a developing ylide. In this regard, it has been found that higher yields of Hammick adducts are obtained upon thermolysis of α-trimethylsilyl esters of π-deficient carboxylic acids, including pyridines, pyridazines, pyrimidines, and pyrazines (equation 2) [19].

Lastly, we cannot leave the area of thermal decarboxylations without at least some mention of the lability of 2-pyridine acetic acid and related species. As indicated, these compounds undergo loss of carbon dioxide via a cyclic transition state with great ease, roughly akin to β-ketoacids:

Decarboxylation of 2-pyridine acetic acids:

Analogous to decarboxylaton of β-ketoacids. Slower with 4-pyridine acetic acids, and fails with 3-pyridine acetic acids (resonance stabilization not possible).

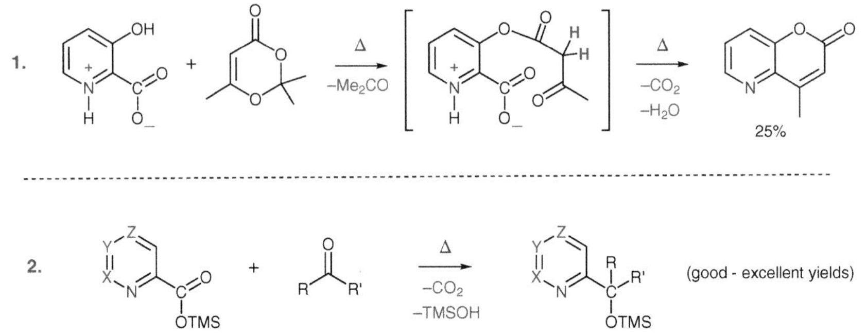

pyridine: X,Y,Z = CH. pyridazine: X = N; Y,Z = CH. pyrimidine: Y = N; X,Z = CH. pyrazine: Z = N; X,Y = CH).

Figure 9.11 Variants of the Hammick reaction.

Much of π-deficient heterocycle synthesis is about displacing good leaving groups from the α- and γ-positions, offering a ready means of generating complexity. Indeed, Chapter 6 was devoted nearly exclusively to this topic. A suitably placed halogen also opens the door to a myriad of cross-coupling reactions. But how are these leaving groups introduced? In contrast to benzene chemistry, electrophilic halogenation is not an option, at least in the absence of powerful activating groups.

In Figure 7.6 we showed that pyridine N-oxide undergoes facile reaction with $POCl_3/NEt_3$ to afford an excellent yield of 2-chloropyridine. This conversion follows a well-traveled mechanistic pathway, involving activation, nucleophilic addition, and ultimately elimination. As it happens, the same reagent will transform 2-pyridone to the corresponding chloride derivative, and by an analogous mechanism (top Figure 9.12). Again it is worth emphasizing that the lactam carbonyl in 2-pyridone is highly nucleophilic, due to amide resonance of the type indicated. Thus, it is a ready partner in displacing Cl^- from $POCl_3$ in a preliminary activation step. The resultant intermediate then undergoes nucleophilic addition at the electron deficient 2-position by Cl^-, generating a tetrahedral adduct that collapses with loss of $HOPOCl_2$. By similar means we can transform 4-pyridone to 4-chloropyridine (middle Figure 9.12), but note that there is no pathway available for preparing 3-chloropyridine from 3-hydroxypyridine (bottom Figure 9.12). It is not that the phosphorylation step fails—we can still convert the hydroxyl functionality into a far better leaving group. It is simply that the β-position is not activated toward nucleophilic addition.

An alternative route to 2-chloropyridine involves diazotization of 2-aminopyridine, which you will recall exists almost entirely as the amine tautomer (top Figure 9.13). When carried out with HONO/HCl, the resultant diazonium salt undergoes rapid displacement with Cl^- to generate the target compound.

f) Hydroxyl, amino and sulfide groups:

Reaction of hydroxy pyridines with POCl₃

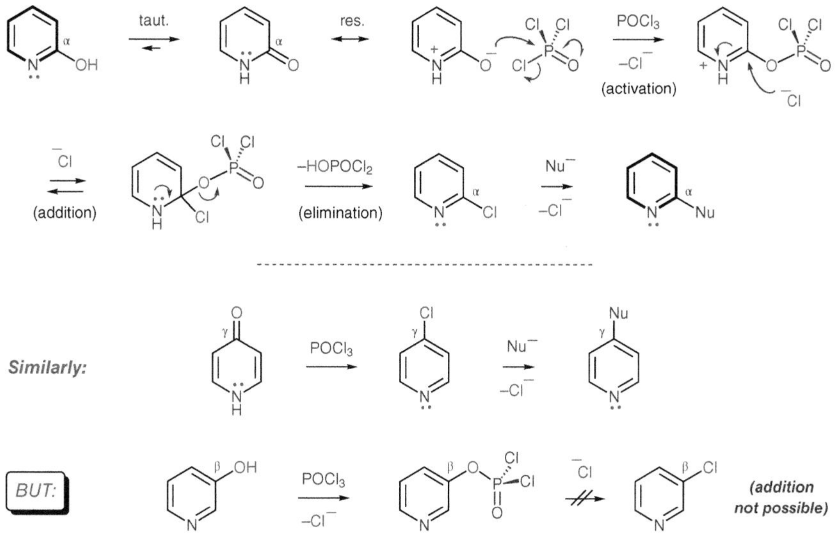

Figure 9.12 Manipulation of existing substituents.

Reaction of aminopyridines with HONO/HCl

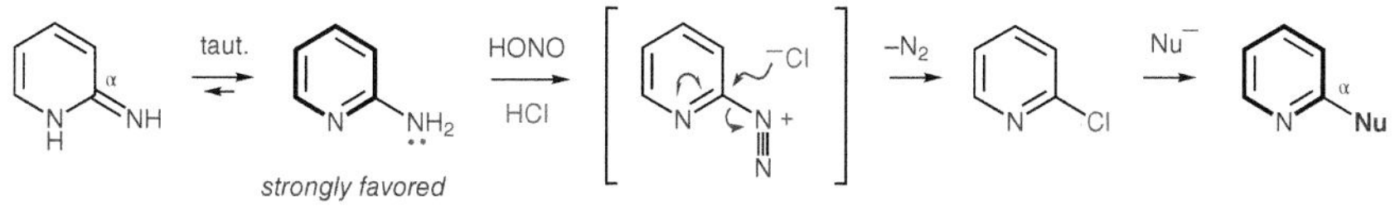

Reaction of hydroxypyridines with P₄S₁₀

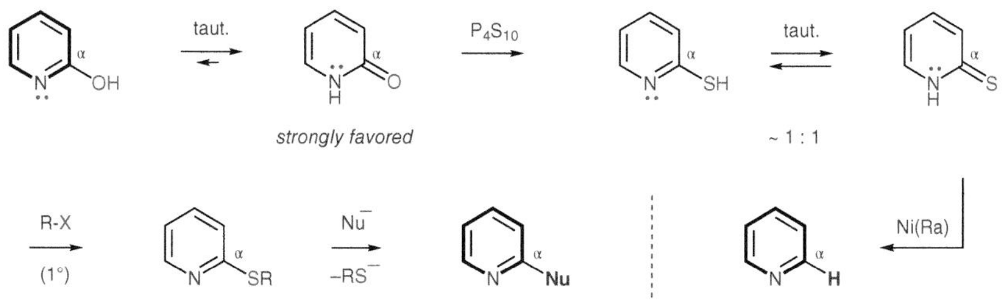

Figure 9.13 Manipulation of existing substituents.

Finally, in certain cases, we might wish to convert the carbonyl group in 2-pyridone, or a related species, to the corresponding thiolactam, which is readily accomplished with P_4S_{10} (middle Figure 9.13). In solution, 2(1H)-pyridinethione is present as an approximate 1:1 tautomeric mixture with 2-pyridinethiol, with the exact equilibrium depending on concentration, temperature, and the nature of the solvent. This, of course, stands in marked contrast to 2-pyridone, which exists nearly exclusively in the lactam form. In any event, when would such a transformation be useful? As with most mercaptans, sulfur can be extruded with Raney nickel, returning the parent pyridine nucleus. More importantly, however, alkylation with primary alkyl halides produces the corresponding thioimidates (bottom Figure 9.13), which are excellent substrates for both nucleophilic displacement, as well as transition metal catalyzed cross-coupling reactions [20]. In this last regard, they have the advantage of greater stability and ease of handling as compared to iminoyl chlorides.

To close this chapter, let us review a synthesis of quinine by Taylor and Martin that nicely illustrates many of the principles discussed so far (Scheme 9.4) [21]. Our starting material is 4-hydroxy-6-methoxyquinoline, which we know to be most stable in the keto form indicated. The final target, quinine, incorporates a quinuclidine ring appended by a methylene group to

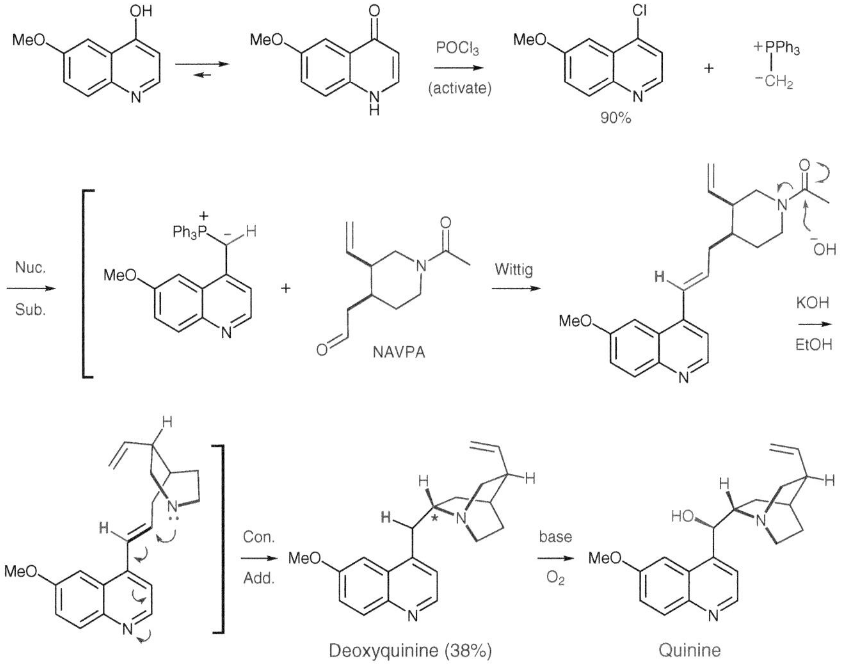

Scheme 9.4 Putting it all together: synthesis of quinine.

the quinoline skeleton. We begin as usual with an activation step, involving in this case introduction of a good leaving group into the 4-position. This was carried out in excellent yield employing the reagent POCl$_3$, to give 4-chloro-6-methoxyquinoline [22]. Note that the chloride group in this compound occupies a position in direct conjugation to the carbon-nitrogen double bond, and is thus subject to facile nucleophilic displacement. Now, settle back for a moment to take in this scheme, for we are about to accomplish many transformations in a "single pot" without isolation of intermediates.

The first of these involved formation of a new carbon-carbon bond, by nucleophilic substitution with the simple Wittig reagent derived from methyl iodide and triphenylphosphine. The product of this reaction, however, is more acidic than the starting reagent (why?), and thus undergoes immediate de-protonation to form the new Wittig reagent shown in brackets. Next to add was the piperidine aldehyde that we will abbreviate as NAVPA (the formal chemical name is rather lengthy), and condensation was allowed to proceed. In trial experiments, the major product of this reaction was shown by NMR to be the *E*-alkene pictured in the right-middle of the scheme, but remember, nothing is being isolated here. Rather, the crude product was subjected to alkaline hydrolysis to cleave the N-acetyl group, producing a reactive intermediate that underwent spontaneous intramolecular conjugate addition to the electron deficient vinyl group. What an elegant concept! Thus was formed the quinuclidine ring, shown in blue, for which there was precedent in earlier studies by Uskokovic et al [23]. The overall result was a 38% yield of deoxyquinine admixed with its epimer at the starred (*) position. Furthermore, the end game was now at hand, since deoxyquinine had previously been converted to quinine by base-induced hydroxylation with molecular oxygen [23]. This last reaction, it should not go un-noticed, was only feasible because of the relatively high acidity of the proton highlighted in red, wherein the conjugate base is stabilized by resonance into the quinoline ring:

Deoxyquinine

9.1 Summary

We have now reached a milestone of sorts in our study of heterocyclic chemistry, having focused much of our attention on π-deficient heteroaromatic rings. Along the way we have gained considerable insight into their chemical reactivity and physical properties, and delved into synthetic methodology. It is no longer a surprise that the principle chemistry of these compounds involves nucleophilic addition, or that ring substituents may have special properties. Also, we understand the importance of aiming for a "thermodynamic well" when synthesizing such ring systems de Novo. Many of these concepts have been illustrated with examples from the literature, carried out by some of the most accomplished practitioners in the field.

However, this is only part of the story. Beginning with Chapter 10 we will devote like attention to the chemistry of π-excessive heterocycles, which as their name implies, undergo facile electrophilic aromatic substitution. How different from their π-deficient cousins, where the site of highest electron density is invariably the heteroatom! We will also take up the topic of synthesizing heterocycles from other heterocycles. But before proceeding, spend some time with the practice problems below, which again are illustrative of the remarkable transformations that π-deficient ring systems can undergo.

Problems for Practice [24]

References

1 Fall, Y.; Van Bac, N.; Langlois, Y. *Tetrahedron Lett.* **1986**, *27*, 3611–3614.
2 See, for example,(a) DeStevens, G.; Halamandaris, A.; Strachen, P.; Donoghue, E.; Dorfman, L.; Huebner, C. F. *J. Med. Chem.* **1963**, *6*, 357–361. (b) Wischmann, K. W.; Logan, A. V.; Stuart, D. M. *J. Org. Chem.* **1961**, *26*, 2794–2796. (c) Ziegler, K.; Zeiser, H. *Justus Liebigs Ann. Chem.* **1931**, *485*, 174–192. (d) Woodward, R. B.; Kornfeld, E. C. in *Organic Syntheses*, Coll. Vol. III, John Wiley & Sons, Inc., New York, New York, **1955**, p. 413. (e) Pasquinet, E. Rocca, P.; Godard, A.; Marsais, F.; Quéguiner, G. *J. Chem. Soc., Perkin Trans. 1* 1998, 3807–3812. (f) Zelinski, R. P.; Benilda, M. *J. Am. Chem. Soc.* **1951**, *73*, 696–697.
3 Jerchel, D.; Heck, H. E. *Justus Liebigs Ann. Chem.* **1958**, *613*, 171–179.
4 Ahrens, F. B. *Chem. Ber.* **1896**, *29*, 2996–2999.
5. (a) Dunn, A. D. *Org. Prep. Proc. Int.* **1999**, *31*, 120–123. (b) Clarke, K.; Goulding, J.; Scrowston, R. M. *J. Chem. Soc. Perkin Trans. 1*, **1984**, 1501–1505.
6 Kagan, E. S.; Ardashev, B. I. *Chemistry of Heterocyclic Compounds* **1967**, *3*, 559–560.
7 Larson, R. D.; Corley, E. G.; King, A. O.; Carroll, J. D.; Davis, P.; Verhoeven, T. R.; Reider, P. J.; Labelle, M.; Gauthier, J. Y.; Xiang, Y. B.; Zamboni, R. J. *J. Org. Chem.* **1996**, *61*, 3398–3405.
8 Klumpp, D. A. *Synlett* **2012**, *23*, 1590–1604.
9 Cope, A. C.; Dryden, H. L. Jr.; Overberger, C. G.; D'Addieco, A. A. *J. Am. Chem. Soc.* **1951**, *73*, 3416–3418.
10 Zhou, J.; List, B. *J. Am. Chem. Soc.* **2007**, *129*, 7498–7499.
11 Danishefsky, S.; Cavanaugh, R. *J. Am. Chem. Soc.* **1968**, *90*, 520–521.
12 Danishefsky, S.; Cain, P.; Nagel, A. *J. Am. Chem. Soc.* **1975**, *97*, 380–387.
13 For a review, see(a) Jung, M. E. *Tetrahedron* **1976**, *32*, 3–31. (b) Ramachandran, R.; Newman, M. S. in *Organic Syntheses*, Coll. Vol. *V*., John Wiley & Sons, Inc., New York, New York, **1973**, p. 486. (c) Corey, E. J.; Ohno, M.; Mitra, R. B.; Vatakencherry, P. A. *J. Am. Chem. Soc.* **1964**, *86*, 478–485.
14 Cohen, T.; Schambach, R. A. *J. Am. Chem. Soc.* **1970**, *92*, 3189–3190.
15 Dunn, G. E.; Lee, G. K. J.; Thimm, H. *Can. J. Chem.* **1972**, *50*, 3017–3027.
16 Haake, P.; Mantecón, J. *J. Am. Chem. Soc.* **1964**, *86*, 5230–5234. The rate data given is for the corresponding N-methyl derivatives, which likely will correlate with the zwitterionic species.
17 Kulka, M. *J. Am. Chem. Soc.* **1946**, *68*, 2472–2473.
18 (a) Bohn, B.; Heinrich, N.; Vorbrüggen, H. *Heterocycles*, **1994**, *37*, 1731–1746. (b) Dyson, P.; Hammick, D. Ll. *J. Chem. Soc.* **1937**, 1724–1725.
19 Effenberger, F.; König, J. *Tetrahedron* **1988**, *44*, 3281–3288.
20 (a) Srogl, J.; Liu, W.; Marshall, D.; Liebeskiind, L. S. *J. Am. Chem. Soc.* **1999**, *121*, 9449–9450. (b) Ghosh, I. Jacobi, P.A. *J. Org. Chem.* **2002**, *67*, 9304–9309.
21 Taylor, E. C.; Martin, S. F. *J. Am. Chem. Soc.* **1974**, *96*, 8095–8102.

22 Riegel, B.; Albisetti, C. J., Jr.; Lappin, G. R.; Baker, R. H. *J. Am. Chem. Soc.* **1946**, *68*, 2685–2686.

23 Gutzwiller, J.; Uskokovic, M. *J. Am. Chem. Soc.* **1970**, *92*, 204–205.

24 Problems for practice 3:(a) Taylor, E. C.; Knopf, R. J.; Cogliano, J. A.; Barton, J. W.; Pfleiderer, W. *J. Am. Chem. Soc.* **1960**, *82*, 6058–6064. (b) Clark, J.; Gelling, I.; Southon, I. W.; Morton, M. S. *J. Chem. Soc. C* **1970**, 494–498. (c) Shaw, E. *J. Org. Chem.* **1962**, *27*, 883–885. (d) Biffin, M. E. C.; Brown, D. J.; Porter, Q. N. *J. Chem. Soc. C*, **1968**, 2159–2162.

10

π-Excessive Heterocycles: General Properties

What constitutes an aromatic π-excessive heterocycle? Mainly two criteria—they must satisfy Hückel's rule, and have higher average π-electron density per atom than found in benzene (as a standard, the π-electron density per carbon in benzene is 1.0). Typically, such heterocycles incorporate a five-membered ring, and as we have already seen (Chapter 2), they have quite specific orbital requirements. By way of review, these are illustrated at the top in Figure 10.1 for furan, thiophene, and pyrrole, which can be considered the parent members of this class.

Electronically, furan (X=O) and thiophene (X=S) share much in common, since the heteroatom in each case bears two electron pairs, one of which contributes to the aromatic sextet. To accomplish this, both oxygen and sulfur must be sp^2-hybridized, affording five p-orbitals over which to delocalize six π electrons. The remaining free electron pair is non-bonding, and occupies an sp^2-orbital in the plane of the ring (i.e., orthogonal to the π-system). The case for pyrrole is somewhat different, in that nitrogen bears an N-H bond in addition to a single electron pair. Once again, however, aromaticity is achieved via sp^2-hybridization, with the electron pair residing in a p-orbital as part of the π-system. To complete the picture, nitrogen is bonded to hydrogen employing an orthogonal sp^2-s σ-bond.

Not surprisingly, aromaticity decreases with increasing electronegativity of the heteroatom, due to greater localization of the π-electron cloud (middle Figure 10.1). Thus, furan, with a resonance energy (RE) of 16 kcal/mol, is the least aromatic of the parent ring systems, as compared to a calculated RE of 24-27 kcal/mol for cyclopentadienyl anion (cf. also Figure 2.2 in Chapter 2). The relatively modest RE of furan is reflected in both its stability and reactivity profile. For example, furan is unstable toward hot mineral acids, much as an acyclic enol ether. And, in many of its reactions furan behaves as a typical

Introductory Heterocyclic Chemistry, First Edition. Peter A. Jacobi.
© 2019 John Wiley & Sons Ltd. Published 2019 by John Wiley & Sons Ltd.

1. Orbital requirements:

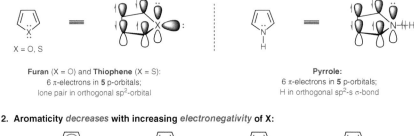

Furan (X = O) and **Thiophene** (X = S):
6 π-electrons in **5** p-orbitals;
lone pair in orthogonal sp²-orbital

Pyrrole:
6 π-electrons in **5** p-orbitals;
H in orthogonal sp²-s σ-bond

2. Aromaticity *decreases* with increasing *electronegativity* of X:

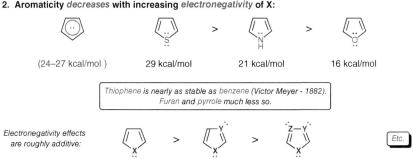

(24–27 kcal/mol)   29 kcal/mol   21 kcal/mol   16 kcal/mol

Thiophene *is nearly as stable as* benzene *(Victor Meyer - 1882).*
Furan *and* pyrrole *much less so.*

*Electronegativity effects
are roughly additive:*   > > Etc.

Figure 10.1 π-Excessive heterocycles: general properties

diene, undergoing, for example, facile Diels-Alder cyclization with dimethyl acetylenedicarboxylate:

At the other end of the aromaticity scale resides thiophene, whose RE of 29 kcal/mol is approaching that of benzene (RE = 36 kcal/mol). As expected, this material is acid stable and shows no inclination to participate in Diels-Alder cyclizations at atmospheric pressure. In fact, the close similarity between benzene and thiophene prompts a brief historical aside [1].

Our story unfolds on an autumn day in 1882, and features a youthful Viktor Meyer, of Zürich Polytechnic—we can picture him passionately engaged in one of his lectures on benzene chemistry. He had recently taken over these classes on the passing of his good friend Professor Wilhelm Weith, and was in the midst of demonstrating a well-known color test for benzene. This involved treatment with isatin and sulfuric acid, which produced a blue dye known as indophenine (the so-called von Baeyer indophenine test; however, the structure of indophenine was not yet established). Meyer had practiced this reaction many times and was confident of the outcome. However, to his great astonishment, and presumably embarrassment, on this afternoon no color developed. What had gone astray? To make a long story short, the key lay in the source of benzene. At the time, nearly all benzene was derived by distillation of coal tar, and it was extremely difficult to obtain pure samples from the natural source. But on this occasion his laboratory assistant, a soon-to-become famous Traugott Sandmeyer, had substituted synthetic benzene derived by decarboxylation of calcium benzoate. The difference between the synthetic sample, and the coal tar-derived material, was that the latter was invariably contaminated with thiophene. Within a year Meyer succeeded in isolating pure thiophene, and we now know that the indophenine test is specific for this ring system [2]:

thiophene   isatin   indophenine

As for other ring systems, aromaticity generally decreases with an increasing number of heteroatoms in the ring, due to increased localization of π-electron density (bottom Figure 10.1).

Would we expect π-excessive heterocycles to exhibit basic properties? It depends. Recall that basicity is the norm for π-deficient heterocycles such as pyridine (pK_a = 5.2), wherein the electron pair on nitrogen is localized in an sp^2-orbital, orthogonal to the aromatic sextet:

Consequently, pyridine undergoes protonation with relatively little disruption to aromatic stability, aside from the fact that there is increased π-cloud localization in the pyridinium conjugate acid. In contrast, pyrrole is an exceedingly weak base (pK_a = −3.8), since protonation completely destroys aromaticity (top Figure 10.2). In fact, under forcing conditions protonation occurs on carbon, which is kinetically favored at C-2. Thiophene and furan are weaker still, even though in principle they can undergo protonation on the heteroatom without completely surrendering aromaticity. However, in practice this is not observed, and protonation once again takes place on carbon.

And what if the ring incorporates two heteroatoms, in particular in a 1,3-relationship? Does basicity increase or decrease? The answer at first may seem surprising, since we know that substituting nitrogen for carbon decreases basicity in the π-deficient class:

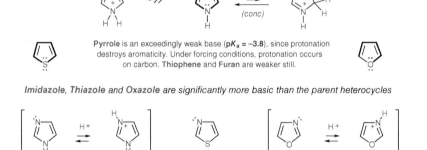

$pK_a = 5.2$ > $pK_a = 1.3$ > $pK_a < 0$

——— decreasing basicity ———→

In Chapter 4 we ascribed this difference to the inductive effect of the extra nitrogen atoms, which due to their electronegativity destabilize the conjugate acid derived by protonation (cf. Figure 4.2). That is, as in any acid-base equilibria, the more stable the conjugate acid, the stronger the base from which it is derived.

3. Basicity:

Pyrrole is an exceedingly weak base (**$pK_a = -3.8$**), since protonation destroys aromaticity. Under forcing conditions, protonation occurs on carbon. **Thiophene** and **Furan** are weaker still.

Imidazole, Thiazole and Oxazole are significantly more basic than the parent heterocycles

$pK_a = 6.9$ $pK_a = 2.5$ $pK_a = 0.8$

Protonation does not destroy aromaticity, and the conjugate acid is stabilized by resonance.

NOTE: This is the opposite trend seen with π-deficient rings, where the inductive effect decreases basicity (cf. Fig. 13).

Figure 10.2 π-Excessive heterocycles: general properties

But consider the case of imidazole, which, with a pK_a of 6.9, is many orders of magnitude stronger as a base than pyrrole ($pK_a = -3.8$) (middle Figure 10.2). It is even considerably stronger than pyridine ($pK_a = 5.2$), and approaching the base strength of typical trialkylamines ($pK_a \sim 10-11$). How do we rationalize this? Actually, comparing the basicity of imidazole and pyrrole is somewhat akin to comparing apples and oranges, since their electronic structures are quite different. In pyrrole, the electron pair contributed by nitrogen is part of the aromatic sextet, and therefore is not available for protonation (vide supra). In imidazole, though, the second ring nitrogen contains an electron pair that resides in an orthogonal sp^2-orbital, so protonation does not destroy aromaticity (bottom Figure 10.2). Not only that, but the conjugate acid is stabilized by resonance involving identical contributing structures, which is particularly favorable:

By way of comparison, thiazole, with a pK_a of 2.5, is still much more basic than thiophene, but less so than imidazole. And oxazole is weaker yet (pK_a = 0.8). In neither case will resonance stabilization of the conjugate acid be as effective as for imidazole, and with oxazole, there is also the destabilizing influence of the very electron withdrawing oxygen atom (cf. middle Figure 10.2). So, of the three common two-heteroatom containing five-membered ring systems, imidazole is far and away the strongest base, and it is an excellent catalyst for acyl transfer reactions, substitutions, and, of course, also proton transfers. Reflecting this fact, the imidazole ring of histidine is a common participant in many biologically important transformations.

This takes us to acid stability, which, amongst the parent π-excessive heterocyclic rings, varies directly with resonance energy. Thus, thiophene, with an RE of 29 kcal/mole, is comparable to benzene in its stability to hot aqueous acid, while furan (RE = 16) undergoes ready hydrolytic ring opening (top Figure 10.3). We will have more to say about this latter transformation in our future discussions on latent functionality. However, for the present, suffice it to say that furan and its derivatives are valuable precursors to a wide range of 1,4-dicarbonyl derivatives, which have considerable synthetic utility. Pyrrole, on the other hand, occupies a special position in that it is relatively reluctant to participate in ring opening reactions. But this is only partly due to its greater RE of 21 kcal/mol. Rather, under strongly acidic conditions, the preferred reaction pathway involves oligomerization and polymerization, often with the formation of dark, resinous tars. Under carefully controlled conditions, however, two products of trimerization have been isolated in ~40% combined yield (middle Figure 10.3) [3]. This involved treatment of pyrrole for 30 seconds with ice cold 25% aqueous HCl, which afforded a 2:1 mixture of trans- and cis-isomers [4].

The mechanism for this trimerization is of interest, in that the pyrrole ring functions as both a nucleophile and, in its protonated form, as a cationic Lewis acid (bottom Figure 10.3). That is, it will readily accept an electron pair to form a new carbon-carbon bond. Note that protonation of pyrrole might take place at either C-2 or C-3, and as alluded to above, C-2 is the kinetically favored point of attack. However, C-3 protonation provides the more potent electrophilic species, which suffers rapid nucleophilic addition by a second molecule of pyrrole (moving counter-clockwise from the start in Figure 10.3). Simple loss of a

4. Acd Stability:

stable to hot aqueous acid

Pyrrole *undergoes trimerization under strongly acidic conditions:*

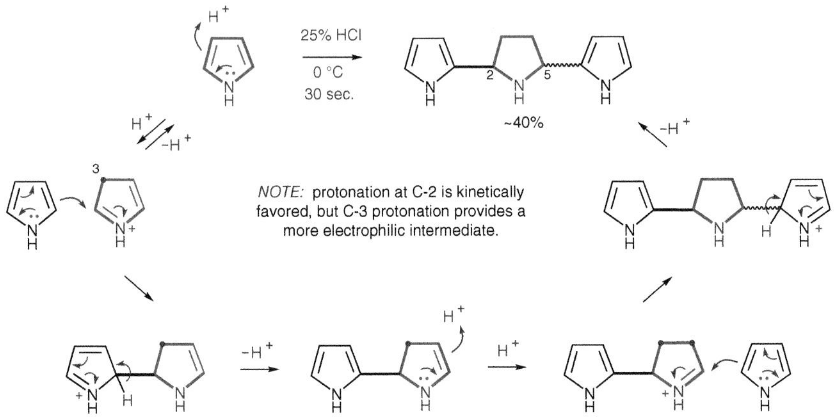

25% HCl

0 °C

30 sec.

~40%

NOTE: protonation at C-2 is kinetically favored, but C-3 protonation provides a more electrophilic intermediate.

Figure 10.3 π-Excessive heterocycles: general properties

proton then effects re-aromatization of the pyrrole ring, and gives the neutral intermediate shown at the bottom center of the figure. And where from here? Once again, protonation affords a reactive iminium ion intermediate, which is rapidly captured by a third molecule of pyrrole. And finally, deprotonation leads to the observed products, which are 2,5-disubstituted pyrrolidines.

Two of the key steps in the trimerization of pyrrole thus involve electrophilic aromatic substitution, following a mechanistic pathway analogous to that observed in benzene chemistry. Indeed, pyrrole is comparable in this regard to very electron rich benzene derivatives, undergoing facile substitution reactions with a wide variety of electrophiles. Although somewhat less reactive, thiophene and furan behave in similar manner, with the ring carbons exhibiting significant nucleophilic character (top Figure 10.4). As previously described, this is due to resonance involving the electron pair contributed by the heteroatom, which is delocalized over all positions in the five-membered ring. In fact, in the case of pyrrole (X=NH), recall that the dipole moment actually points away from nitrogen and toward carbon:

vs

1.80 D

1.57 D

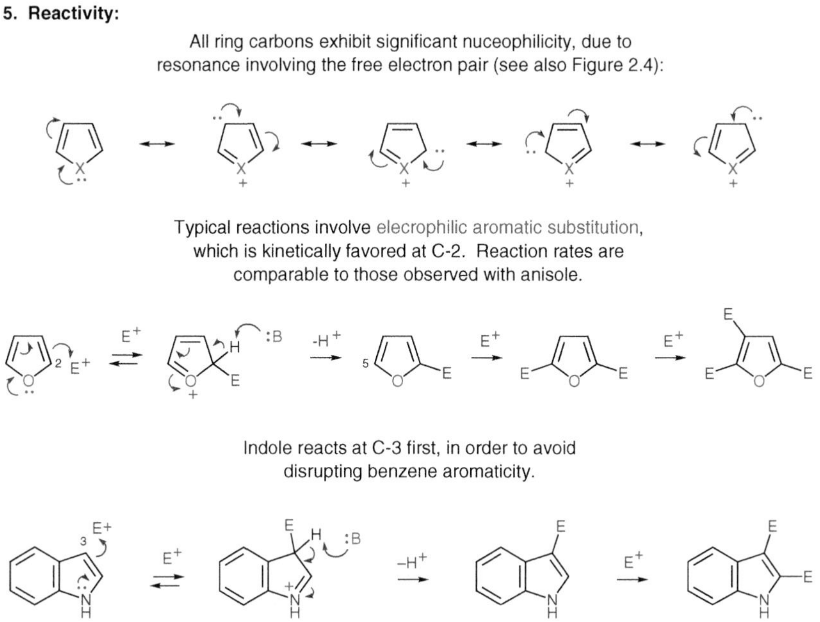

5. Reactivity:

All ring carbons exhibit significant nuceophilicity, due to resonance involving the free electron pair (see also Figure 2.4):

Typical reactions involve elecrophilic aromatic substitution, which is kinetically favored at C-2. Reaction rates are comparable to those observed with anisole.

Indole reacts at C-3 first, in order to avoid disrupting benzene aromaticity.

Figure 10.4 π-Excessive heterocycles: general properties.

So by now we are comfortable with the fact that π-excessive heterocycles are nucleophilic at carbon, but which positions are most so? As illustrated for furan (middle Figure 10.4), initial electrophilic attack is almost exclusively at C-2 versus C-3, pointing to a substantially lower energy of activation. On the surface of it this might seem curious, since there is little difference in calculated electron density between the two sites (cf. Figure 2.4). Rather, it is generally accepted that relative nucleophilicity in these ring systems has less to do with electron density, and more with transition state stabilization in the rate determining step. That is, as bonding begins for C-2 attack, the developing positive charge is spread over four atoms, including oxygen (left pathway in Figure 10.5). In contrast, C-3 attack leads to a carbocation intermediate that is stabilized only by the free electron pair on oxygen, with no contribution from the remaining double bond. The result is a relatively large difference in transition state enthalpy ($\Delta\Delta H$), which is reflected in the regiochemical outcome of reaction. In any event, it is only after both α-positions are occupied that an electrophile will add to C-3 and C-4 (middle Figure 10.4).

Finally, moving on to the bottom of Figure 10.4, we observe a somewhat different situation for fused ring systems. In the case of indole, initial electrophilic attack is at C-3, by which path electron donation from nitrogen need not disrupt the aromaticity of the benzene ring (curly arrows in the figure). However,

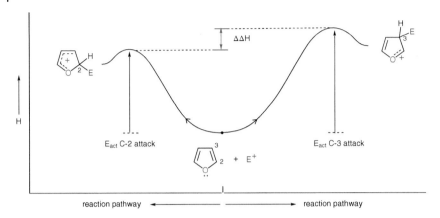

Figure 10.5 C-2 versus C-3 electrophilic aromatic substitution.

this barrier is not so great that a second electrophile is unable to add to C-2. In contrast, the less electron rich benzo[*b*]furan prefers C-2 substitution, via the intermediacy of a benzylic cation.

References

1 Cameron, M. D. *J. Chem. Ed.* **1949**, *26*, 521–524.
2 Tormos, G. V.; Belmore, K. A.; Cava, M. P. *J. Am. Chem. Soc.* **1993**, *115*, 11512–11515.
3 Potts, H. A.; Smith, G. F. *J. Chem. Soc.* **1957**, 4018–4022.
4 Zhao, Y.; Beddoes, R. L.; Joule, J. A. *J. Chem. Research (S)*, **1997**, 42–43.

11

π-Excessive Heterocycles: De Novo Syntheses

Our discussion of the synthesis of π-excessive heterocycles will be divided into two sections, consisting of (1) Preparation "de Novo" with all substituents present, and (2) Introduction of new substituents (top Figure 11.1). As with π-deficient heterocycles, no attempt has been made to present a comprehensive account of synthetic methodology. For that purpose, the appropriate reviews can be consulted. Rather, our main objective is to offer a "sampling" of the basic principles of π-excessive heterocyclic synthesis. We begin with the parent members of this class, furan, pyrrole, and thiophene.

In the category of de Novo syntheses, it is worth emphasizing that product stability may provide an opportunity for exercising thermodynamic control, just as with π-deficient heterocycles. That is, if at all possible, we want to "sink into" the thermodynamic well represented by aromaticity, with minimal adjustments to oxidation level. Note, for example, that succindialdehyde is in the same oxidation state as furan, and we can draw a straightforward mechanism interconverting the two (middle Figure 11.1). In the forward direction this involves cyclodehydration, initiated by acid-catalyzed enolization of one of the aldehyde groups. Protonation of the neighboring aldehyde then affords an intermediate that is strongly activated toward intramolecular cyclization, producing the hemiacetal shown at the bottom left of the figure. We are now only three steps removed from generating the aromatic furan ring, involving (1) proton transfer (p.t.), (2) dehydration, and (3) loss of a proton from C-3. Conversely, hydrolysis of furan would be initiated by C-3 protonation followed by capture of water at C-2. So which is favored? Clearly it depends on the reaction conditions, wherein an anhydrous medium would favor furan formation. In any event, in the forward direction this is the parent member of a class of reactions known as the Paal-Knorr furan synthesis, reported independently by the German chemists Carl Paal and Ludwig Knorr in 1884 [1]. We shall see below that this reaction is quite general for substituted derivatives. However, it has seen little utility for synthesizing furan itself, in part because succindialdehyde undergoes competitive polymerization under acidic conditions. Also,

Introductory Heterocyclic Chemistry, First Edition. Peter A. Jacobi.
© 2019 John Wiley & Sons Ltd. Published 2019 by John Wiley & Sons Ltd.

1. **Preparation "de Novo" with all substituents present.** *Product stability may provide an opportunity for exercising thermodynamic control.*
2. Introduction of new substituents.

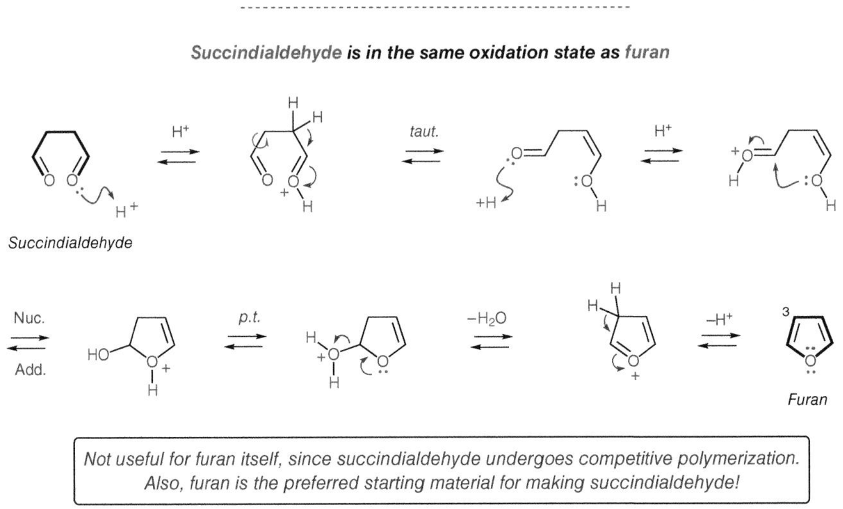

----

*Succindialdehyde **is in the same oxidation state as** furan*

Succindialdehyde

Furan

> *Not useful for furan itself, since succindialdehyde undergoes competitive polymerization. Also, furan is the preferred starting material for making succindialdehyde!*

Figure 11.1 Synthesis of π-excessive heterocycles.

furan is abundantly derivable from natural sources, and indeed, is the preferred starting material for synthesizing succindialdehyde.

Aside from its non-applicability to the parent heterocycle, the Paal-Knorr synthesis is one of the most versatile of all furan syntheses (top Figure 11.2). Yields are generally high, and it is limited only by the availability of the starting 1,4-dicarbonyl compounds, or their synthetic equivalents. And what of the parent ring system, which we note above is "abundantly derivable from natural sources?" On a commercial scale, furan is produced by gas phase decarbonylation of furfural, which in turn is derived from such agricultural waste products as oat hulls or corncobs (bottom Figure 11.2; the name furfural has its Latin roots in the noun furfur, roughly translated as "bran," or the hard outer layers of cereal grain). The procedure is remarkably simple, and takes advantage of the fact that the polysaccharides present in these last materials are a rich source of pentoses, via acid hydrolysis. The Quaker Oats company is credited with developing the first economically viable synthesis of furfural, initially as a means of increasing the market position of oat hulls as a livestock feed [2]. Mechanistically, this transformation requires loss of three molecules of water to provide the aromatic furan ring, corresponding to the three hydroxyl groups at C2–C4. However, the precise sequence of events is not known with certainty, and probably varies with conditions. A reasonable pathway is provided

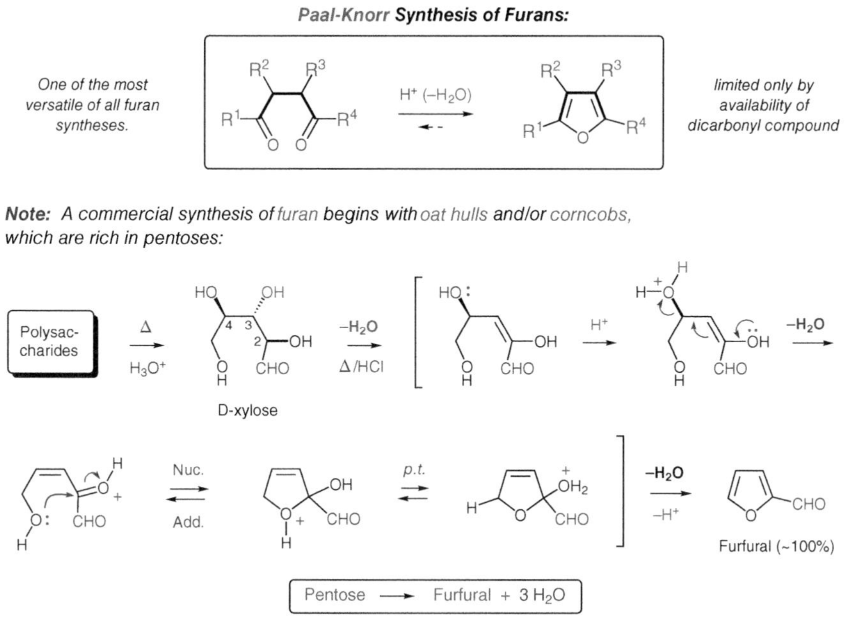
Paal-Knorr Synthesis of Furans:
One of the most versatile of all furan syntheses.
H⁺ (–H₂O)
limited only by availability of dicarbonyl compound
Note: A commercial synthesis of furan begins with oat hulls and/or corncobs, which are rich in pentoses:
Polysac-charides
H₃O⁺
D-xylose
–H₂O
Δ/HCl
H⁺
–H₂O
Nuc.
Add.
p.t.
–H₂O
–H⁺
Furfural (~100%)
Pentose ⟶ Furfural + 3 H₂O

Figure 11.2 De Novo synthesis.

in the figure [3], but the most important feature is that no oxidants or reductants are required.

Coming back again to Paal and Knorr, the mid-1880s were an incredibly productive period for both researchers. Closely related to the Paal-Knorr synthesis of furans are the Paal thiophene synthesis and the Paal-Knorr synthesis of pyrroles (top Figure 11.3). Both of these syntheses employ 1,4-dicarbonyl compounds as starting materials, and it was Paal in 1885 who first described their conversion to thiophenes with P_4S_{10} [4]. Although more modern reagents have since supplanted the use of P_4S_{10}, this strategy has found great utility in the synthesis of a diverse range of thiophene derivatives, including polythiophenes and materials for use in liquid crystal displays (bottom Figure 11.3) [5].

Also in 1885, and in nearly back-to-back papers, Paal and Knorr published their experiments on the pyrrole synthesis that bears their name (top right Figure 11.3) [4, 6]. In addition to ammonia (R=H), a wide variety of primary amines has been employed. The mechanism for this transformation most likely proceeds through the intermediacy of the bis-hemiaminal shown in brackets at the top of Figure 11.4, although the timing of ring closure may vary depending upon substrates and reaction conditions. A useful variant makes use of cyclic ketals of the type shown in the bottom left of the figure, wherein the corresponding 1,4-dicarbonyl compounds are generated in situ.

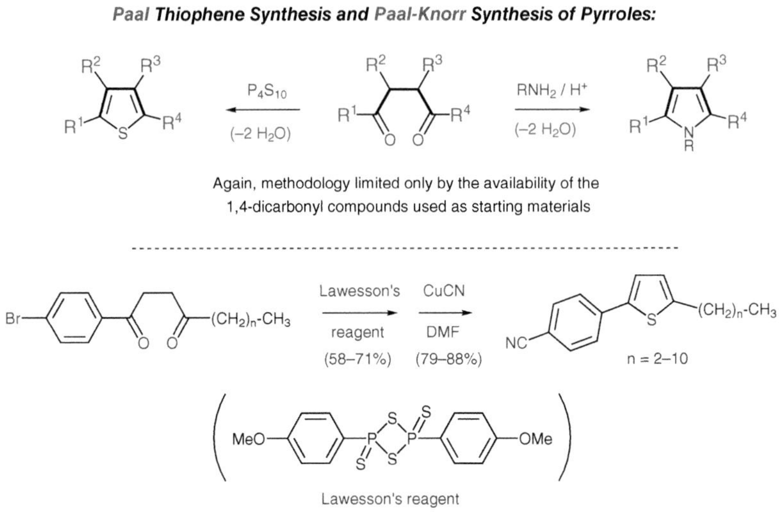

Paal *Thiophene Synthesis* and *Paal-Knorr* Synthesis of Pyrroles:

Again, methodology limited only by the availability of the
1,4-dicarbonyl compounds used as starting materials

Figure 11.3 De Novo synthesis.

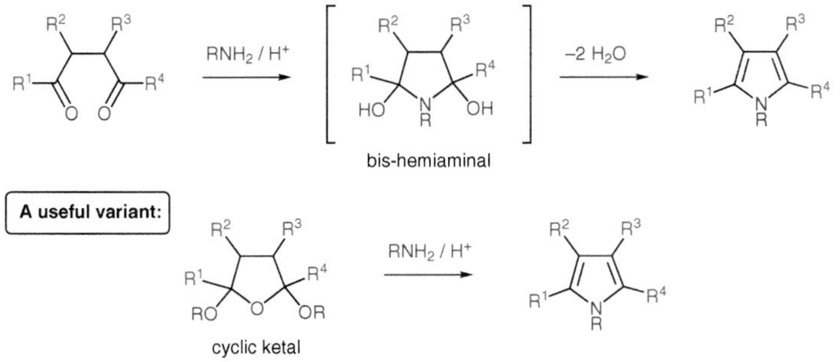

Figure 11.4 Mechanism of pyrrole synthesis.

This last modification was put to good use in a novel synthesis of porphobilinogen, a deceptively simple-appearing pyrrole that is the biological precursor to nearly all naturally occurring tetrapyrroles (Scheme 11.1) [7]. It is
also of interest as a pro-drug in tumor photodynamic therapy (PDT). The
synthesis began with the commodity chemical furfurylamine (R=H), which
was converted to the allylic alcohol shown by a three step sequence consisting of alkylation (R=dibenzosuberyl), in situ acylation, and aldol condensation with acrolein (86% overall yield). Upon heating with methyl orthoacetate
(MOA), this last material gave a 72% yield of the 7-oxonorbornene derivative
at the top right of the scheme, by initial Johnson orthoester Claisen rearrangement followed by intramolecular Diels-Alder cyclization. And why the

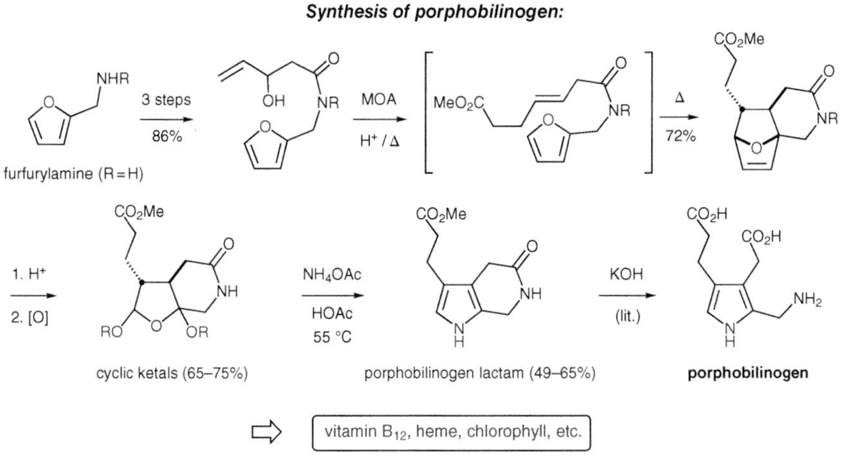

Synthesis of porphobilinogen:

Scheme 11.1 De Novo synthesis.

dibenzosuberyl group? An interesting feature of this transformation is that with R=H the Diels-Alder reaction failed completely. The bulky dibenzosuberyl group functions by providing a steric bias for the conformation leading to cyclization [8]:

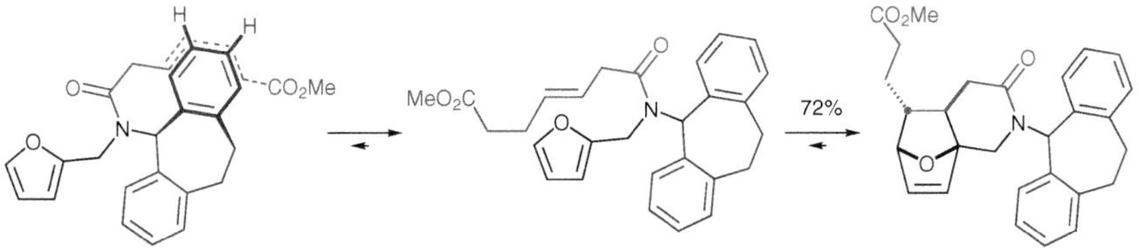

In any event, by this stage the entire carbon skeleton of porphobilinogen had been fashioned, and it was time to expose the latent 1,4-dicarbonyl precursor. This was initiated by acid hydrolysis of the dibenzosuberyl substituent, followed by oxidative cleavage of the carbon-carbon double bond with O_3 and then $I_2/PhI(OAc)_2$. In this manner was produced a 65–75% yield of the cyclic ketals shown in the bottom left of the scheme (R=OAc, OCHO), which were directly converted to the known porphobilinogen lactam methyl ester employing ammonium acetate in aqueous acetic acid. Hydrolysis according to the literature procedure then gave porphobilinogen in eight steps, and ~20% overall yield from furfurylamine.

Finally, Diels-Alder chemistry and ozonolysis also played a key role in synthesizing the Paal-Knorr intermediate shown in brackets in Scheme 11.2 (TIPS=triisopropylsilyl). In this case aminolysis utilizing ammonium carbonate provided the target pyrrole in greater than 50% overall yield [9].

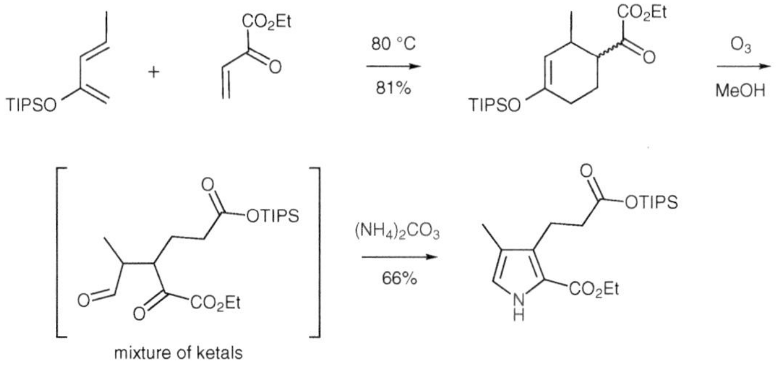

Scheme 11.2

This concludes our discussion on synthesizing π-excessive heterocycles via the intermediacy of 1,4-dicarbonyl derivatives. However, we have not heard the last from our (now) Professor Knorr, who went on to publish extensively in heterocyclic chemistry. Undoubtedly, though, his most lasting contribution resides in the pyrrole synthesis that is closely associated with his name [10]:

Knorr *Pyrrole Synthesis:*

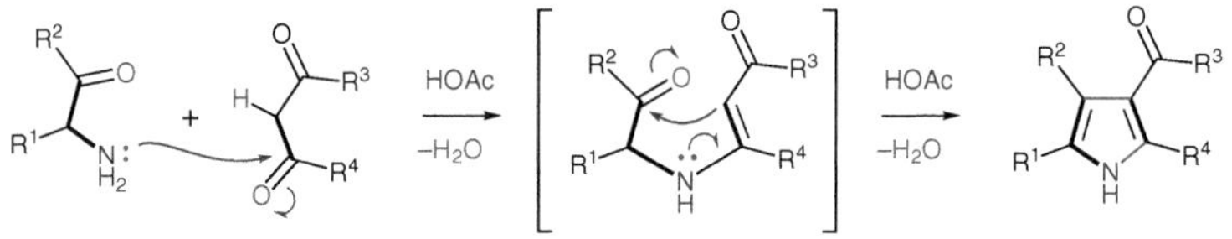

Perhaps the most versatile of all pyrrole syntheses.

The Knorr pyrrole synthesis is perhaps the most versatile of all pyrrole syntheses, combining an α-aminoketone with a β-dicarbonyl compound to provide the title compounds in generally good-excellent yields. The initial step is believed to be enamine formation, followed by ring closure and aromatization with loss of water (see above). A notable strength of this approach is that there is a wide degree of latitude for substituents R^1–R^4. Perhaps the biggest challenge lies in synthesizing the α-aminoketone starting materials, which are susceptible to self-condensation and are generally prepared in situ.

Thus, in its original version, Knorr carried out his synthesis in "one pot," beginning with two equivalents of ethyl acetoacetate dissolved in glacial acetic acid (Scheme 11.3) [10]. Half of the starting material was then converted to the α-aminoketone derivative shown in the bottom left of the scheme, by sequential treatment with one equivalent of sodium nitrite, followed by two equivalents of zinc dust. This led to rapid formation of what has ever since been known simply as "Knorr's pyrrole." Published decades later as an Organic Syntheses procedure, the reported yield is 57–64% on >200 g scale [11].

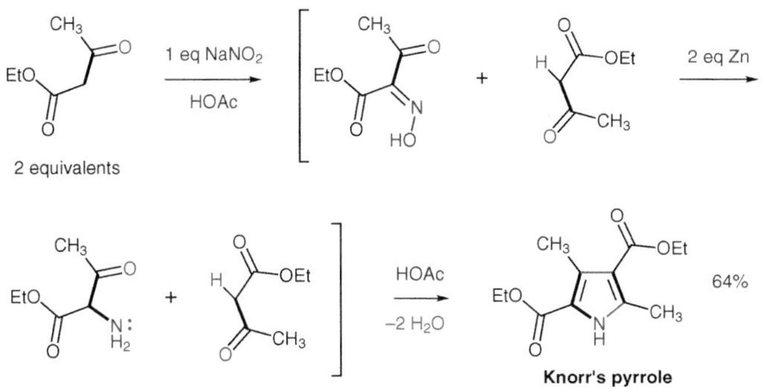

Scheme 11.3 De Novo synthesis.

This strategy was nicely exploited in an efficient synthesis of piquindone (Ro 22-1319), an antipsychotic agent that is a potent antagonist of dopamine (Scheme 11.4) [12]. Although never marketed, piquindone exhibited moderate efficacy in treating schizophrenia, and showed early promise in countering Tourette's syndrome. Coffen et al. began their synthesis with the readily available tetrahydropyridine derivative arecolone, which with dimethyl malonate was cleanly converted into the diketo compound shown in the brackets. In fact, this conversion was so clean that purification was unnecessary, and the crude product was treated directly with 2-aminopentan-3-one (red) to afford the target pyrrole in 62% overall yield. As further testament to the robustness of this synthesis, it was readily adaptable to multi-kilogram scale.

It remains to outline a useful modification of the Knorr pyrrole synthesis, wherein tetrasubstituted pyrroles can be prepared directly from alkylated β-diketones (Scheme 11.5) [13]. In this case the requisite α-aminoketone has its origin in a β-keto acid, which undergoes smooth decarboxylative nitrosation

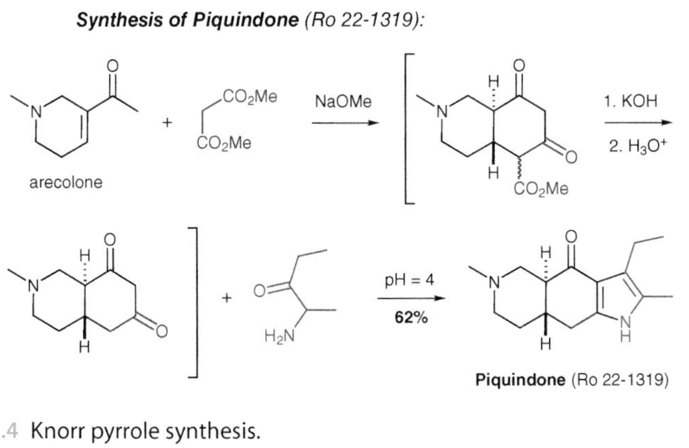

Scheme 11.4 Knorr pyrrole synthesis.

A useful modification: *Tetrasubstituted pyrroles can be prepared directly from alkylated β-diketones:*

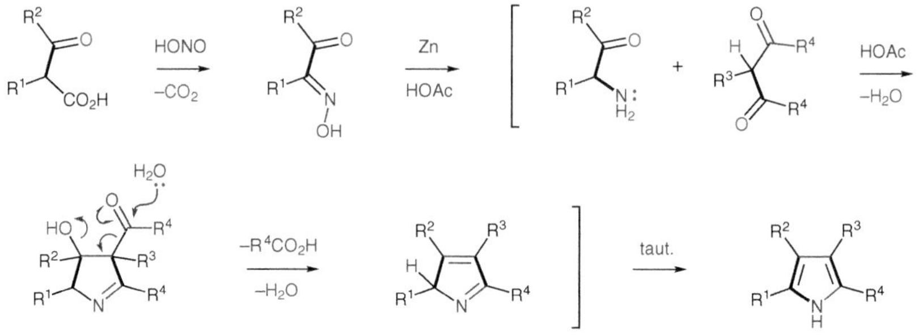

Scheme 11.5 Knorr pyrrole synthesis.

on treatment with nitrous acid (HONO). Note an important difference here as compared to the "original version" (Scheme 11.3), in that the R^1-substituent need not be an activating group to enable oxime formation. As before, though, the amino group is generated in situ by reduction with Zn in acetic acid. Next comes the all-important step involving condensation with a symmetrical β-diketone derivative, in which the central carbon is occupied by an R^3-substituent. On first inspection, one might question how such an intermediate could lead to an aromatic pyrrole ring, since the C3-position is blocked by both an acyl and alkyl group. However, once again thermodynamics prevails, and aromaticity is achieved by loss of the elements of R^4CO_2H and H_2O (see curly arrows), followed by tautomerization.

Remember Professor Hantzsch? Earlier, we described his ground breaking synthesis of pyridines, first published in 1881 and still perhaps the most general strategy of its class (Scheme 5.2). Fast forward now to 1890, and Hantzsch's name is again prominently featured in the context of a new heterocycle synthesis, this time involving pyrroles (top Figure 11.5) [14]. In a remarkably simple protocol, a β-ketoester, an α-chloroaldehyde or ketone, and aqueous ammonia are mixed together and allowed to react with gentle warming. No particular attention is paid to the order of addition. However, mechanistic studies have shown that the first step in this synthesis involves enamine formation between ammonia and the keto portion of the β-ketoester, followed by in situ C-alkylation with the α-chlorocarbonyl derivative (structures in brackets). Finally, cyclodehydration gives a 2,3-disubstituted or 2,3,5-trisubstituted pyrrole in moderate to good yields.

And what if one lowers the concentration of ammonia, or substitutes a base such as potassium hydroxide or pyridine? Feist (1902) [15], and later Bénary (1911) [16], explored the first of these options in the early years of the new century, demonstrating that β-dicarbonyl compounds undergo reaction with α-chloroaldehydes or ketones to provide di- or trisubstituted furans

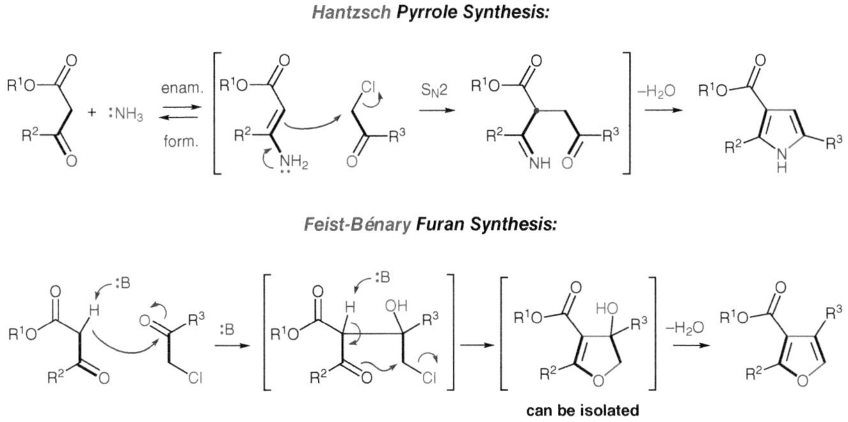

Hantzsch **Pyrrole Synthesis:**

Feist-Bénary **Furan Synthesis:**

can be isolated

Figure 11.5 De Novo synthesis.

(bottom Figure 11.5). However, pyrroles were still a significant byproduct. As pertains to alternative bases, pyridine has emerged as a more satisfactory choice, since it eliminates the problem of competing pyrrole formation. Mechanistically, there are a couple of points of interest here, starting with the fact that the initial adduct does not suffer immediate dehydration, as is the usual pathway with Knoevenagel condensations. Rather, enolization and intramolecular O-alkylation intercede, producing a dihydrofuran that under carefully controlled conditions can be isolated. More about this below. Typically, though, elimination of a molecule of water is effected in situ to provide the aromatic furan ring. As an aside, the Hantzsch pyrrole synthesis is occasionally portrayed as a modification of the Feist-Bénary furan synthesis, involving higher concentrations of ammonia. However, giving credit where credit is due, the historical record shows the reverse.

A number of efforts have been made to adapt the so-called "interrupted" Feist-Bénary reaction to natural product synthesis. In a noteworthy example, Calter et al. developed a novel route to the bicyclic core of 7-deoxyzaragozic acid, beginning with sodium malondialdehyde (blue) and the simple brominated β-ketoester derivative shown in red (top Figure 11.6) [17]. Remarkably, only four steps were required to convert the dihydrofuran product of "interrupted" condensation to the target compound (the numbering corresponds to that of zaragozic acid itself). The Calter group also reported the first catalytic, asymmetric version of this class of reaction, achieving ee's as high as 98% employing a quinine derived catalyst (bottom Figure 11.6) [18].

For now, we will put aside further discussion of de Novo furan syntheses, although this topic will resurface later in the context of "heterocycles from other heterocycles." Also, we are nearing the end of our discussion on pyrroles, with only two to three related examples left to consider. To set the stage, let us

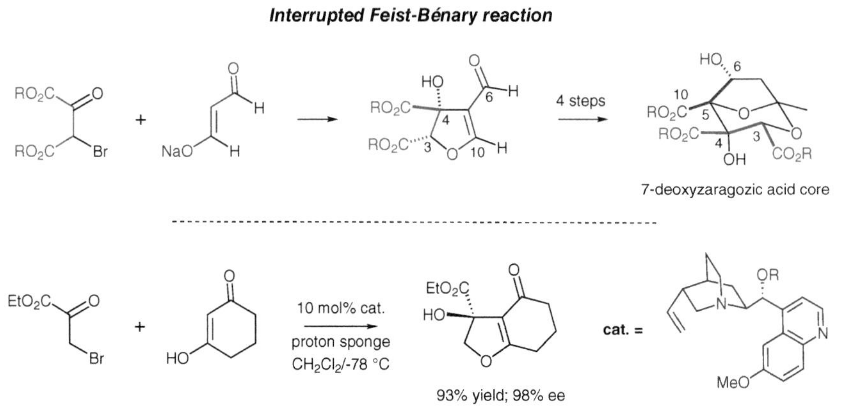

Interrupted Feist-Bénary reaction

7-deoxyzaragozic acid core

93% yield; 98% ee

Figure 11.6 Interrupted Feist-Benary reaction.

start with the Barton-Zard pyrrole synthesis, often cited as the epitome of a "3 + 2" approach to pyrroles (Scheme 11.6) [19]. This unusually versatile strategy begins with aliphatic nitro compounds, which, due to their relatively acidic nature, undergo facile base-promoted condensation with aldehydes (the Henry reaction). Typically, the resultant β-hydroxynitro adducts are captured with acetic anhydride, which renders them particularly prone to E1cB elimination to afford nitroalkenes. And here begins the really interesting chemistry (in brackets). Not surprisingly, nitroalkenes are extraordinarily reactive as Michael acceptors, in the present case undergoing rapid conjugate addition with the lithium anion derived from stabilized isocyanide derivatives. This sets into motion a series of steps leading to aromaticity, consisting of 5-*endo-dig* cyclization, followed by proton transfer, and ultimately, elimination of nitrite anion to generate a 2,3,4-trisubstituted pyrrole.

Numerous workers have put this methodology to good use synthesizing novel tetrapyrroles and related species, wherein the requisite pyrrole building

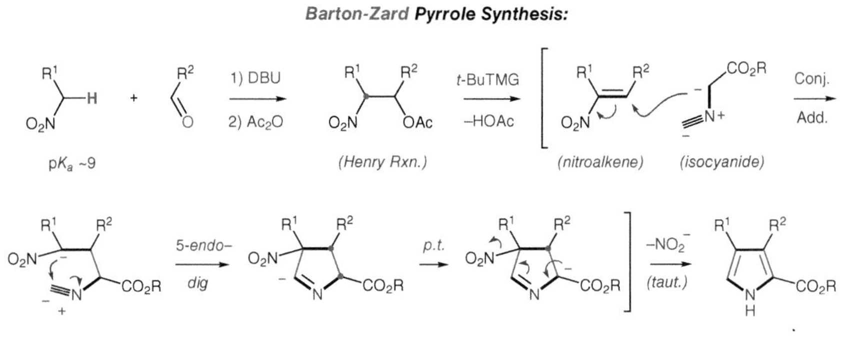

Barton-Zard Pyrrole Synthesis:

Scheme 11.6 De Novo synthesis.

blocks were otherwise difficult to access. In addition, notable variants to this strategy have been developed by van Leusen and Montforts, in the first instance employing tosylmethyl isocyanide as the three atom nucleophilic component [20], and in the second utilizing vinylsulfones as the Michael acceptor [21]:

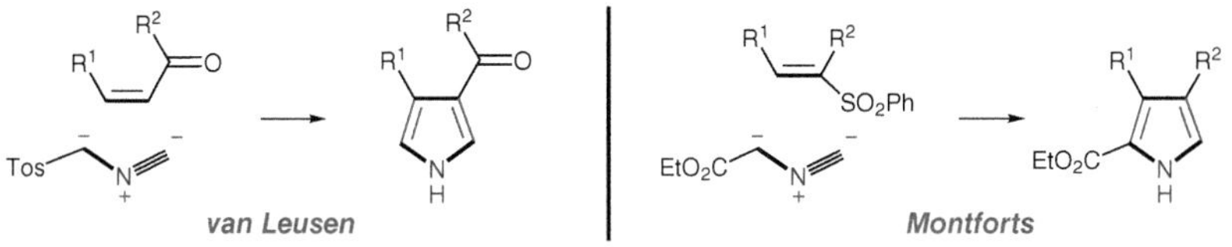

van Leusen   *Montforts*

The Montforts methodology is particularly well suited for synthesizing fused-ring pyrroles, since the vinylsulfone precursors are conveniently prepared from cycloalkenes. This is achieved by a three step sequence consisting of (1) electrophilic addition of PhSCl, (2) oxidation, and (3) elimination of chloride. A representative example is illustrated in Scheme 11.7, beginning with the Diels-Alder adduct derived from diethyl fumarate and butadiene. The desired vinylsulfone was obtained in ~70% overall yield, and was directly converted to the target pyrrole employing ethyl isocyanoacetate with *t*-BuOK (43%) [21].

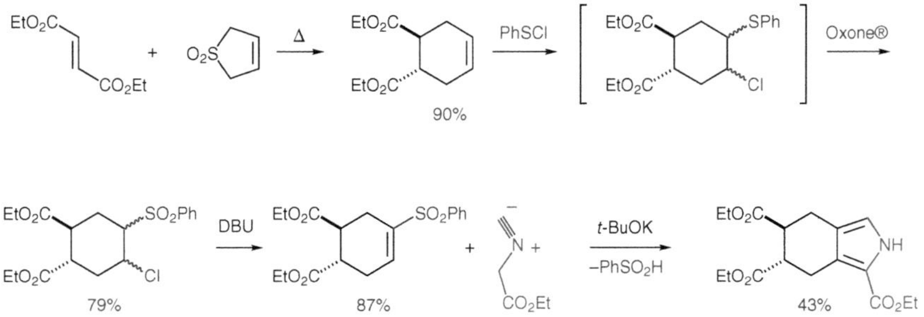

Scheme 11.7

Finally, lest sulfur be neglected, we close this section with the Hinsberg thiophene synthesis, long taken for granted as a *bis* aldol-like condensation (Scheme 11.8) [22]. Interestingly, though, the primary product of this reaction has one of the ester groups hydrolyzed, regardless of how carefully water is excluded. It is only after base workup that the corresponding dicarboxylic acid is obtained, which undergoes facile decarboxylation to produce 3,4-disubstituted thiophenes. Related condensations afford furans, pyrroles, and selenophenes, in each case in partially hydrolyzed form (Scheme 11.8).

So what is going on here? Although first described in 1910, it remained until 1965 for Wynberg and Kooreman to unravel the detailed mechanism for the Hinsberg thiophene synthesis [23]. Their solution was based upon isotopic labeling studies, utilizing ^{18}O-enriched benzyl as the dicarbonyl component,

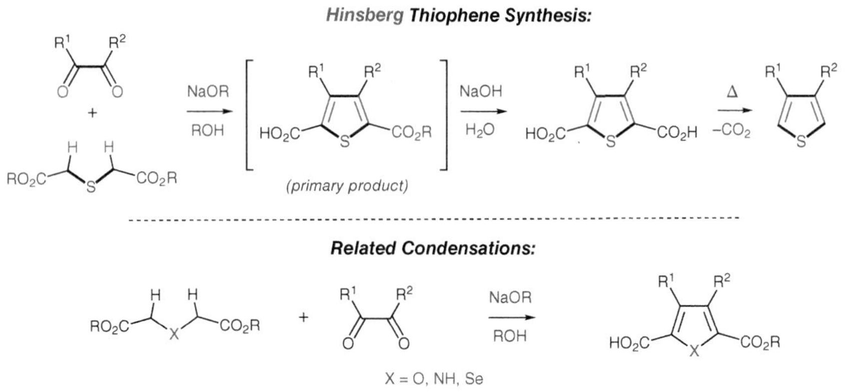

Scheme 11.8 De Novo synthesis.

and diethyl thiodiacetate as the active methylene substrate (Scheme 11.9). The thiophene product was obtained in 93% yield upon brief exposure to *t*-BuOK in *t*-BuOH. And here is where the ^{18}O labeling paid dividends. Namely, were a *bis*-aldol pathway followed, all of the label should have been lost to dehydration in generating the aromatic thiophene ring. However, the experimental observation was strikingly different, in that exactly half of the ^{18}O was found in the 2-carboxy substituent.

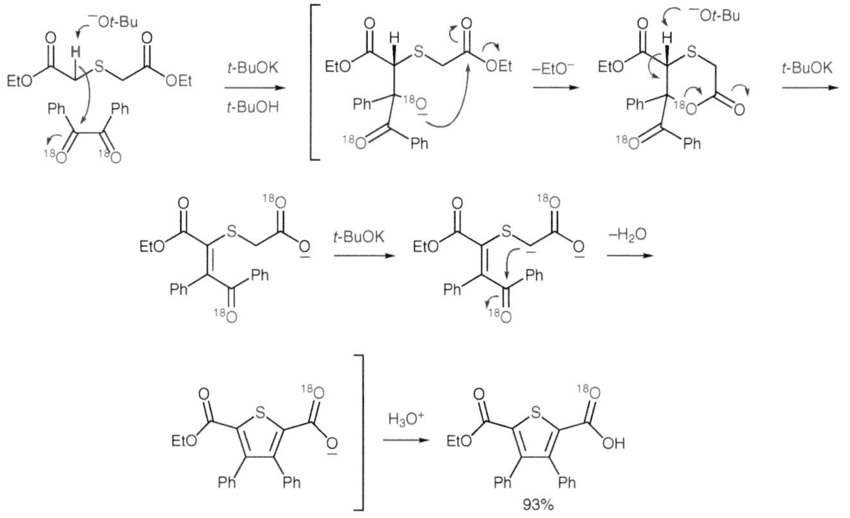

Scheme 11.9

These results are best explained by a mechanism reminiscent of a Stobbe condensation, wherein initial reaction with a benzyl carbonyl group takes place as indicated. However, the intermediate alkoxy anion is immediately intercepted by the ester group five bonds removed, to produce a six membered ring lactone (the second structure in brackets). Base-induced ring opening then generates the open chain intermediate shown at the bottom left of the scheme, in which the ^{18}O label has been transferred to the carboxy group. The remaining steps leading to product are straightforward, involving aldol-like condensation to achieve aromatization, followed by acid workup.

11.1   Synthesis of 1,3-Azoles

Are you getting the feeling that a lot of ground breaking science came to the fore in the late nineteenth and early twentieth century? If so you are correct. By way of historical context, U.S. citizens were still debating the American Civil War, Teddy Roosevelt made his famous ride up San Juan Hill in 1898, and the Wright brothers sent imaginations soaring with their 1903 flight at Kitty Hawk, North Carolina. Add to this, physicists were celebrating Einstein's "miracle papers" of 1905, which forever changed our views on space, time, mass, and energy. And their chemist colleagues were isolating and determining the structures of complex natural products without the aid of modern spectroscopic techniques.

   Into this arena arrived (the later to be knighted) Robert Robinson, who, it can be fairly said, ushered in a new age of organic synthesis. Readers will perhaps recognize his name from the annulation reaction outlined in Scheme 9.2, and we will hear more of him in future chapters. To summarize a long and distinguished career, Sir Robert was awarded the Nobel Prize in Chemistry in 1947, "for his investigations on plant products of biological importance, especially the alkaloids." However, as a young man, he is probably best remembered for the 1909 oxazole synthesis that bears his name [24a], along with that of Siegmund Gabriel (Figure 11.7) [24b].

   By Robinson's time, of course, Paal-Knorr type syntheses were well established methodology, involving cyclodehydration of 1,4-dicarbonyl derivatives to give furans, pyrroles, and thiophenes (cf. Figures 11.2 and 11.3). Therefore,

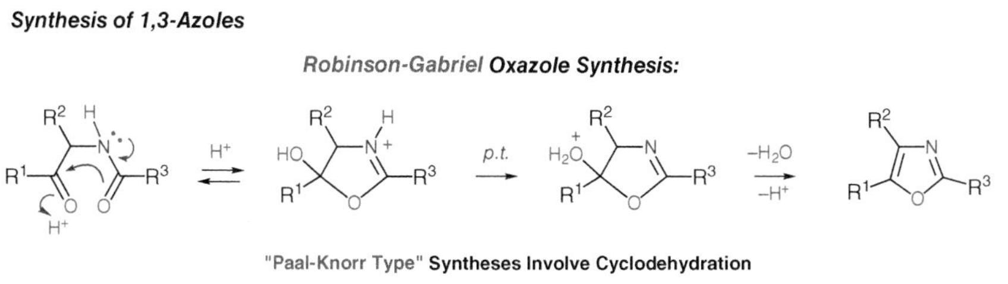

Synthesis of 1,3-Azoles

*Robinson-Gabriel **Oxazole Synthesis:***

"Paal-Knorr Type" Syntheses Involve Cyclodehydration

Figure 11.7 De Novo synthesis.

it comes as no surprise that a similar approach might be applied to the synthesis of oxazoles, which are the first of the 1,3-azoles we shall consider (azoles can be generally defined as 5-membered ring heterocycles containing nitrogen and at least one other heteroatom). The starting materials for the Robinson-Gabriel synthesis are 2-acylamidoketones, which undergo cyclodehydration employing a variety of dehydrating agents (Figure 11.7). Classically, though, this was accomplished with H_2SO_4 [24a]. Mechanistic studies have demonstrated that it is the amide carbonyl that functions as the nucleophile in the initial ring closure, presumably with strong participation of the free electron pair on nitrogen [25]. Subsequent proton transfer (p.t.) then activates the hydroxyl group to elimination, with aromatization providing the driving force.

A particular strength of the Robinson-Gabriel methodology is that the starting 1,4-dicarbonyl derivatives are oftentimes derivable by a Dakin-West reaction, wherein an amino acid is converted directly into a 2-acylamidoketone:

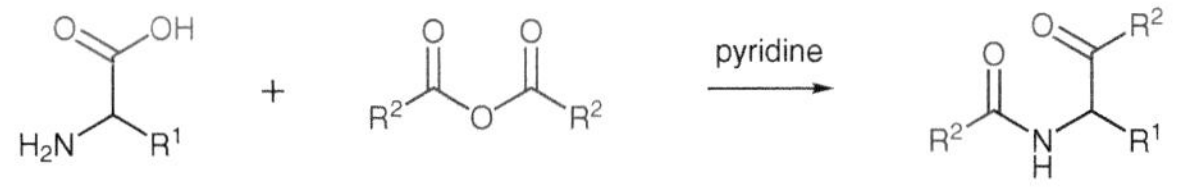

Dakin-West reaction

Typical reaction conditions involve warming with an acid anhydride in pyridine as solvent. Although the mechanism for this conversion is complex, yields can be quite high. This is nicely illustrated by the synthesis of the oxazole containing PPARα/γ agonist depicted at the bottom right of Scheme 11.10, which was obtained in two steps, and >80% overall yield, beginning with the readily available amino acid derivative shown at the top [26]. The first of these steps involved Dakin-West reaction with acetic anhydride (Ac$_2$O) in pyridine (95%), which was followed by cyclodehydration utilizing the reagent combination H_2SO_4/Ac$_2$O in ethyl acetate (87%). Adding to the impressiveness of this route, the synthesis was readily adapted to multi-kilogram scales.

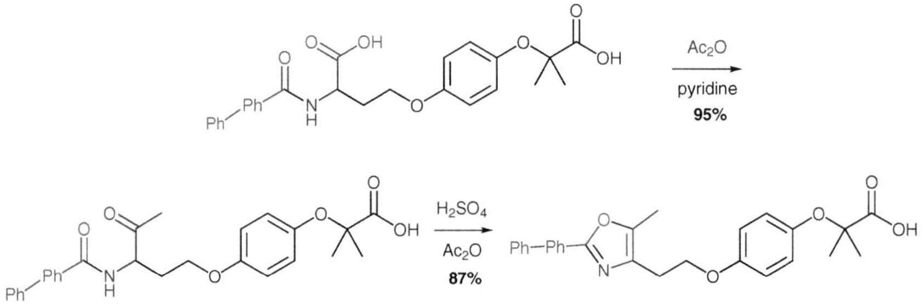

Scheme 11.10

As with the Barton-Zard pyrrole synthesis (cf. Scheme 11.6), oxazoles are also conveniently derived utilizing a so-called "3 + 2" approach. One of the best known of these is the Schöllkopf oxazole synthesis, in which a carboxylic acid derivative is treated with an α-metalated isocyanide (top, Figure 11.8; X = Cl, OR, etc.) [27]. The initial product of this reaction is an α-isocyanoketone, which under basic conditions is in equilibrium with the 2-oxazole anion derived by 5-*endo-dig* cyclization (structures in brackets). Acid workup then affords the corresponding mono- or disubstituted oxazole in generally good-excellent yields. An interesting application of this methodology is provided by the synthesis of the furanosesquiterpene petasalbine, in which a key transformation involved reaction of lithiomethyl isocyanide with a 6-membered ring lactone (bottom Figure 11.8) [28]. The desired oxazole alcohol was obtained in 51% yield, and was converted in three steps to the target natural product. As to the nature of these steps, we will postpone this discussion until a later section describing the chemical reactivity of oxazoles. For now, suffice it to say that these ring systems are exceptionally reactive as dienes in Diels-Alder reactions.

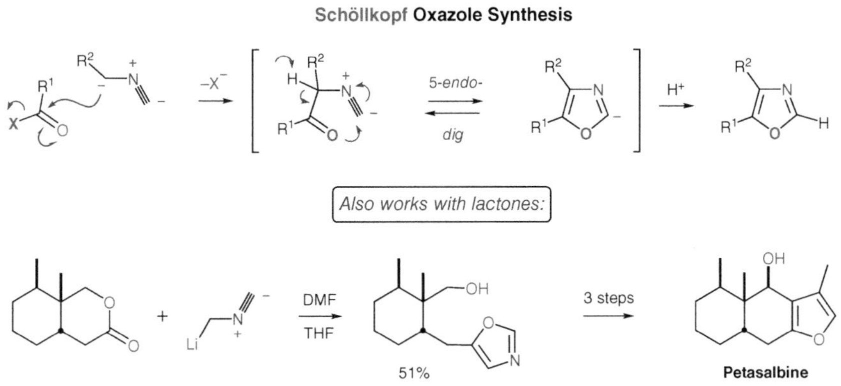

Figure 11.8 De Novo synthesis.

Would you be surprised to learn that most oxazole syntheses have their counterpart in thiazole chemistry? Probably not, given the close structural resemblance of these species. Thus, the Robinson-Gabriel oxazole synthesis finds its direct analogy in the Gabriel thiazole synthesis (1910) [29a], in which 2-acylamidoketones are treated with P_4S_{10}, or its more modern equivalent, Lawesson's reagent (Figure 11.9) [29b]. In addition, thionoesters react with metalated isocyanides in much the same fashion as in the Schöllkopf oxazole synthesis, as do thionolactones [30].

And what would a heterocycle synthesis be without the name Hantzsch attached to it? Already we have highlighted his pyridine and pyrrole syntheses (vide supra), and to those we now add the Hantzsch thiazole synthesis (top,

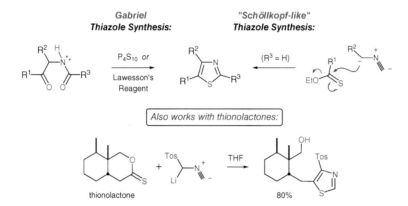

Figure 11.9 De Novo synthesis.

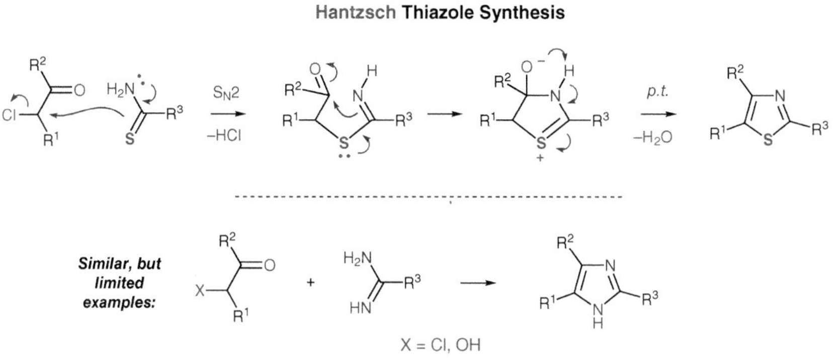

Hantzsch Thiazole Synthesis

Figure 11.10 De Novo synthesis.

Figure 11.10) [31]. Although over 130 years old, it is still probably the most versatile of thiazole syntheses, in part because of the ready availability of the starting materials. These consist of an α-chloroaldehyde or ketone in combination with a thioamide, the latter being highly nucleophilic at sulfur. Simple S_N2 displacement then joins the fragments together, with final ring closure being effected by cyclodehydration. In principle, this approach also seems well suited for synthesizing imidazoles, substituting amidine derivatives for thioamides (bottom Figure 11.10). However, there are only a limited number of such examples in the literature, and side reactions can be a problem.

A more reliable route to imidazoles is provided by the van Leusen imidazole synthesis, outlined at the top of Figure 11.11 [32]. The mechanism for this synthesis has a "déjà vu" feel to it, in that a metalated tosylmethyl isocyanide again plays a key role as a three atom component (see also the van Leusen

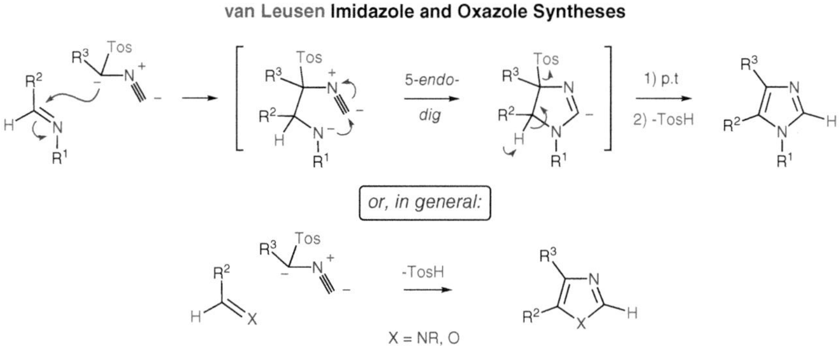

van Leusen Imidazole and Oxazole Syntheses

Figure 11.11 De Novo synthesis.

pyrrole synthesis in Chapter 9). In this case, however, the two atom component consists of an imine, derived by in situ condensation of an aldehyde with a primary amine. Nucleophilic addition now generates an amine anion, which undergoes rapid ring closure by 5-*endo-dig* cyclization (structures in brackets). Elimination of TosH then affords the aromatic imidazole ring. Finally, keeping pace with Hantzsch, van Leusen added a third name reaction to his credit, demonstrating the wide applicability of this methodology to oxazole synthesis (bottom Figure 11.11) [33].

11.2   Synthesis of 1,2-Azoles

At the risk of going off topic, let us return briefly to an earlier chapter on heterocycle synthesis, in which we described pyrimidines as among the easiest of heterocyclic ring systems to prepare (cf. Figure 5.3). Most often this involves a "3 + 3" approach, wherein an N-C-N fragment reacts with an electrophilic C-C-C component to generate the aromatic ring. A case in point is the condensation of an amidine with a 1,3-dicarbonyl derivative to afford pyrimidines directly, by a pathway involving bis-imine formation:

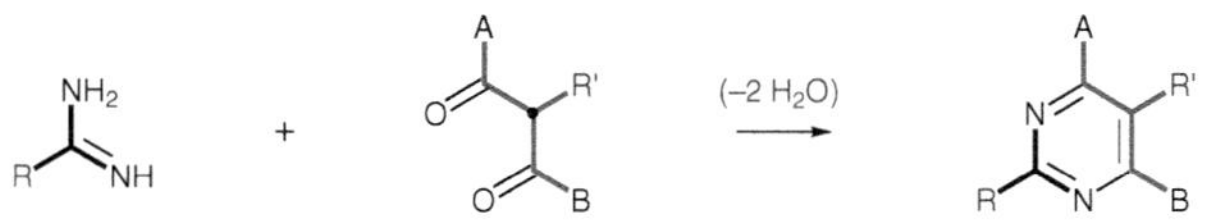

So what is the connection? Simply that condensations of this type find their counterpart in many 1,2-azole syntheses. Thus, nearly all pyrazole and isoxazole syntheses begin with an intact N-N or N-O bond, mirroring the role of amidines in pyrimidine synthesis. Also, electrophilic C-C-C components are in many cases key players. For example, aliphatic and aromatic hydrazines

Synthesis of pyrazoles and isoxazoles

Nearly all syntheses begin with an intact N-N or N-O bond:

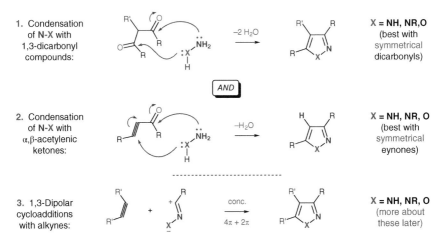

1. Condensation of N-X with 1,3-dicarbonyl compounds:

$-2 H_2O$

X = NH, NR,O (best with symmetrical dicarbonyls)

AND

2. Condensation of N-X with α,β-acetylenic ketones:

$-H_2O$

X = NH, NR, O (best with symmetrical eynones)

3. 1,3-Dipolar cycloadditions with alkynes:

conc.

$4\pi + 2\pi$

X = NH, NR, O (more about these later)

Figure 11.12 De Novo synthesis.

undergo facile condensation with 1,3-dicarbonyl derivatives to afford aromatic pyrazoles directly (X=NH, NR), with the best results obtained with symmetrical dicarbonyls (Figure 11.12, equation 1) [34]. Analogously, hydroxylamine affords the corresponding isoxazoles (X=O) [35].

Numerous variants to this "3 + 2" strategy exist, and there is considerable leeway in the choice of the electrophilic C-C-C fragment. For example, β-ketonitrile derivatives with hydrazine afford 3(5)-aminopyrazoles, while β-ketoesters produce pyrazolones. In addition, as shown in equation 2, α,β-acetylenic ketones are suitable reaction partners, being in the proper oxidation state for direct conversion to pyrazoles and isoxazoles [34, 35]. Here we should point out that there is some ambiguity in the mechanistic pathway for this last ring forming process, which is pH dependent. Under acidic conditions, it is likely the initial step involves hydrazone or oxime formation, followed by intramolecular conjugate addition. However, the reverse order may be favored under alkaline conditions.

Lastly, this brings us to Equation 3, one of the most versatile of all strategies for preparing pyrazoles and isoxazoles. The key step in both cases involves a 1,3-dipolar cycloaddition to an alkyne, a class of concerted $4\pi + 2\pi$ reaction that we will discuss in greater detail in Chapter 15. For now, we will only note that with X=O the reactive species are known as nitrile oxides, which must be generated in situ and afford high yields of isoxazoles [36]. Alternatively, with X=NR the derived nitrile imines are excellent precursors to N-substituted pyrazoles [36a].

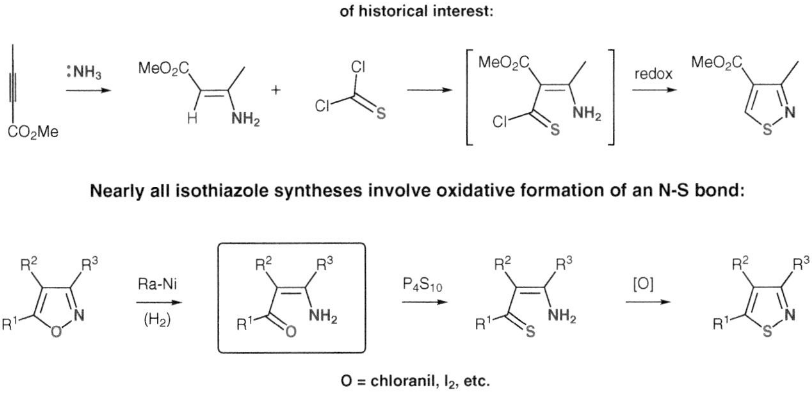

of historical interest:

Nearly all isothiazole syntheses involve oxidative formation of an N-S bond:

O = chloranil, I$_2$, etc.

Figure 11.13 De Novo synthesis.

Notably absent from our discussion thus far are the isothiazoles (X=S in Figure 11.12), which present a new set of challenges. Chiefly this is due to the lack of availability of intact N-S synthons, such as thiohydroxylamine (H$_2$N-SH) or nitrile sulfides (RCNS). Reflecting this fact, the first synthesis of an isothiazole ring was only achieved in 1963, and even then in serendipitous fashion (top Figure 11.13) [37]. The investigator was R. B. Woodward, and the intention was to prepare the simple thiono derivative shown in brackets. Unexpectedly, though, this species underwent spontaneous redox cyclization, producing an isothiazole that would serve as a template for a remarkable synthesis of colchicine. More about this later. At this juncture we need only underscore that, building upon Woodward's observation, most isothiazole syntheses involve oxidative formation of an N-S bond, according to the general strategy shown at the bottom of Figure 11.13 [38]. Interestingly, the starting enamides are frequently best prepared by Raney nickel cleavage of the N-O bond in the corresponding isoxazoles (cf. structure in box) [39]. Sulfonation with P$_4$S$_{10}$ then affords the requisite β-aminoenethiones, which undergo in situ ring closure using a variety of mild oxidants (O = chloranil, I$_2$, etc.).

11.3   Fischer Indole Synthesis

Ask yourself "who was the greatest organic chemist of the nineteenth century?" Put the same question to any second semester organic student and they will likely (or should) answer Emil Fischer. Fischer received the Nobel Prize in Chemistry in 1902, "in recognition of the extraordinary services he has rendered by his work on sugar and purine syntheses." However, this citation barely does justice to the impact he had on the field. Spend an afternoon with the

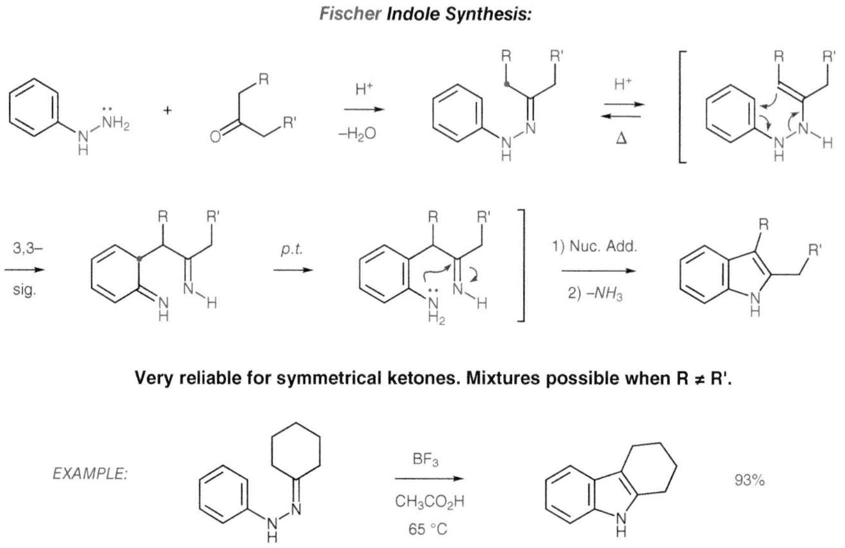

Fischer *Indole Synthesis*:

Very reliable for symmetrical ketones. Mixtures possible when R ≠ R'.

EXAMPLE:

$$BF_3$$
$$CH_3CO_2H$$
$$65\ °C$$

93%

Figure 11.14 De Novo synthesis.

"Fischer Proof" of glucose structure (1891), and you will come to appreciate this as one of the greatest intellectual achievements of the era. And who has not employed Fischer projections, or pondered the implications of his "lock and key" mechanism of enzyme activation? Add to this his many contributions to heterocyclic chemistry, of which we shall consider just one—the indole synthesis that bears his name. Initially reported in 1883 [40], it is one of the most recognizable name reactions in all of organic chemistry (Figure 11.14).

In its most simple representation, a phenylhydrazone is converted to an indole in a single step. In actuality, though, Fischer himself was well aware of the complexity of this reaction, writing in the last paragraph of his paper that "Dieser Vorgang ist so merkwürdig, dass wir einstweilen es nicht wagen, eine Erklärung desselben zu geben [41]." What he was referring to, of course, was not the initial formation of a phenylhydrazone, a reaction that he had discovered some years earlier. Rather, it was the unprecedented sequence of steps that transpired upon heating these derivatives in acid, accompanied by vivid changes in color, and a "noticeable amount of ammonia."

Not surprisingly, the mechanism of indole formation has been studied extensively, with the initiating step involving tautomerization of the hydrazone to an ene-hydrazine (Figure 11.14, first structure in brackets) [42]. Two aspects of this ene-hydrazine deserve special mention, the first being that it possesses a relatively weak N-N bond; and second, it has two double bonds that are now ideally situated to undergo a 3,3-sigmatropic shift (curly arrows). In this fashion the terminus of the enamine carbon migrates to the *ortho* position of the benzene ring, concomitant with N-N bond scission. With most of the "heavy lifting" now accomplished, simple proton transfer (p.t.) generates an aromatic aniline ring, in which the amino group is in juxtaposition to a highly electrophilic imine (last structure in brackets). Acid catalyzed condensation, with loss of ammonia, then produces the aromatic indole product.

In general, this strategy is very reliable for aldehydes and symmetrical ketones, as evidenced by the high yielding transformation shown at the bottom of Figure 11.14 [43]. However, mixtures are possible when R≠R'. An exception to this complication is with ketones bearing only one set of α-protons, which can be quite complex in nature. Gribble et al. provided an excellent example of this last circumstance during their studies of indole-fused oleanolic acid derivatives. Note in this case that the intermediate hydrazone need not be isolated [44]:

This draws to a close one aspect of our discussion of π-excessive heterocyclic synthesis. But before leaving this chapter, it is worth providing a glimpse of things to come. In our discussions thus far we have limited ourselves to "de Novo" syntheses involving non-heterocyclic starting materials, whether the products be π-deficient or π-excessive in nature. However, more than once we have alluded to the possibility of preparing heterocycles from other heterocycles. When would such a strategy be beneficial? To begin with, the starting materials must be readily available, and preferably from a renewable resource. Given this qualification it is natural that our attention might turn to furan, which, as we have already described, is abundantly derived from agricultural waste products (cf. Figure 11.2). Indeed, since furan is cheap, and easily functionalized, it is frequently employed as a starting material for other ring systems (Figure 11.15). In the examples shown, advantage is taken of the fact that furan rings are in the same oxidation state as 1,4-dicarbonyl derivatives, which, in turn, are versatile precursors to pyrroles and thiophenes (cf. also Figure 11.3).

In principle, then, a highly functionalized furan could be converted to a like-substituted pyrrole by acid hydrolysis, followed by condensation with ammonia or a primary amine. In analogous fashion, thiophenes would result upon treatment with P_4S_{10}. However, there is a problem with this scenario, in that the derived 1,4-dicarbonyl derivatives are frequently unstable to the strongly acidic conditions necessary to open the furan ring. A more general solution involves initial de-aromatization employing the reagent combination of bromine in

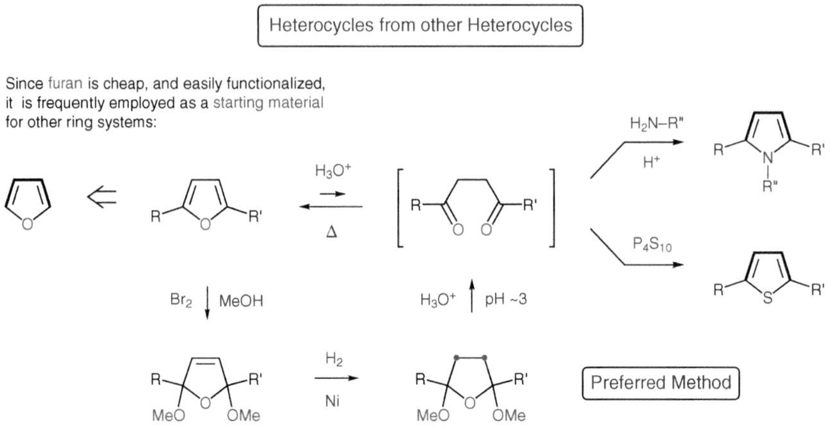

Heterocycles from other Heterocycles

Since furan is cheap, and easily functionalized, it is frequently employed as a starting material for other ring systems:

Figure 11.15 De Novo synthesis.

methanol, which effects rapid oxidation of furans to 2,5-dihydro-2,5-dimethoxyfurans (bottom left, Figure 11.15). We shall take up the mechanism for this transformation in the following chapter. For now, suffice it to say that we have converted a relatively stable furan ring into an acid-labile *bis* cyclic acetal, which upon catalytic hydrogenation returns us to the oxidation state of the starting furan. But what a change in reactivity! The derived 2,5-dimethoxytetrahydrofuran now suffers facile hydrolysis in the pH range of two to three, to afford excellent yields of the 1,4-dicarbonyl derivatives shown in brackets. Typically, though, these are not isolated, but rather are carried through directly to the desired pyrrole or thiophene product.

References

1 (a) Paal, C. *Chem. Ber.* **1884**, *17*, 2756–2767. (b) Knorr, L. *Chem. Ber.* **1884**, *17*, 2863–2870.
2 Brownlee, H. J.; Miner, C. S. *Ind. Eng. Chem.* **1948**, *40*, 201–204
3 Feather, M. S.; Harris, D. W.; Nichols, S. B. *J. Org. Chem.* **1972**, *37*, 1606–1608.
4 Paal, C. *Chem. Ber.* **1885**, *18*, 367–371.
5 Brettle, R.; Dunmur, D. A.; Marson, C. M.; Pinol, M.; Toriyama, K. *Chem. Lett.* **1992**, 613–616.
6 Knorr, L. *Chem. Ber.* **1885**, *18*, 299–311.
7 Jacobi, P. A.; Li, Y. *J. Am. Chem. Soc.* **2001**, *123*, 9307–9312.
8 Jung, M. E.; Gervay, J. *J. Am. Chem. Soc.* **1991**, *113*, 224–232.
9 Jacobi, P. A.; Cai, G. *Tetrahedron Lett.* **1991**, *32*, 1765–1768.
10 Knorr, L. *Chem. Ber.* **1884**, *17*, 1635–1642.

11 Fischer, H. in *Organic Syntheses*, Coll. Vol. II, John Wiley & Sons, Inc., New York, New York, **1943**, p. 202.

12 Coffen, D. L.; Hengartner, U.; Katonak, D. A.; Mulligan, M. E.; Burdick, D. C.; Olson, G. L.; Todaro, L. J. *J. Org. Chem.* **1984**, *49*, 5109–5113.

13 (a) Johnson. A. W.; Markam, E.; Price, R.; Shaw, K. B. *J. Chem. Soc.* **1958**, 4254–4257. (b) Johnson, A. W.; Price, R. in *Organic Syntheses*, Coll. Vol. V., John Wiley & Sons, Inc., New York, New York, **1973**, p. 1022.

14 (a) Hantzsch, A. *Chem. Ber.* **1890**, *23*, 1474–1476. For a review, see (b) Roomi, M. W.; MacDonald, S. F. *Can. J. Chem.* **1970**, *48*, 1689–1697.

15 Feist, F. *Chem. Ber.* **1902**, *35*, 1545–1556.

16 Bénary, E. *Chem. Ber.* **1911**, *44*, 493–496.

17 Calter, M. A.; Zhu, C.; Lachicotte, R. J. *Org. Lett.* **2002**, *4*, 209–212.

18 Calter, M. A.; Korotkov A. *Org. Let.* **2011**, *13*, 6328–6330.

19 Barton, D. H. R.; Kervagoret, J.; Zard, S. Z. *Tetrahedron* **1990**, *46*, 7587–7598.

20 Van Leusen, A. M.; Siderius, H.; Hoogenboom, B. E.; van Leusen, D. *Tetrahedron Lett.* **1972**, *13*, 5337–5340.

21 Haake, G.; Struve, D.; Montforts, F.-P. *Tetrahedron Lett.* **1994**, *35*, 9703–9704.

22 Hinsberg, O. *Chem. Ber.* **1910**, *43*, 901–906.

23 Wynberg, H.; Kooreman, H. J. *J. Am. Chem. Soc.* **1965**, *87*, 1739–1742.

24 (a) Robinson, R. *J. Chem. Soc.* **1909**, *95*, 2167–2174. (b) Gabriel, S. *Chem. Ber.* **1910**, *43*, 134–138.

25 Wasserman, H. H.; Vinick, F. J. *J. Org. Chem.* **1973**, *38*, 2407–2408.

26 Godfrey, A. G.; Brooks, D. A.; Hay, L. A.; Peters, M.; McCarthy, J. R.; Mitchell, D. J. *Org. Chem.* **2003**, *68*, 2623–2632.

27 Schöllkopf, U. *Angew. Chem. Int. Ed. Engl.* **1977**, *16*, 339–348, and references cited therein.

28 Jacobi, P. A.; Walker, D. G. *J. Am. Chem. Soc.* **1981**, *103*, 4611–4613.

29 (a) Gabriel, S. *Chem. Ber.* **1910**, *43*, 1283–1287. (b) Sanz-Cervera, J. F.; Blasco, R.; Piera, J.; Cynamon, M.; Ibáñez, I.; Murguía, M.; Fustero, S. *J. Org. Chem.* **2009**, *74*, 8988–8996.

30 Jacobi, P. A.; Frechette, R. F. *Tetrahedron Lett.* **1987**, *28*, 2937–2940.

31 Hantzsch, A.; Weber, J. H. *Chem. Ber.* **1887**, *20*, 3118–3132.

32 Van Leusen, A. M.; Wildeman, J.; Oldenziel, O. H. *J. Org. Chem.* **1977**, *42*, 1153–1159.

33 Van Leusen, A. M.; Hoogenboom, B. E.; Siderius, H. *Tetrahedron Lett.* **1972**, *13*, 2369–2372.

34 Fusco, R. in *Pyrazoles, Pyrazolines, Pyrazolidines, Indazoles and Condensed Rings*, Wiley, R. H., Ed., John Wiley & Sons, Inc., New York, New York, **1967**.

35 Grünanger, P; Vita-Finzi, P. in *Isoxazoles*, Part One, Taylor, E. C.; Weissberger, A., Eds., John Wiley & Sons, Inc., New York, New York, **1991**.

36 (a) Caramella, P.; Grünanger, P. in *1,3-Dipolar Cycloaddition Chemistry*, Vol. 1, Padwa, A., Ed., John Wiley & Sons, Inc., New York, New York, **1984**. See also (b) Padwa, A. in *1,3-Dipolar Cycloaddition Chemistry*, Vol. 2, Padwa, A., Ed.,

John Wiley & Sons, Inc., New York, New York, **1984**. (c) Jager, V.; Colinas, P. A. in *Synthetic Applications of 1,3-Dipolar Cycloaddition Chemistry Toward Heterocycles and Natural Products*, Padwa, A.; Pearson, W. H., Eds., John Wiley & Sons, Inc., New York, New York, **2002**.

37 Woodward, R. B. in *Harvey Lectures Series 59 (1963–64)*, Academic Press, Inc., New York, New York, **1965**, pp. 31–47.

38 Wooldridge, K. R. H. *Adv. Heterocycl. Chem.* **1972**, *14*, 1–41.

39 McGregor, D. N.; Corbin, U.; Swigor, J. E.; Cheney, L. C. *Tetrahedron* **1969**, *25*, 389–395.

40 Fischer, E.; Jourdan, F. *Chem. Ber.* **1883**, *16*, 2241–2245.

41 Rough translation: "This process is so strange that for the time being we dare not give an explanation of it."

42 (a) Allen, C. F. H.; Wilson, C. V. *J. Am. Chem. Soc.* **1943**, *65*, 611–612. (b) Carlin, R. B.; Fisher, E. E. *J. Am. Chem. Soc.* **1948**, *70*, 3421–3424. (c) Douglas, A. W. *J. Am. Chem. Soc.* **1979**, *101*, 5676–5678. (d) Hughes, D. L. *Org. Prep. Proc. Int.* **1993**, *25*, 607–632

43 Snyder, H. R.; Smith, C. W. *J. Am. Chem. Soc.* **1943**, *65*, 2452–2454.

44 Finlay, H. J.; Honda, T.; Gribble, G. W. *Arkivoc* **2002**, *12*, 38–46.

12

π-Excessive Heterocycles: Introduction of New Substituents

And by what means do we functionalize π-excessive heterocycles? Must we again take a "cleansing rinse," as in the case with their π-deficient cousins (cf. Chapter 6)? Not necessarily. It turns out that we are treading in much more familiar territory here, since most methodology finds its parallel in benzene chemistry. Thus, as previously described, the chemical reactivity of thiophene, furan, and pyrrole is similar to that of anisole, in that they are ready partners in a variety of electrophilic substitution reactions (cf. Chapter 2). However, as introduced in Figure 12.1, the reagents involved may require modification.

Let us begin with thiophene (top left Figure 12.1), which shares many properties in common with benzene (remember our distraught Viktor Meyer?). Consequently, to use an overworked phrase, we can play chemical "hard-ball" with this substrate, overriding the fact that it is the least nucleophilic of the parent π-excessive rings (but still considerably more so than benzene). Thus, sulfonation is cleanly achieved upon standing at room temperature (RT) in concentrated H_2SO_4 [1], conditions under which benzene is stable, but which would rapidly decompose either furan or pyrrole (as an aside, can you suggest now how Meyer eventually separated thiophene from benzene?). So what recourse do we have for these latter ring systems? Fortunately, alternative reagents have been devised that accomplish the same purpose but in the absence of powerful acids. One such reagent is the crystalline complex derived by treating sulfur trioxide (SO_3) with pyridine, which can be employed in an inert solvent such as dichloroethane (DCE):

These are conditions that are compatible even with the easily polymerized furan, which affords a 41% yield of furan-2-sulfonic acid at RT (top center

Introductory Heterocyclic Chemistry, First Edition. Peter A. Jacobi.
© 2019 John Wiley & Sons Ltd. Published 2019 by John Wiley & Sons Ltd.

2. Introduction of new substituents; Electrophilic Substitution. *Reactivity is similar to anisole, BUT....*

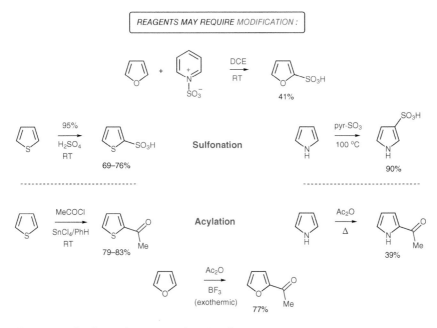

Figure 12.1 Synthesis of π-excessive heterocycles.

Figure 12.1) [2]. Interestingly, the main by-product in this reaction is furan-2,5-disulfonic acid (15%), which is formed almost exclusively at higher temperatures. Moving on to pyrrole, one observes a divergent reaction pathway with the same reagent (top right Figure 12.1). At 100 degrees Celsius, this ring system undergoes clean sulfonation, and for many years it was assumed that the product was the kinetically favored pyrrole-2-sulfonic acid. However, re-investigation of this reaction demonstrated that the actual product, formed in 90% yield, is the thermodynamically more stable pyrrole-3-sulfonic acid [3]. With hindsight this result is perhaps not so surprising, given that aromatic sulfonation is a readily reversible process:

Turning now to the bottom left of Figure 12.1, thiophene affords a 79–83% yield of 2-acetylthiophene upon Friedel-Crafts acylation with acetyl chloride/SnCl$_4$ at RT [4]. Remarkably, this reaction is conducted in benzene, providing further insight into the relative nucleophilicity of these two ring systems. Little, if any, of the 3-isomer is formed, and a similar protocol can be employed with a wide variety of acid chlorides. In most cases, $SnCl_4$ is preferred over $AlCl_3$ as a Lewis acid catalyst, since it causes less decomposition. The acid sensitive (but more reactive) furan requires still milder conditions, giving a 77% yield of 2-acetylfuran with the reagent combination of Ac_2O/BF_3 (bottom center Figure 12.1; the reaction exotherm is controlled with ice cooling) [5]. Comparable yields are obtained with propionic and butyric anhydrides (81% and 93%, respectively), and once again, there is no mention of the corresponding 3-isomers. Finally, the highly reactive pyrrole affords a 39% yield of 2-acetylpyrrole simply upon heating in acetic anhydride, this time accompanied by ~9% of the 3-isomer (bottom right Figure 12.1; older literature cites a virtually quantitative yield of the 2-isomer) [6].

Let us next consider nitration, which in benzene chemistry is effected with a mixture of concentrated sulfuric and nitric acids. Consult any first year organic text, and you will find that the purpose of H_2SO_4 is to protonate the weaker acid HNO_3 (top Figure 12.2). Elimination of a molecule of water then yields the active electrophile nitronium ion ($+NO_2$). Could our robust thiophene ring stand up even to these conditions? It is best not to try, since at a minimum you will see tar formation—under some conditions escalating to "explosive violence" (this last apparently due to nitrosation side reactions, in which HONO is generated autocatalytically during the course of reaction) [7]. Much more satisfactory is the reagent derived by dissolving fuming nitric acid in acetic anhydride, which affords the mixed anhydride acetyl nitrate (bottom Figure 12.2).

Both thiophene and pyrrole undergo smooth nitration with acetyl nitrate, affording 50–85% yields of the corresponding 2-nitro derivatives at 0–10 degrees Celsius (top Figure 12.3; in the case of pyrrole, ~13% of the 3-nitro isomer is also formed) [8]. Mechanistically, there has been little speculation on the course of these reactions, but it is unlikely that $+NO_2$ is the reactive species. Firstly, benzene is essentially inert to acetyl nitrate, which would not be the

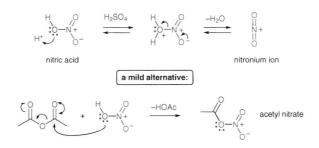

Figure 12.2 Generation of acetyl nitrate.

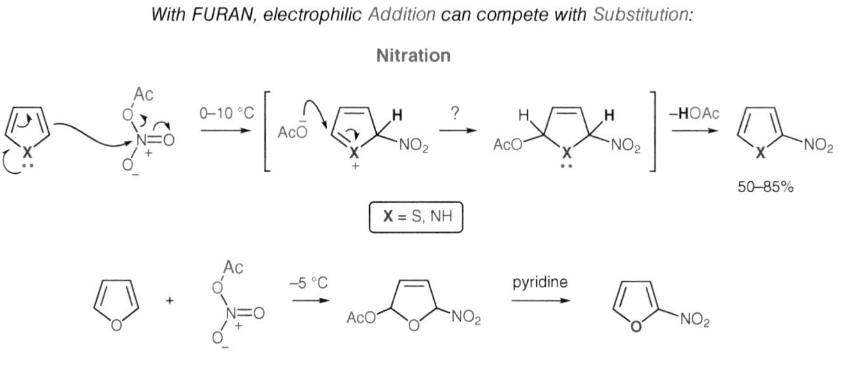

With FURAN, electrophilic Addition can compete with Substitution:

Nitration

X = S, NH

Figure 12.3 Introduction of new substituents.

case were +NO$_2$ present. Also, nitronium ion cannot be detected by Ramon or IR spectroscopy in freshly prepared solutions of this reagent [9]. This leaves the mixed anhydride itself, which in either protonated or neutral form might undergo direct nucleophilic attack by the π-excessive ring (curly arrows). This would give the first intermediate shown in brackets, which parenthetically would be the same as that obtained with +NO$_2$. From here, though, the pathway forward is less clear. Simple re-aromatization by loss of the proton highlighted in bold is one possibility, following the classical mechanism observed in benzene chemistry. However, a second possibility exists, in that the initial cation might be intercepted by acetate anion, in this manner producing a 2,5-dihydro-2-acetoxy-5-nitro derivative (second structure in brackets). 1,4-Elimination of HOAc would then lead to the observed products. Some credence for this pathway is provided by the fact that such intermediates are actually isolable in the case of nitration of furan with acetyl nitrate (bottom Figure 12.3) [10]. Aromatization is then achieved by treatment with a weak base such as pyridine.

Not surprisingly, thiophene, pyrrole and furan are exceptionally reactive toward halogenation, in most cases requiring no added catalyst (top Figure 12.4). At 25 degrees Celsius, for example, thiophene is about 10^7 times more reactive than benzene in chlorination, and 10^9 more reactive in bromination (and remember, thiophene is the least nucleophilic of the trio!) [11]. With Cl$_2$ and Br$_2$, 2,5-bis-halogenation predominates, although modified reagents, and low temperatures, can diminish this reaction pathway. Iodine is less reactive, but with HgO affords 72–75% yields of 2-iodothiophene at RT (equation 1 in Figure 12.4; again utilizing benzene as solvent) [12]. A small amount of 2,5-diiodothiophene is also formed.

At the other end of the reactivity scale sits pyrrole, where tetrahalogenation is the norm, even with I$_2$ (equation 2) [13]. Furan occupies an "in-between" position, and in fact is unreactive toward iodine. Still, though, it is difficult to control

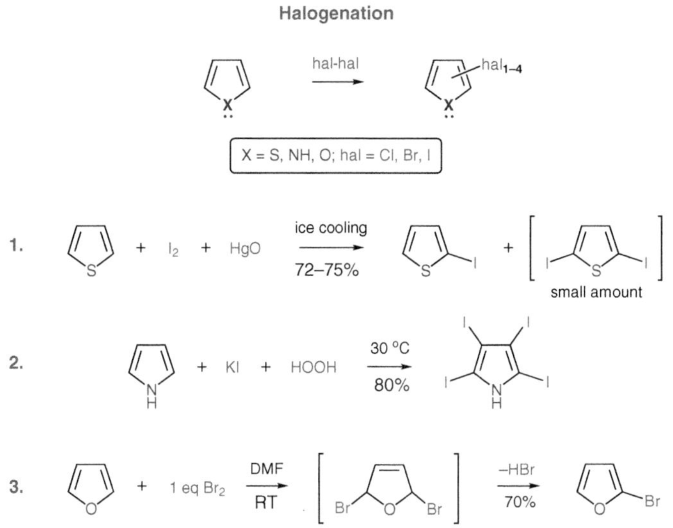

Halogenation

Figure 12.4 Introduction of new substituents.

polyhalogenation with Cl_2 and Br_2. In the case of bromination, this outcome can be mitigated by employing the much milder reagent dioxane dibromide, a stable 1:1 complex that has found considerable utility in brominating activated aromatic systems [14]. Alternatively, the combination of Br_2 in dimethylformamide (DMF) has proven to be highly effective. With this solvent, one obtains a 70% yield of 2-bromofuran at room temperature, and it has been postulated that the reaction involves initial 1,4-addition (structure in brackets) [15]. Low temperature NMR studies utilizing molecular Br_2 in CS_2 tend to support this postulate [16]. However, as the authors note, the spectral observation of such intermediates does not constitute proof that bromination occurs via an addition-elimination mechanism. It could also be that these adducts represent a side equilibrium.

What is not in doubt, however, is that bromination of furan in MeOH takes place by electrophilic addition. In this case the relatively stable 2,5-dihydro-2,5-dimethoxyfuran is isolated in 75-79% yield as a mixture of cis and trans isomers (top Scheme 12.1) [17]. As to a mechanism, some literature sources posit the intermediacy of 2,5-dibromo-2,5-dihydrofuran, as in equation 3 of Figure 12.4. However, given the nucleophilicity of MeOH, and its overwhelming concentration, the sequence outlined in Scheme 12.1 seems more likely. Here, the initially formed cation is captured not by Br⁻, but directly with MeOH. This is followed by methanolysis of the 2-bromo substituent (curly arrows). In analogous fashion, bromination in HOAc affords a 70% yield of 2,5-diacetoxy-2,5-dihydrofuran (bottom Scheme 12.1) [18].

Coming back to pyrrole, this ring system is sufficiently nucleophilic to undergo diazo coupling with benzenediazonium salts, which are generally

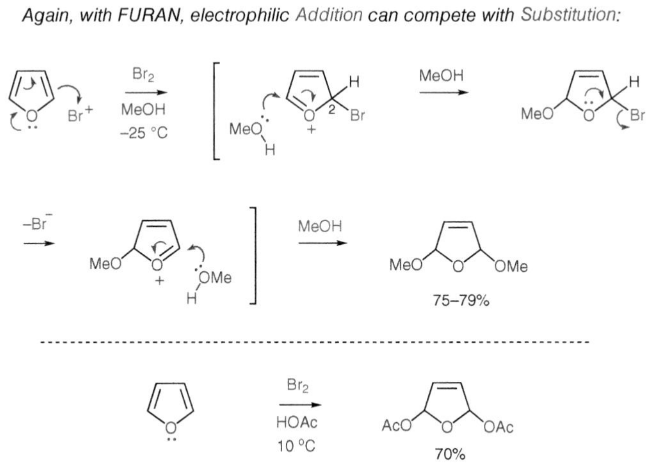

Again, with FURAN, electrophilic Addition can compete with Substitution:

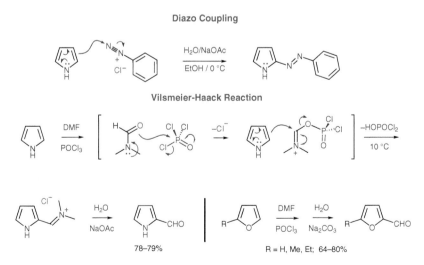

Scheme 12.1 Introduction of new substituents.

considered to be relatively poor electrophiles. The "parent" reaction of this class was first reported in 1886, in connection with a study on dyestuffs (top Scheme 12.2) [19]. And who discovered this reaction? The lead author was one Otto Fischer, a cousin of Emil Fischer, with whom he had many fruitful collaborations. Indeed, it has been suggested that most of Otto's contributions to the field centered around the pioneering work of Emil. In any event, our dyestuff discoverer is not to be confused with a distinguished German chemist of a later

Diazo Coupling

Vilsmeier-Haack Reaction

Scheme 12.2 Introduction of new substituents.

generation, Ernst Otto Fischer. Ernst (almost universally called E.O., and no relation to Emil) went on to share the 1973 Nobel Prize in Chemistry with Sir Geoffrey Wilkinson for their work on organometallic compounds, including ferrocene. But we digress!

Pyrrole and its derivatives are also ready participants in a formylation process known as a Vilsmeier-Haack reaction, which is limited to only the most reactive of aromatic rings (middle Scheme 12.2). Historically, the so-called Vilsmeier reagent has been generated in situ from a variety of formylated secondary amines and $POCl_3$. More often than not, though, dimethylformamide (DMF) plays the duel role of solvent as well as formyl transfer agent, as in the example shown [20]. And what is the active formylating agent? What we can say with certainty is that $POCl_3$ undergoes an exothermic reaction with DMF, in which a chloride ion is displaced by the relatively nucleophilic amide carbonyl group (curly arrows in brackets). From here two pathways are available. In the most direct of these, the resultant positively charged imidate ester undergoes nucleophilic acyl substitution, with pyrrole functioning as the nucleophile, and $HOPOCl_2$ serving as the leaving group (second structure in brackets). Alternatively, depending upon the substrates, this step may be preceded by formation of the corresponding iminoyl chloride. In either case, with pyrrole itself the initial product is a 2-substituted iminium salt (bottom left Scheme 12.2), which undergoes rapid hydrolysis to pyrrole-2-carboxaldehyde upon aqueous workup (78–79% yield) [20]. In similar fashion, but under somewhat more vigorous conditions, furan and its 2-alkyl derivatives undergo selective formylation at the unoccupied 5-position (bottom right Scheme 12.2) [21]. Lastly, there are relatively few examples of thiophene or its derivatives undergoing a Vilsmeier-Haack reaction with the reagent combination DMF/$POCl_3$. However, the parent ring does afford a 68% yield of thiophene-2-carboxaldehyde on warming with $POCl_3$ and N-methylformanilide [22]:

Somewhat higher yields are obtained with alkyl substituted derivatives, while 2-nitrothiophene and 2-acetylthiophene are completely unreactive. These last results, of course, are not unexpected, given the strongly deactivating nature of these groups.

Return now for a moment to our earlier discussion on the Mannich condensation, "which in its many variants is one of the most important bond forming processes in alkaloid chemistry." In Chapter 9 we saw how reactions of this category can be utilized for functionalizing α-methyl groups in π-deficient heterocycles (cf. Figures 9.3 and 9.4). Can this reaction also be employed for functionalizing π-excessive heterocycles? The answer is yes, but with some qualifications.

The active electrophile in such reactions is an iminium salt, classically formed by interaction of formaldehyde with a secondary amine under mildly acidic conditions (Figure 12.5; later we shall see that there is a wide latitude of choices for both the amine and carbonyl components). As indicated, in many cases the iminium electrophile has only to be generated in situ, and the equilibrium need not lie far to the right. For discussion purposes we will refer to these as *conditions a*, which are suitable for highly nucleophilic ring systems such as pyrrole (X = NH). Indeed, pyrrole and many of its derivatives undergo facile Mannich condensations with a wide range of aldehydes and amines, generally proceeding at room temperature (RT) in moderate to excellent yield. Remarkably, even strongly deactivated pyrrole rings undergo successful condensation, albeit under more vigorous conditions (top Figure 12.6) [23].

By way of comparison, furan (X = O) is much less reactive under *conditions a*, at the least requiring an activating alkyl substituent G, along with elevated temperatures. However, it undergoes ready condensation with preformed iminium salts at RT in MeCN (*conditions b*, bottom right Figure 12.6) [24]. Finally, bringing up the rear is thiophene (X = S), which even with preformed iminium salts requires heating to reflux in MeCN (bottom left Figure 12.6) [25].

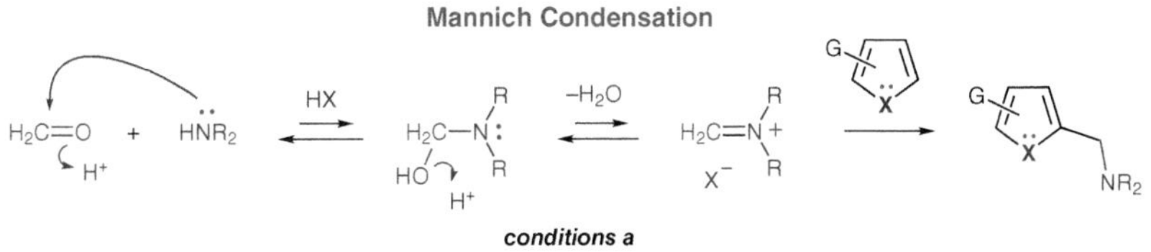

Figure 12.5 Introduction of new substituents.

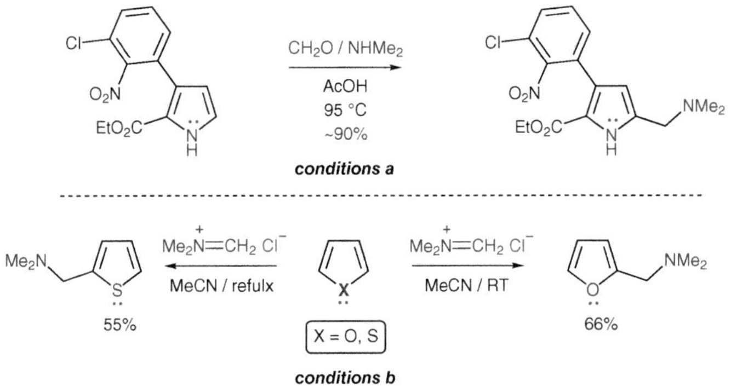

Figure 12.6 Conditions for Mannich condensations.

And of what utility are such Mannich condensations with π-excessive hetero-cycles? Oftentimes the goal is to introduce an α-methyl substituent incorporat-ing a good leaving group. For this purpose, the dimethylamino group is further activated by quaternization, with the resultant salt undergoing facile displace-ment. Note that in most cases this substitution likely follows an S_N1 pathway, rendered all the more favorable by electron donation from the heteroatom:

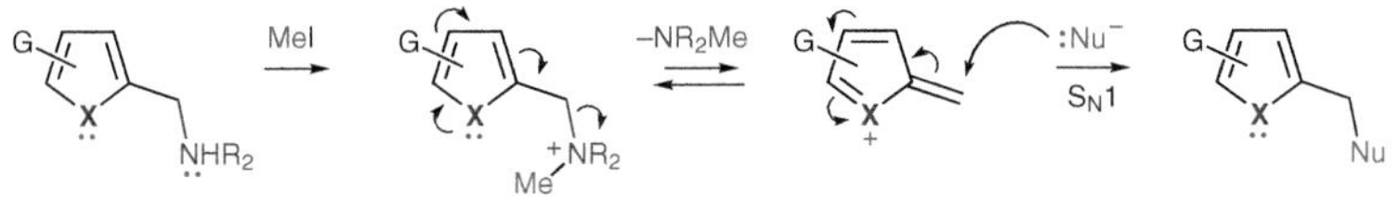

Nor is the Mannich condensation limited to monocyclic π-excessive hetero-cycles. Indole and its derivatives are particularly reactive, with the important proviso that reaction now occurs at the β-position (cf. bottom Figure 10.4). This reactivity was nicely exploited in an early synthesis of d,l-tryptophan, in which *conditions a* were utilized in the first key bond forming step (Scheme 12.3) [26]. The product of this reaction is a naturally occurring alkaloid known as gramine, which without purification, was converted in 80% overall yield to the corresponding methiodide salt (this on ~250 g scale). What remained was to carry out a nucleophilic displacement with a synthetic equivalent of glycine, highlighted in red in the final product. Ultimately this was accomplished in 63–70% yield employing the sodium salt of acetamidomalonic ester. The remaining steps were now routine, involving ester hydrolysis, decarboxylation, and cleavage of the acetamide group. This sequence was carried out essentially in one "pot," providing d,l-tryptophan in ~45% overall yield from indole [26].

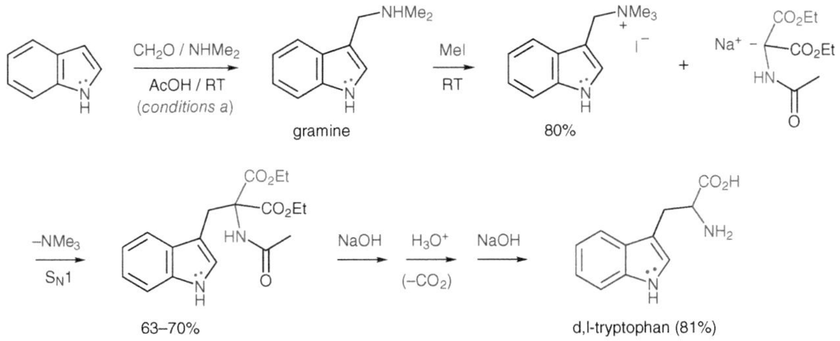

Scheme 12.3

So where from here? To this point we have covered many of the reactions that can be categorized as electrophilic in nature, which for the most part are applicable to 1,2- and 1,3-azoles as well. However, as we shall see shortly, reac-tions of this class are not the only means available for functionalizing π-excessive heterocycles, and we can draw other parallels to anisole chemistry. But before moving on, let us consider one additional natural product synthesis that is widely regarded as a classic in the field, and which draws together a number of principles we have discussed in the past two chapters.

Some readers will already be familiar with the 1963 publication describing this synthesis (at least in an historical sense), with its famous opening exclamation of "STRYCHNINE! [27a]" How Woodwardian! It is doubtful whether the scientific literature could accommodate such a writing style in the present day. However, this was Robert Burns Woodward, and if Emil Fischer was the preeminent organic chemist of the nineteenth century, then many would argue that RBW held that mantel in the twentieth. Of our target, it was said at the time that "For its molecular size it is the most complex substance known:"

strychnine

This can be argued, but according to Woodward, "... with six nuclear asymmetric centers and seven rings constituted from only twenty-four skeletal atoms, the case is a good one."

Woodward chose as his starting materials phenyl hydrazine and methyl veratryl ketone, two items of commerce that he had every reason to believe would participate in a successful Fischer indole synthesis. And well they did, affording a 54% yield of 2-veratrylindole simply upon warming with polyphosphoric acid (top right of Scheme 12.4). Now, here we might pause for a moment and ask ourselves "why the veratryl group? This has no obvious place in the strychnine skeleton." And for the uninitiated this is certainly a legitimate question to raise. But it is not by happenstance that Woodward's biographers have referred to him as "an architect and artist in the world of molecules [27b]," for in this one stroke he has not only formed rings I and II, but he has incorporated nearly all of the elements necessary for constructing rings III and IV. More about this below.

First, though, let us focus on ring V, which required functionalization at the indole β-position. This was initiated by Mannich condensation utilizing *conditions a*, which afforded a 92% yield of the dimethylaminomethyl derivative shown in the second row of the scheme. Quaternization with methyl iodide then set the stage for displacement with ⁻CN (97%), which when followed by LAH reduction produced a >80% overall yield of the β-aminoethyl derivative shown at the right in row 3. All of this chemistry is by now familiar to us, given our discussions of the preceding pages. Nor is there anything out of the ordinary with the following step, which involved imine formation with ethyl glyoxylate

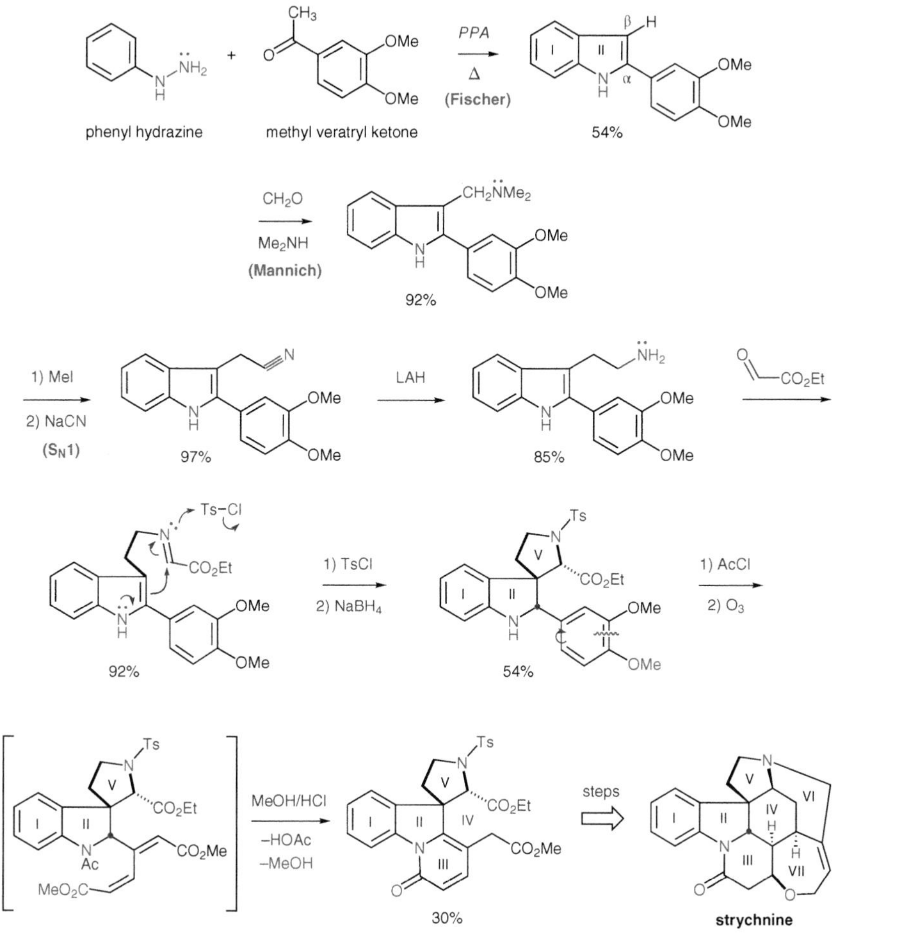

Scheme 12.4 Putting it all together.

(92%). This takes us down to the left structure of row four, where things get a bit more interesting. Woodward's plan was to introduce the spirocyclic ring V according to the general electron flow indicated by curly arrows, which you will recognize as an intramolecular Mannich-like condensation. But how to drive this step forward? After all, we *are* sacrificing the aromaticity of a pyrrole ring, and the reaction is clearly reversible. The solution turned out to be surprisingly simple, involving capture of the initial adduct with one equivalent of *p*-toluenesulfonyl chloride (TsCl). With the retro-Mannich reaction blocked, reduction of the resultant indolenine with NaBH$_4$ then gave a 54% overall yield of the key intermediate shown to the right of row four. Three rings down, four to go.

Now is a good time to return to our veratryl ring, whose sole purpose thus far has been to shield the indole α-position from electrophilic attack. However, Woodward had much more in mind. To put it in his own words, "Our veratryl group had now served admirably its function of directing the processes involved in the elaboration of ring V, and at this point we undertook to examine whether it could be made to perform its second function." The second function, of course, pertained to the further unveiling of the strychnine skeleton, and here must be noted a special property of the veratryl ring. While aromatic, Woodward correctly predicted that oxidative cleavage between the two strongly electron donating methoxy groups should be facile. To start the sequence, the basic aniline-like nitrogen was protected by acylation, followed by ozonolysis at low temperature. And what of the two methyl ester groups freed by such an operation (structure in brackets)? One (in red) was destined to become part of ring III in strychnine, following bond rotation and lactam formation. The other (in black) provided a convenient handle for elaboration of ring IV. Left to explore are the remaining steps leading to strychnine, which the reader is encouraged to peruse in the original literature. You will not be disappointed! By any measure this was an ingenious use of latent functionality, and cemented Woodward's reputation as a master of organic synthesis (in his youth, a "Wunderkind"). In 1965 he was awarded the Nobel Prize in Chemistry for "his outstanding achievements in the art of organic synthesis." He most certainly would have shared another for his "Woodward-Hoffmann Rules," were it not for his untimely death in 1979 at the relatively young age of 62.

For now, this will close our discussion on electrophilic aromatic substitution, with its many parallels to anisole chemistry. We have seen how with simple monocyclic rings initial attack is favored at the α-position, and for much the same reasons that anisole prefers *ortho-* and *para*-attack (left Figure 12.7) [28]. These are the positions that allow for the most effective stabilization of a developing positive charge, by resonance donation of an electron pair from the heteroatom. However, yet to be described is a fascinating area of aromatic chemistry wherein a C-H bond is directly converted into a carbon-metal bond (right Figure 12.7) [29]. Taking again anisole as our model, it is interesting to note that the acidity of the *ortho* proton is far greater than that in benzene itself (pK_a 39 vs. pK_a 43). Primarily this is due to the inductive electron withdrawing effect of the methoxy group, with its ability to delocalize a developing negative charge. Since inductive effects drop off rapidly with distance, it comes as little surprise that the *para* lithio derivative is ~3.6 kcal/mole less stable than its internally coordinated *ortho* isomer, as determined by calorimetric quenching studies. However, with respect to kinetic acidity, other factors are at play as well.

In hydrocarbon solvents, anisole and *n*-BuLi enter into a rapid equilibrium, in which lithium becomes coordinated to the methoxy group (for simplicity, the lithiating agent is depicted as monomeric, but the actual species is

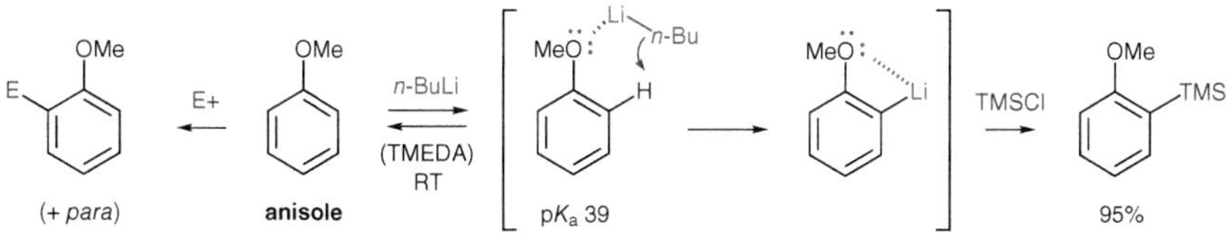

Figure 12.7 Lithiation of anisole.

undoubtedly oligomeric). According to the "coordination only mechanism," the effects of this complexation are threefold [29b]. First, the C-Li bond of *n*-BuLi is polarized to a greater extent, rendering the alkyl residue even more basic. Second, the *ortho* C-H bond is now more acidic, due to the stronger inductive electron withdrawing effect of the complexed methoxy group. And third, there is the entropic advantage of abstracting the *ortho* proton via a six-membered ring transition state (the rate determining step). Alternatively, in the presence of tetramethylethylenediamine (TMEDA), the less Lewis acidic *n*-BuLi/ TMEDA complex may function simply as a superior base, removing the most acidic proton (the "acid-base mechanism") [29b]. In any event, taken in the aggregate, these phenomena constitute the basis for a process known as "directed *ortho* metalation" (DoM), for which there are many directing groups possible [29c]. For our present comparison, however, the methoxy group functions admirably, and proton abstraction is highly efficient. Thus, upon quenching with trimethylsilyl chloride (TMSCl), there is obtained a 95% yield of *o*-trimethylsilyl anisole (right Figure 12.7) [30].

Given that inductive effects are an important stabilizing factor in *ortho* metalations, what can we expect for so-called α-metalations, involving proton abstraction from an sp^2-hybridized carbon that is directly attached to a heteroatom [31]? The simplest such system is methyl vinyl ether, which undergoes rapid and quantitative lithiation upon treatment with *t*-BuLi in THF between -65 and -5 degrees Celsius (Figure 12.8) [32]. These are far milder conditions than required for *o*-lithiation of anisole, and such transformations have considerable synthetic utility. Thus, the resulting lithiated species serves as an acyl anion equivalent, affording high yields of α-substituted vinyl ethers on capture with a wide range of electrophiles (second structure in brackets). Typically, though, these are not isolated, but rather are directly hydrolyzed to the corresponding methyl ketone [32].

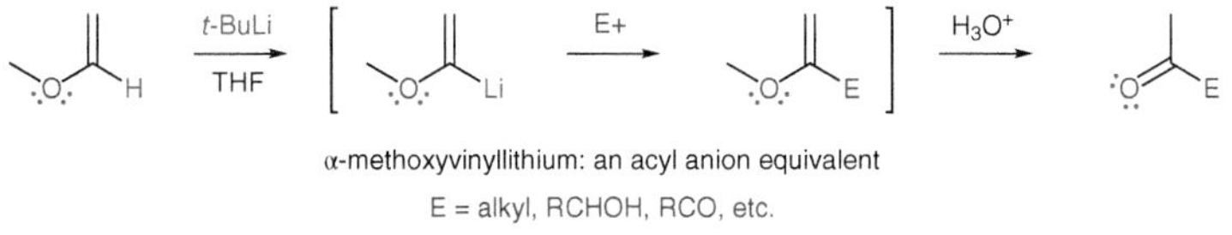

α-methoxyvinyllithium: an acyl anion equivalent

E = alkyl, RCHOH, RCO, etc.

Figure 12.8 Lithiation of methyl vinyl ether.

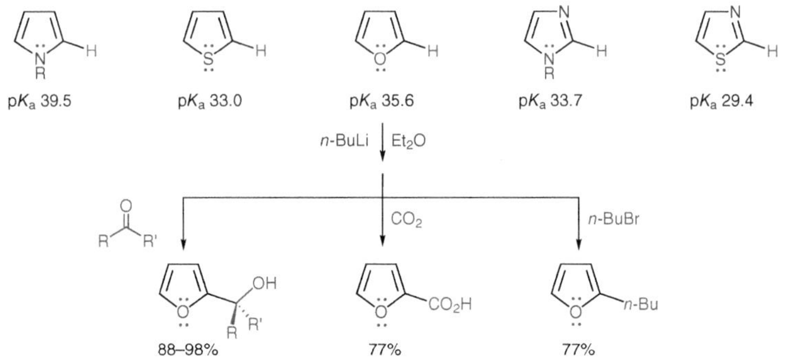

Figure 12.9 Lithiation of π-excessive heterocycles.

So how do these results carry over to π-excessive heterocycles, which incorporate a similar bonding motif? Simply put, if an α-proton is present we expect it to exhibit markedly enhanced acidity, in direct analogy to the case with methyl vinyl ether (cf. Figure 12.8). Furthermore, the derived lithiated species should be strongly nucleophilic. Buttressing this analogy, a sampling of relevant pK_a values is given in Figure 12.9, along with some representative conversions undergone by furan [33]. Note that, with the exception of pyrrole, all of these ring systems are several orders of magnitude stronger as carbon acids than anisole, and all undergo similar transformations as furan. Amongst the mono-heteroatom ring systems, thiophene in particular is readily deprotonated (pK_a 33.0), a result that has been attributed to *d*-orbital participation by sulfur [34]. Not surprisingly, N-alkylpyrroles (pK_a 39.5) are slowest to lithiate.

With these examples, we are coming near the end of our chapter on "Introduction of New Substituents," but we would be remiss in not mentioning one further natural product synthesis, this relating to our present subject material. Our target molecule is *cis*-jasmone (Scheme 12.5), a volatile fragrance

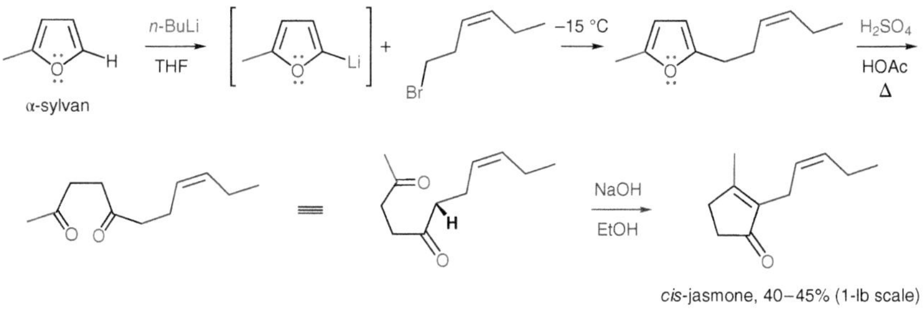

Scheme 12.5

factor isolated from the essential oils of jasmine flowers. Commercially it has found some use in perfumes and cosmetics. And while certainly no strychnine as regards to complexity, it does present a challenge with respect to scalability. The solution outlined is that of G. Büchi and H. Wüest, and was carried out in three steps on 1-lb scales [35].

Büchi's starting material was α-sylvan (2-methylfuran), an item of commerce manufactured by vapor phase hydrogenation of furfural (which you will recall from Figure 11.2 is itself derived from various agricultural waste products). As with furan, α-sylvan undergoes facile lithiation at the vacant 5-position, producing the nucleophilic intermediate shown in brackets. This was quenched at -15 degrees Celsius with *cis*-1-bromohex-3-ene, after which all atoms were in place corresponding to the final objective. In fact, so clean was this reaction that the product was carried forward without purification. It remained mainly to unmask the 1,4-dicarbonyl derivative at the core of the furan skeleton, which was accomplished by hydrolysis employing aqueous $HOAc/H_2SO_4$. The product as drawn at the bottom left of the scheme emphasizes its 2,5-disubstituted furan lineage. However, when redrawn as in the bottom center, one sees clearly that we are only an aldol condensation away from producing *cis*-jasmone. This was carried out utilizing NaOH/EtOH, with the pleasing result of a 40–45% overall yield.

Once again we have reached a milestone of sorts, having covered many aspects of π-excessive heterocyclic chemistry. We began our discussion with an overview of general properties (Chapter 10), and have progressed through de Novo syntheses (Chapter 11) up to the current chapter, Introduction of New Substituents. Still to come is the topic of non-aromatic 5-membered ring heterocycles, which in many cases are best prepared by 1,3-dipolar cycloadditions (vide infra). First, though, let us delve into some other ring transformations of the aromatic family, starting with Diels-Alder reactions.

References

1 Steinkopf, W.; Ohse, W. *Ann. Chem.* **1924**, *437*, 14–22.
2 Scully, J. F.; Brown, E. V. *J. Org. Chem.* **1954**, *19*, 894–901.
3 Mizuno, A.; Kan, Y.; Fukami, H.; Kanei, T.; Miyazaki, K.; Matsuki, S. *Tetrahedron Lett.* **2000**, *41*, 6605–6609.
4 Johnson, J. R.; May, G. E. in *Organic Syntheses*, Coll. Vol. *II*, John Wiley & Sons, Inc., New York, New York, **1943**, p. 8.
5 Heid, J. V.; Levine, R. *J. Org. Chem.* **1948**, *13*, 409–412.
6 Anderson, Jr. A. G.; Exner, M. M. *J. Org. Chem.* **1977**, *42*, 3952–3955.
7 Butler, A. R.; Hendry, J. B. *J. Chem. Soc. (B)* **1971**, 102–105.
8 (a) Babasinian, V. S. in *Organic Syntheses*, Coll. Vol. II, John Wiley & Sons, Inc., New York, New York, **1943**, p. 466. (b) Cooksey, A. R.; Morgan, K. J.; Morrey, D. P. *Tetrahedron* **1970**, *26*, 5101–5111.

9 Bordwell, F. G.; Garbisch, Jr., E. W. *J. Am. Chem. Soc.* **1960**, *82*, 3588–3598.

10 (a) Clauson-Kaas, N.; Fakstorp, J. *Acta. Chem. Scand.* **1947**, *1*, 210–215. (b) Michels, J. G.; Hayes, K. J., *J. Am. Chem. Soc.* **1958**, *80*, 1114–1116, and references cited therein. (c) Balina, G. Kesler, P.; Petre, J.; Pham, D.; Vollmar, A. *J. Org. Chem.* **1986**, *51*, 3811–3818.

11 Marino, G. *Tetrahedron* **1965**, *21*, 843–848.

12 Minnis, W. in *Organic Syntheses*, Coll. Vol. II, John Wiley & Sons, Inc., New York, New York, 1943, p. 357.

13 Treibs, A.; Kolm, H. G. *Ann. Chem.* **1958**, *614*, 176–198.

14 Chaudhuri, S. K.; Roy, S.; Bhar, M. S.; Bhar, S. *Synth. Commun.* **2007**, *37*, 579–583.

15 Keegstra, M. A.; Klomp, A. J. A.; Brandsma, L. *Synth. Commun.* **1990**, *20*, 3371–3374.

16 Baciocchi, E.; Clementi, S.; Sebastiani, G. V. *J. Chem. Soc., Chem. Commun.* **1975**, 875–876.

17 Burness, D. M. in *Organic Syntheses*, Coll. Vol. *V*, John Wiley & Sons, Inc., New York, New York, **1973**, p. 403.

18 Clauson-Kaas, N. *Acta Chem. Scand.* **1947**, *1*, 379–381.

19 Fischer, O.; Hepp, E. *Chem. Ber.* **1886**, *19*, 2251–2259

20 Silverstein, R. M.; Ryskiewicz, E. E.; Willard, C. in *Organic Syntheses*, Coll. Vol. IV, John Wiley & Sons, Inc., New York, New York, **1963**, p. 831.

21 Traynelis, V. J.; Miskel, Jr., J. J.; Sowa, J. R. *J.Org. Chem.* **1957**, *22*, 1269–1270

22 King, W. J.; Nord, F. F. *J. Org. Chem.* **1948**, *13*, 635–640.

23 Tanaka, K.; Kariyone, K.; Umio, S. *Chem. Pharm. Bull.* **1969**, *17*, 616–621.

24 Heaney, H,; Papageorgiou, G.; Wilkins, R. F. *Tetrahedron Lett.* **1988**, *29*, 2377–2380.

25 Dowle, M. D.; Hayes, R.; Judd, D. B.; Williams, C. N. *Synthesis* **1983**, 73–75.

26 (a) Snyder, H. R.; Smith, C. W.; Stewart, J. M. *J. Am. Chem. Soc.* **1944**, *66*, 200–204. (b) Snyder, H. R.; Smith, C. W. *J. Am. Chem. Soc.* **1944**, *66*, 350–351.

27 (a) Woodward, R. B.; Cava, M. P.; Ollis, W. D.; Hunger, A.; Daeniker, H. U.; Shenker, K. *Tetrahedron* **1963**, *19*, 247–288. (b) *Robert Burns Woodward: Architect and Artist in the World of Molecules*, Benfey, O. T.; Morris, P. J. T., Eds., Chemical Heritage Foundation, Philadelphia, Pennsylvania, **2001**.

28 Olah, G. A.; Olah, J. A.; Ohyama, T. *J. Am. Chem. Soc.* **1984**, *106*, 5284–5290.

29 (a) Gilman, H.; Bebb, R. L. *J. Am. Chem. Soc.* **1939**, *61*, 109–112. (b) Gschwend, H. W.; Rodriquez, H. R. *Organic React.* **1979**, *26*, 1–360. (c) Snieckus, V. *Chem. Rev.* **1990**, *90*, 879–933. (d) Rennels, R. A.; Maliakal, A. J.; Collum, D. B. *J. Am. Chem. Soc.* **1998**, *120*, 421–422, and references cited therein.

30 Slocum, D. W.; Moon, R.; Thompson, J.; Coffey, D. S.; Li, J. D.; Slocum, M. G.; Siegel, A.; Gayton-Garcia, R. *Tetrahedron Lett.* **1994**, *35*, 385–388.

31 For a review, see Lever, Jr., O. W. *Tetrahedron* **1976**, *32*, 1943–1971.

32 Baldwin, J. E.; Höfle, G.; Lever, Jr., O. W. *J. Am. Chem. Soc.* **1974**, *96*, 7125–7127.

33 Ramanathan, V.; Levine, R. *J. Org. Chem.* **1962**, *27*, 1216–1219.

34 Gilman, H.; Shirley, D. A. *J. Am. Chem. Soc.* **1949**, *71*, 1870–1871.

35 Büchi, G.; Wüest, H. *J. Org. Chem.* **1966**, *31*, 977–978.

13

Ring Transformations of π-Excessive Heterocycles: Diels-Alder Reactions

It would be difficult to imagine the field of organic synthesis lacking the most famous name reaction of them all, first described in 1928 by Otto Diels and Kurt Alder (winners of the 1950 Nobel Prize in Chemistry) [1]. The fundamentals of the Diels-Alder reaction are likely well known to readers, providing as it does one of the most reliable means available for synthesizing complex 6-membered rings. It is both regio- and stereoselective, consisting of a concerted 4π + 2π cycloaddition between a conjugated diene and a substituted alkene or alkyne (the dienophile):

Under so-called "normal demand" circumstances, the reaction is accelerated by electron donating groups (EDG) on the diene and electron withdrawing groups (EWG) on the dienophile. And, as we shall see later in this chapter, the basic concept can be applied to heteroatom substituted π-systems as well.

In principle, nearly all π-excessive heterocycles might function as dienes in Diels-Alder reactions, given their conjugated nature and the fact that they are locked into a cisoid geometry. However, one must bear in mind that there is a cost to be paid for such reactivity, in the form of lost aromatic stabilization. Consequently, for the parent ring systems, it is understandable that Diels-Alder reactivity varies inversely with aromaticity (top Figure 13.1). Furan, in fact, is far more reactive than pyrrole or its simple N-alkyl derivatives, and thiophene is a distant third (requiring high pressure).

As shown at the bottom of Figure 13.1, furan reacts reversibly with moderately strong dienophiles, including maleic anhydride. In so doing it follows all of the classic rules one associates with Diels-Alder chemistry, including *endo*-selectivity under kinetic control. That is, in the lowest energy pathway, the dienophile will approach the furan nucleus in such a fashion as to have maximum orbital overlap between the two pi systems (cf. second figure in brackets).

Introductory Heterocyclic Chemistry, First Edition. Peter A. Jacobi.
© 2019 John Wiley & Sons Ltd. Published 2019 by John Wiley & Sons Ltd.

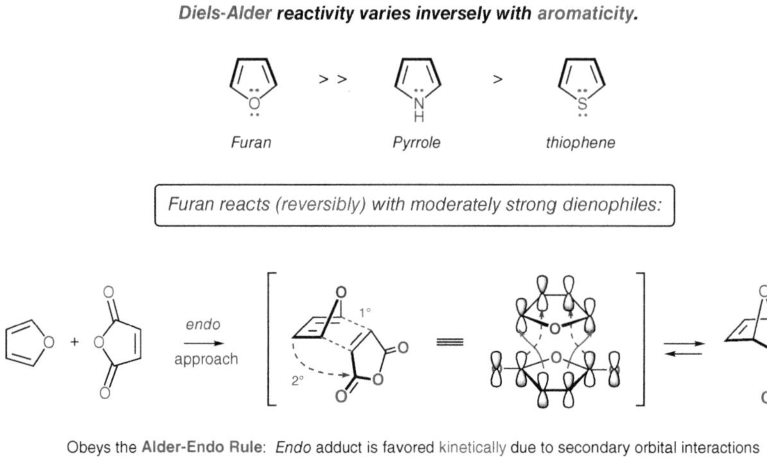

Diels-Alder *reactivity varies inversely with* aromaticity.

| Furan | Pyrrole | thiophene |

Furan reacts (reversibly) with moderately strong dienophiles:

Obeys the **Alder-Endo Rule**: *Endo* adduct is favored kinetically due to secondary orbital interactions

Figure 13.1 Ring transformation of π-excessive heterocycles: Diels-Alder reactions.

Not only does this allow for a strong primary bonding interaction between the HOMO of furan and the LUMO of maleic anhydride (red curly arrows), but the transition state is also stabilized by the secondary orbital interactions shown in blue.

Reflecting this fact, kinetic studies have shown that the *endo*-adduct is initially formed ~500 times faster than the *exo*-adduct at 40 degrees Celsius in MeCN, corresponding to a difference in E_{act} of 3.8 kcal/mole (assuming comparable entropies of activation) [2]. And yet, the *exo*-isomer is found to be ~1.9 kcal/mol more stable by equilibration studies. So which is actually isolated? It turns out there was considerable confusion in the early literature on this point, with Diels and Alder reporting an essentially quantitative yield of the *endo*-adduct (1929) [3]. However, this structural assignment was corrected nearly 20 years later by Woodward and Baer, who proved conclusively that the actual product was *exo* (1948) [4]. In this case thermodynamics prevails, due to the ready reversibility of adduct formation (Figure 13.2). Notice you can establish the same equilibrium whether you begin with furan and maleic anhydride, the *endo*-adduct of furan and maleic anhydride, or the *exo*-adduct. Even the act of attempted purification of the *endo*-adduct causes rapid cycloreversion.

In some cases, then, cycloreversion can be problematic, but in others it can be put to practical advantage. Suppose, for example, you were tasked with synthesizing 3-bromofuran, a species that cannot be prepared by direct electrophilic aromatic substitution. In particular, as we have already seen, carefully controlled *mono*-bromination of furan in DMF gives exclusively 2-bromofuran (cf. Figure 12.4), as would be expected for a kinetically controlled process. What alternatives do we have?

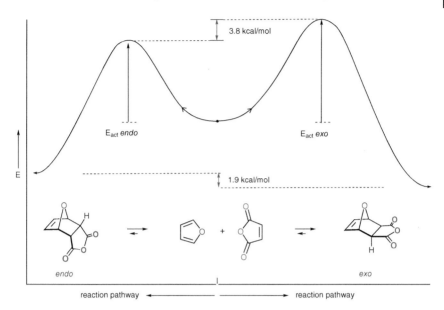

Figure 13.2 Energy diagram for Diels-Alder reaction of furan.

One strategy relies on a transformation known as an Alder-Rickert reaction, a term that over the years has come to be associated with virtually any *retro*-Diels-Alder reaction, including those that simply regenerate starting materials [5]. To be accurate, though, this designation should be reserved for the more specific case where the products of cycloreversion differ from the initial dienophile and diene. A classic example involves the Diels-Alder adduct derived from diethyl acetylenedicarboxylate and cyclohexadiene, which on distillation produces diethyl phthalate and ethylene [5]:

As applied to furan chemistry, the Alder-Rickert reaction is useful for preparing both 3- and 3,4-substituted derivatives (Figure 13.3). Thus, coming back to 3-bromofuran, this material is most efficiently synthesized beginning with the *exo*-Diels-Alder adduct from furan and maleic anhydride, which undergoes clean bromination across the isolated carbon-carbon double bond (top left Figure 13.3) [6a]. The dibromide, of course, is stable to *retro*-Diels-Alder reaction, since there are no π-orbitals to allow concert. However, all of this changes upon dehydrobromination to give the vinyl bromide shown in brackets. So labile is this compound that in practice it need not be isolated, but rather

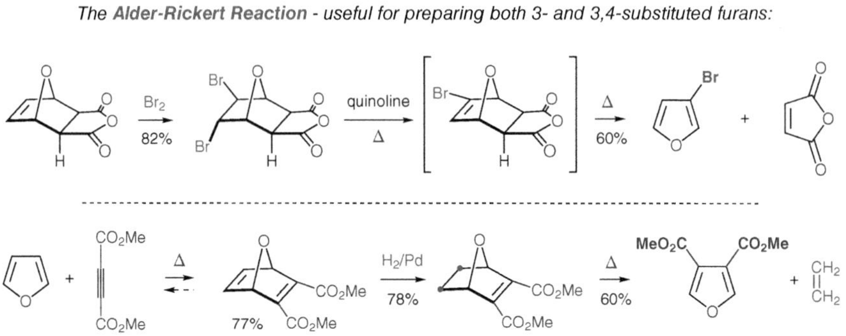

*The **Alder-Rickert Reaction** - useful for preparing both 3- and 3,4-substituted furans:*

Figure 13.3 Ring transformation of π-excessive heterocycles: Diels-Alder reactions.

affords a 60% overall yield of 3-bromofuran upon cycloreversion, together with maleic anhydride [6b].

At the bottom of Figure 13.3 is shown a second example of the utility of Alder-Rickert reactions, this time in conjunction with a synthesis of 3,4-dicarbomethoxyfuran. In this case the starting materials are furan and dimethyl acetylenedicarboxylate, which in refluxing benzene afford a 77% yield of the corresponding Diels-Alder adduct. However, this material is unstable toward cycloreversion to starting materials, and must be directly hydrogenated. The resultant dihydro derivative can now undergo *retro*-Diels-Alder reaction in only one direction, expelling ethylene to give a 60% yield of the desired diester [7].

Sometimes the Alder-Rickert reaction can lead to a scrambling of substituents, as for example in the Diels-Alder reaction between 3,4-dicyanofuran and hexafluoro-2-butyne (Figure 13.4) [8]. Imagine that you carried out this reaction expecting to obtain a normal 4π + 2π adduct **A**. What a surprise you would get, since a careful search of the crude reaction mixture would reveal none of the product in brackets. In its stead would be an approximately 1:1 mixture of 7-oxonorbornadiene derivatives, one incorporating four cyano groups (**B**: 35%), and the other having four trifluoromethyl substituents (**C**: 30%). For the most part, the mechanism for this outcome is clear-cut, commencing with a rapid cycloreversion of the initial Diels-Alder adduct in an Alder-Rickert sense (bottom Figure 13.4). In this manner would be generated dicyanoacetylene and 3,4-bistrifluoromethylfuran, bringing together all of the components necessary to produce the observed mixture. However, still to explain is the complete absence of the "normal" adduct **A**, in particular if thermodynamic equilibrium is attained. Possibly **A** undergoes partial decomposition by an alternative reaction pathway (a significant amount of "carbonaceous" material is reported) [8]. In any event, it would be interesting to explore this reaction computationally.

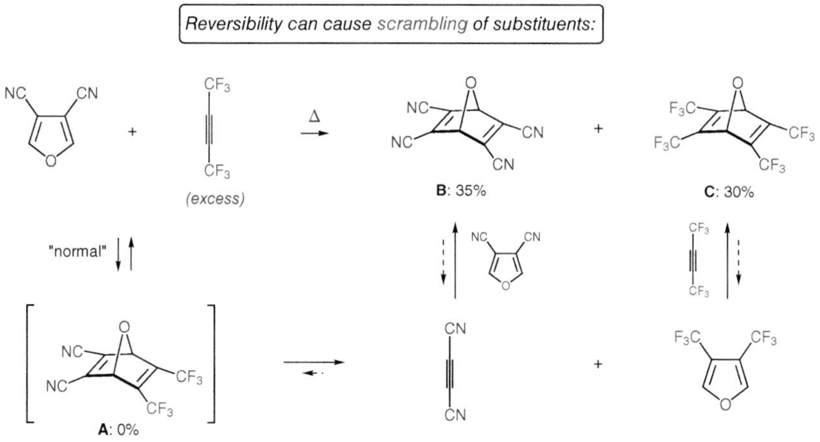

Figure 13.4 Ring transformation of π-excessive heterocycles: Diels-Alder reactions.

So what other "Diels-Alder" surprises does furan have in store? The short answer is many, of which time and space will limit our discussion to but a few. To start, conjugate addition is often a competing process, especially with unsymmetrical, highly polarized, dienophiles. The overall transformation can be viewed as a special case of electrophilic aromatic substitution, reflecting the nucleophilic character of furan:

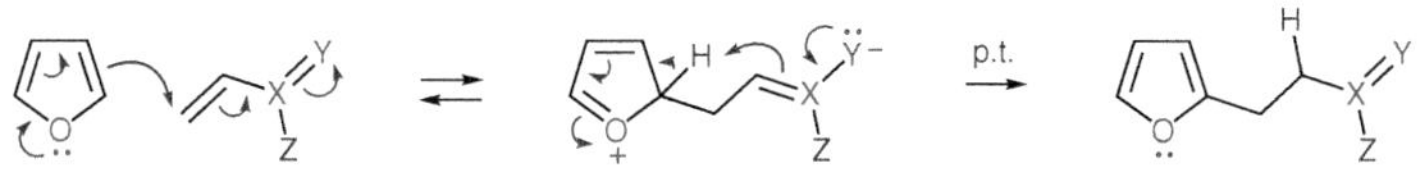

Such is the case with the attempted Diels-Alder reaction of furan with acrolein, which under purely thermal conditions yields little if any of the product of cycloaddition (top Figure 13.5). Rather, the preferred route is that in which aromaticity is preserved, leading to a 51% yield of the *bis*-Michael adduct shown to the right (14% of mono-adduct is also isolated) [9]. It is only with powerful Lewis acid catalysis that a 4π + 2π adduct is formed, and then only at low temperature [10]. If the reaction is allowed to warm, rapid cycloreversion ensues.

A similar pathway prevails with 2-acetylbenzoquinone, which affords a single conjugate addition product with furan (bottom right Figure 13.5; note that addition takes place at the most electrophilic carbon, that in conjugation with both the benzoquinone carbonyl and the acetyl group) [11]. In this case, presumably, the primary adduct is the corresponding hydroquinone or its di-keto tautomer, which undergoes in situ oxidation with excess 2-acetylbenzoquinone. Supporting this premise is the isolation of 2-acetylhydroquinone shown at the bottom right. None of the Diels-Alder adduct corresponding to 4π + 2π cycloaddition was detected.

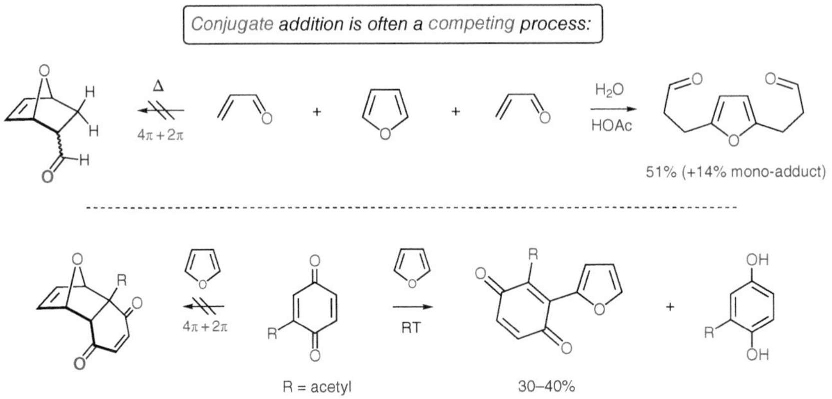

Conjugate *addition is often a* competing *process:*

$$4\pi + 2\pi$$

$$\xrightarrow[\text{HOAc}]{\text{H}_2\text{O}}$$

51% (+14% mono-adduct)

$$4\pi + 2\pi$$

RT

R = acetyl

30–40%

Figure 13.5 Ring transformation of π-excessive heterocycles: Diels-Alder reactions.

An interesting situation presents itself when the furan nucleus contains an appended alkene, in that more than one Diels-Alder adduct may be possible. The simplest such compound is 2-vinylfuran, where an approaching dienophile has two choices (top Figure 13.6). The more "traditional" pathway leads off to the left in the figure, and involves overlap with the two endocyclic double bonds of the furan ring. However, the experimental observation is that none of this product is formed, or at least isolated. Rather, the author reports a 79% yield of the Diels-Alder adduct shown to the immediate right, involving participation of the exocyclic double bond [12]. This material was remarkably stable to cycloreversion, surviving boiling in water. So should this result surprise us? In a sense it seems to have surprised the author of this 1943 paper, who describes his structural assignment as "*provisoirement*" (author's italics).

To be sure, there seems little doubt that the proposed adduct was at least formed initially, as the author presents convincing data for the skeletal type (and remember, these were pre-NMR days). Having said that, though, the

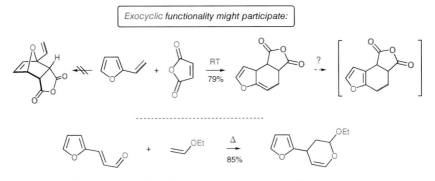

Exocyclic *functionality might participate:*

RT
79%

?

Δ
85%

Figure 13.6 Ring transformation of π-excessive heterocycles: Diels-Alder reactions.

available evidence would also support the more stable aromatic structure shown in brackets, for which there is ample precedent. For example, the same reaction conducted with 2-vinylthiophene yields the analogous aromatic adduct, whose structure was secured by NMR studies [13]. In the final analysis, then, it may well be that both types of furan adducts are formed reversibly, with the equilibrium being drained to the right by re-aromatization of the furan ring (i.e., sinking into the "thermodynamic well"). With the advantage of modern instrumentation, this would be an experiment worth repeating.

A more straightforward example is provided by β-(2-furyl)acrolein, which in principle offers three sites where dienophile addition might occur (bottom Figure 13.6). With ethyl vinyl ether, however, there is formed an 85% yield of the hetero-Diels-Alder adduct shown to the right, in this case proceeding via inverse electron demand [14].

Finally, under the category of Diels-Alder complications, it cannot be over-emphasized that steric factors play an enormous role. In fact, such effects are at least partly responsible for one of the most famous "failed" reactions in organic chemistry (Figure 13.7). To get a sense of this failure let us return briefly to 1929, the year in which Diels and Alder published the second paper in their series on new cycloaddition chemistry (ultimately they would co-author nineteen on this topic, with an additional nine coming from Diels and other contributors) [3]. The title of this paper was "Synthesen in der hydro-aromatischen Reihe, II. Mitteilung: Über Cantharidin," of which the last two words stand out (About Cantharidin). What is cantharidin, and why did Diels and Alder attach such importance to this molecule that it was one of the first syntheses they attempted on inventing their reaction? Partly it was due to the rich history associated with this potent vesicant principle, first isolated in

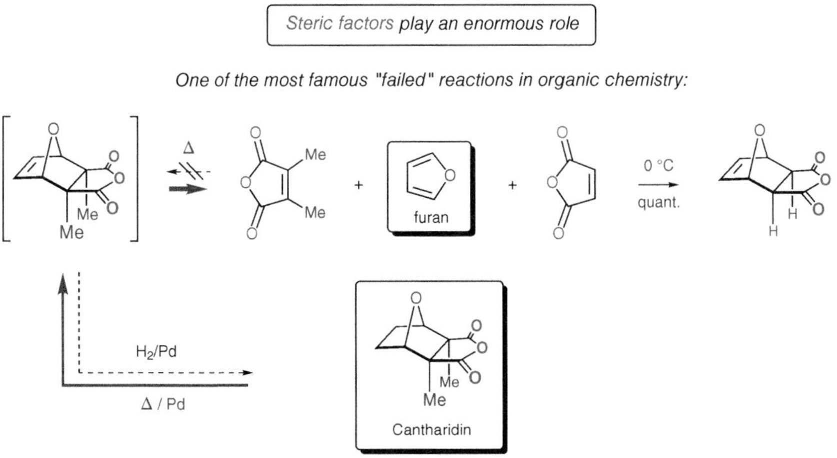

Figure 13.7 Ring transformation of π-excessive heterocycles: Diels-Alder reactions.

crystalline form in 1810. Add to this its notoriety as a supposed aphrodisiac (with its mysterious folk name, Spanish Fly); but in the end it was simply a substance that cried out for exploitation of a $4\pi + 2\pi$ cycloaddition.

The structure of cantharidin is shown in the shadow box at the bottom of Figure 13.7, and we see immediately why this target was so enticing. It is only a hydrogenation step away from the hypothetical *exo*-Diels-Alder adduct between furan and dimethylmaleic anhydride, shown in brackets at the top left of the figure. And recall that in preliminary studies furan and maleic anhydride combined to give an essentially quantitative yield of what we now know to be the *exo*-isomer (top right of figure). So how much difference could those two methyl groups make? Actually, all the difference in the world! As Diels and Alder found out, and generations of organic chemists have since confirmed, no amount of coaxing, nor a legion of catalysts, nor ultra-high pressures, could induce the desired transformation. It is simply thermodynamically not accessible. A telling experiment involves heating naturally derived cantharidin over a palladium catalyst, under conditions known to effect dehydrogenation (heavy blue arrows). No trace of the bracketed structure is obtained, but rather its products of cycloreversion.

The consequences of this failed Diels-Alder reaction were profound, turning what should have been a two-step synthesis into a much more complex endeavor. Indeed, efforts along these lines occupied the attention of some of the world's most accomplished synthetic chemists for many decades. Among these, a youthful Gilbert Stork emerged as one of the first to achieve success, publishing in 1951 an elegant, if somewhat circuitous stereoselective synthesis (Scheme 13.1) [15]. Building upon known chemistry, Stork began his synthesis with the Diels-Alder reaction between furan and dimethyl acetylenedicarboxylate, the adduct of which was immediately reduced to the corresponding unsaturated diester shown at the top right of the scheme (cf. also Figure 13.3). Initially one might suspect that the carbomethoxy groups were destined to

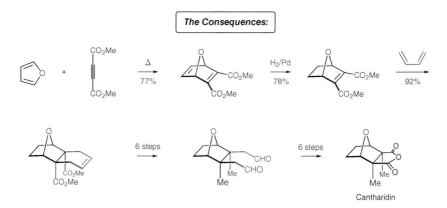

Scheme 13.1 Ring transformation of π-excessive heterocycles: Diels-Alder reactions.

become the *exo*-anhydride functionality in cantharidin, which has a certain appeal if *endo*-dimethylation could be achieved. However, the prospects for such a transformation were slim, as Stork knew from previous workers' results. Instead, the quaternary centers in cantharidin were introduced by yet a second Diels-Alder reaction, wherein butadiene entered exclusively from the sterically less hindered *exo*-face (92% yield). The resultant adduct, shown at the bottom left of the scheme, does have all of the stereochemical features necessary to reach the final target. However, to do so required an additional 12 steps, with multiple adjustments in oxidation level in both the *exo*- and *endo*-substituents. What an enormous impact one failed reaction can have!

And here things stood until 1980, when Dauben et al. re-investigated the cantharidin "problem" using high pressure techniques (Scheme 13.2) [16]. Once again, the "parent" reaction of furan plus dimethylmaleic anhydride failed, even at pressures up to 40 kbars (i.e., nearly 40,000 atmospheres!). This was surprising given the known accelerating effect of increased pressure on reactions having a negative volume of activation. However, far better results were obtained employing a slight variation in dienophile structure, wherein the two methyl groups were conjoined by a sulfur bridge (right Scheme 13.2). Employing this substrate, and at 15 kbar pressure, there was obtained an essentially quantitative yield of two Diels-Alder adducts, with the desired *exo*-isomer greatly predominating (85:15). Dauben attributed this success to both steric and electronic effects, and to this we might add the relief of ring strain in the transition state leading to cycloaddition. In any event, although the two isomers could be separated, it proved expeditious to simply reduce the entire reaction mixture with Raney nickel, by which means cantharidin was produced in 63% overall yield [16].

Interestingly, as a postscript, Grieco et al. in 1990 demonstrated that Dauben's high pressure results could be mimicked by carrying out the Diels-Alder reaction in 5 M LiClO$_4$/Et$_2$O at atmospheric pressure [17]. These conditions failed, however, in a 1996 attempt to synthesize palasonin, a desmethyl analog of

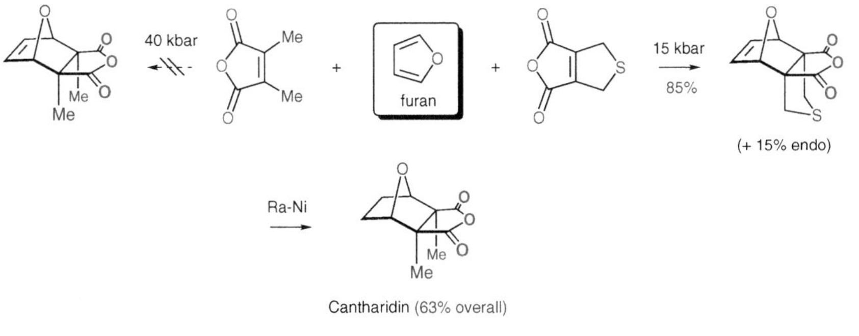

Scheme 13.2 Ring transformation of π-excessive heterocycles: Diels-Alder reactions.

cantharidin isolated from the seeds of *Butea frondosa* (see below) [18]. Detailed kinetic studies showed that the difficulty resided in an unfavorable equilibrium for the initial Diels-Alder cycloaddition, with the reverse reaction having a larger rate constant. Once again, recourse was taken to high pressure, with outstanding results [18]:

Palasonin

Compared to furan, the Diels-Alder chemistry of pyrrole and its simple derivatives is rather limited. Two factors come into play here, the first being that pyrrole has considerably greater aromatic stabilization (RE = 21 kcal/mol versus 16 kcal/mol). Because of this, cycloreversion will always be a competing process, even more so than with furan. In addition, pyrrole is a much stronger nucleophile, opening the door to more favorable reaction pathways. For example, simply stirring pyrrole and dimethyl acetylenedicarboxylate together in Et_2O at room temperature affords a 67% yield of the conjugate addition products shown at the top left of Figure 13.8 [19]. Also obtained is ~6% of product corresponding to N-addition, but none of the hypothetical Diels-Alder adduct shown at the top right. Marginally better results are obtained with N-benzylpyrrole and acetylene dicarboxylic acid, which give an ~8% yield of the Diels-Alder adduct shown in the center of the figure, together with a preponderance of Michael adduct [20]. And, better still are the reactions of N-alkylated pyrroles with benzynes, generally affording moderate-good yields of Diels-Alder products. Typical of this class is the reaction of N-methylpyrrole with 3,5-dichloro-4,6-difluorobenzyne, which gives a 59% yield of the adduct shown at the bottom of Figure 13.8 [21]. Adducts of this type are useful intermediates for synthesizing substituted naphthalene derivatives, via cheletropic extrusion of nitrosomethane from the derived N-oxide.

Finally, as noted at the bottom of Figure 13.8, N-acylpyrroles have enhanced reactivity in Diels-Alder cycloadditions, due to greater bond localization. This reactivity was put to good use in a number of syntheses of (±)-epibatidine, a potent non-opiate analgesic isolated from the Ecuadorian poison frog *Epipedobates tricolor* (Scheme 13.3). Remarkably, this small molecule alkaloid is 200–400 times more potent than morphine in various whole animal tests. The first three steps of a synthesis by Regan and Clayton closely followed literature precedent, initiated by Diels-Alder reaction between tosylacetylene and N-carbomethoxypyrrole [22]. In this manner was formed the requisite bicyclic ring skeleton found in epibatidine, albeit in moderate yield (this same reaction, however, when carried out at twelve kbars, and RT, affords an 81%

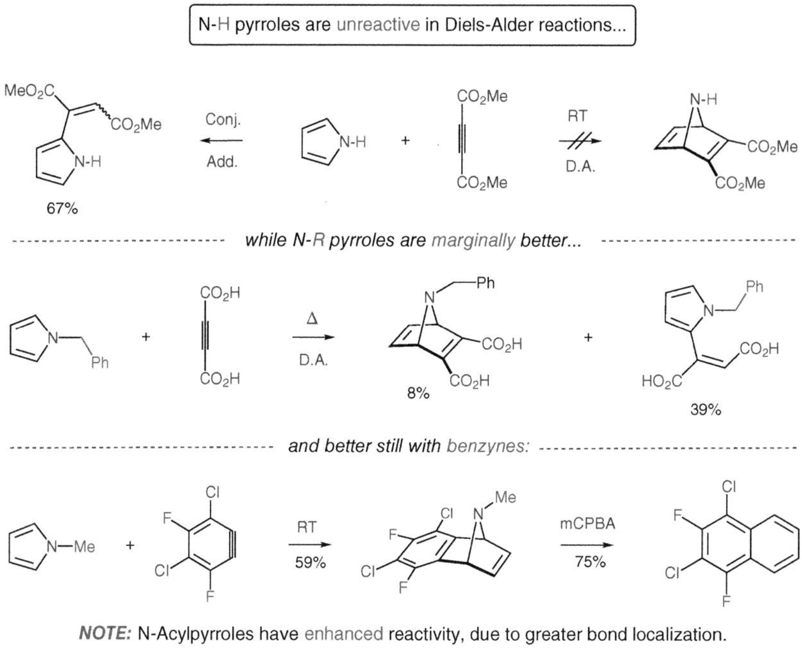

Figure 13.8 Ring transformation of π-excessive heterocycles: Diels–Alder reactions.

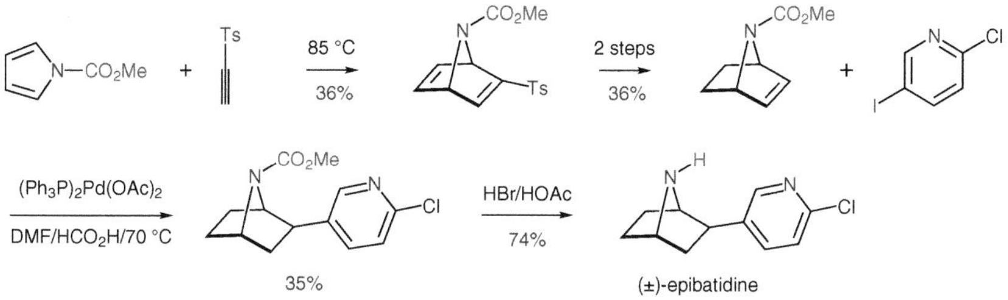

Scheme 13.3 Ring transformation of π-excessive heterocycles: Diels–Alder reactions.

yield of adduct [23]). Next, hydrogenation of the least hindered double bond (H₂/Pd), followed by reductive cleavage of the tosylate group (Na-Hg), afforded a 36% yield of the known 7-azabicyclo[2.2.1]heptene derivative shown at the top right of the scheme. At this point the authors were only two steps removed from their target, the first of which involved a novel reductive Heck-type coupling with 2-chloro-5-iodopyridine. Precedent for this reaction indicated that the *exo*-coupling product should be highly favored, and this in fact was observed. It remained now only to cleave the N-carbomethoxy group, and this was cleanly accomplished utilizing HBr in acetic acid.

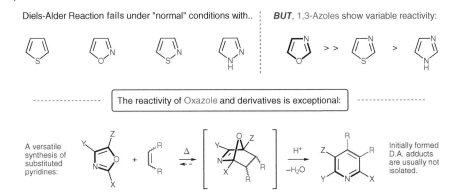

Figure 13.9 Ring transformation of π-excessive heterocycles: Diels-Alder reactions.

Unlike furan and N-acylpyrroles, thiophene is nearly devoid of Diels-Alder reactivity under "normal" conditions, as might be expected based upon its RE of 29 kcal/mol (cf. Figure 13.1) [24]. But what about other π-excessive ring systems, such as the 1,2- and 1,3-azoles? Intuitively, one might expect these azadienes to be ready partners in cycloaddition reactions, given their generally lower aromatic stabilization. However, as indicated at the top left of Figure 13.9, the 1,2-azoles are notoriously unreactive. Why should this be so? The parent reaction of isoxazole with ethylene has been studied at a high level of theory by Houk et al., who unmasked both a positive ΔE (4.0 kcal/mol), as well as a significant activation barrier (28.9 kcal/mol) [25]. The equilibrium for this reaction is therefore predicted to lie far to the left at room temperature or above.

A contributing factor to these results is the high strain energy associated with placing a nitrogen at the bridgehead position in the adduct, which constricts the C-N-C bond angle to 104 degrees, and eliminates any possibility of enamine conjugation (see below). In addition, electrostatic repulsion between the free electron pairs may come into play. And what does this tell us about isothiazoles and pyrazoles, which share a similar lack of reactivity? Although these species were not addressed computationally, the same arguments undoubtedly hold. To date, there appear to be no examples of Diels-Alder reactions between alkene dienophiles and simple π-excessive rings containing two adjacent heteroatoms. For such to occur, one of the heteroatoms would have to occupy an unfavorable bridgehead position:

X = O:, NH, S:

In contrast, 1,3-azoles as a class have a wide range of reactivity, with the oxazole ring being among the most reactive of dienes in Diels-Alder cyclizations (top right Figure 13.9). Thiazoles are considerably less reactive, and there is only a scattering of reports suggesting imidazole as an effective diene. As a point of reference, the reaction of oxazole with ethylene is calculated to be exothermic by 19.7 kcal/mol, with an energy of activation of 16.0 kcal/mole (i.e., ~13 kcal/mol lower than that for isoxazole) [25]. Numbers such as these make for some very interesting chemistry, one aspect of which is depicted at the bottom of Figure 13.9. Here we see that Diels-Alder cyclizations between oxazoles and alkenes are quite general, affording a wide range of adducts of the type shown in brackets. Interestingly, though, such adducts are rarely isolated, finding a ready pathway to aromatization via opening of the oxygen bridge followed by dehydration [26]. Note that with internal alkenes the most common products of aromatization are 3,4-disubstituted pyridines, which have a familiar look about them. Picture, for example, the result with X=H, R=hydroxymethyl, Y=Me, and Z=OH. What you are visualizing is a single step conversion of an oxazole to pyridoxine, aka vitamin B6! And although we cannot quite reach that level of efficiency, we can come pretty close.

One approach is outlined in Scheme 13.4, where the first task was to prepare a suitably substituted oxazole [27]. This was accomplished in two steps, and ~50% overall yield, beginning with commercially available (and inexpensive) (±)-ethyl alanate (top left of scheme). The first of these steps involved N-formylation, which was conveniently effected via the mixed anhydride derived from sodium formate and acetyl chloride (acetic formic anhydride, or AFA). Next to address was the cyclodehydration of this material, leading to 4-methyl-5-ethoxyoxazole (top right of scheme). The reagent of choice here

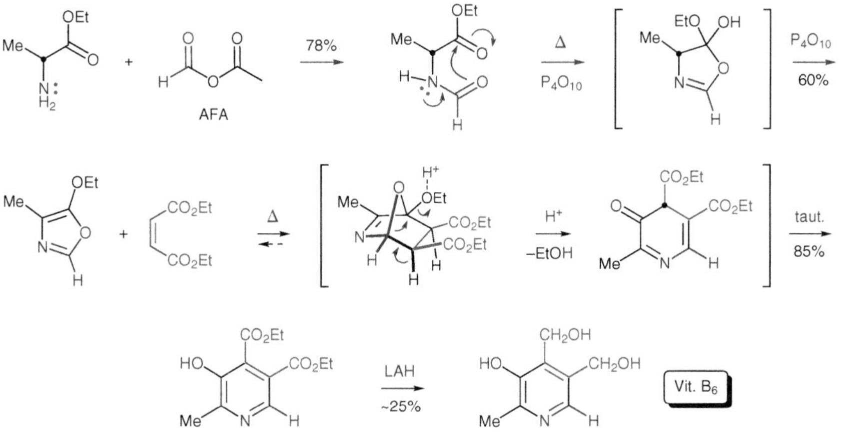

Scheme 13.4 A novel synthesis of pyridoxine.

was P_4O_{10}, for which some interesting mechanistic aspects come into play. There is little doubt as to the initiating step in this ring closure, which takes full advantage of the free electron pair on nitrogen (cf. curly arrows). Thus activated, the nucleophilic formyl group undergoes facile addition to the proximate ethyl ester, producing the cyclic orthoester shown in brackets. And how does this tetrahedral intermediate break down? In the absence of P_4O_{10} the favored reaction pathway would most likely involve ejection of EtOH, to produce a species known as an azlactone. However, P_4O_{10} functions by capturing the hydroxyl group, leading to a formal elimination of the elements of water.

Step three in the synthesis involved Diels-Alder reaction of 4-methyl-5-ethoxyoxazole with diethyl maleate, which was carried out for two hours at 110 degrees Celsius (middle row Scheme 13.4). Without isolation, the resultant adduct was cleaved with ethanolic HCl, affording an 85% yield of the immediate precursor to pyridoxine (bottom left of scheme). A plausible mechanism for this process is provided in the brackets, involving acid-catalyzed fragmentation with loss of ethanol, followed by tautomerization. Lastly, it remained only to reduce the diester groups to the corresponding alcohols utilizing lithium aluminum hydride (LAH), a step that had previously been reported in the literature (~25% yield) [28].

Let us analyze the strengths and weaknesses of this synthesis, beginning with the observation that it requires only four steps from inexpensive starting materials. Other strong points are its highly convergent nature, and the fact that the crucial Diels-Alder reaction takes place in excellent yield. Now for the minus side—the authors have chosen a dienophile in a high oxidation state, necessitating a final low yielding reduction with LAH. Presumably this step could be improved upon. Anyway, as it turns out, there was a good reason for choosing diethyl maleate, in that the more direct route employing cis-butene-1,4-diol gave only modest yields of non-crystalline pyridoxine (top Figure 13.10). So where from here? In the same time period, many workers were investigating different combinations of oxazole dienes and alkene dienophiles, some with notable success. However, one strategy, while in a sense unsuccessful, opened up an entirely new field. The key experiment involved a Diels-Alder reaction between 4-methyloxazole and 2-butyne-1,4-diol, with the expectation that the initially formed adduct might also find a pathway forward to pyridoxine (middle Figure 13.10). This was not an unreasonable expectation, given that the adduct was in the same oxidation state as the final target. But in this case no pyridoxine was formed, or at least not in neutral solvents. Rather, the major product was 3,4-bis(hydroxymethyl)furan (63%), the result of an Alder-Rickert reaction expelling acetonitrile [29]. With hindsight this outcome makes good sense, given the essentially irreversible nature of the last step (i.e., acetonitrile is a feeble dienophile), and the thermodynamic driving force of producing a ring system with greater aromatic stabilization. In fact, additional studies revealed this to be a very general reaction, and it is presently the method of

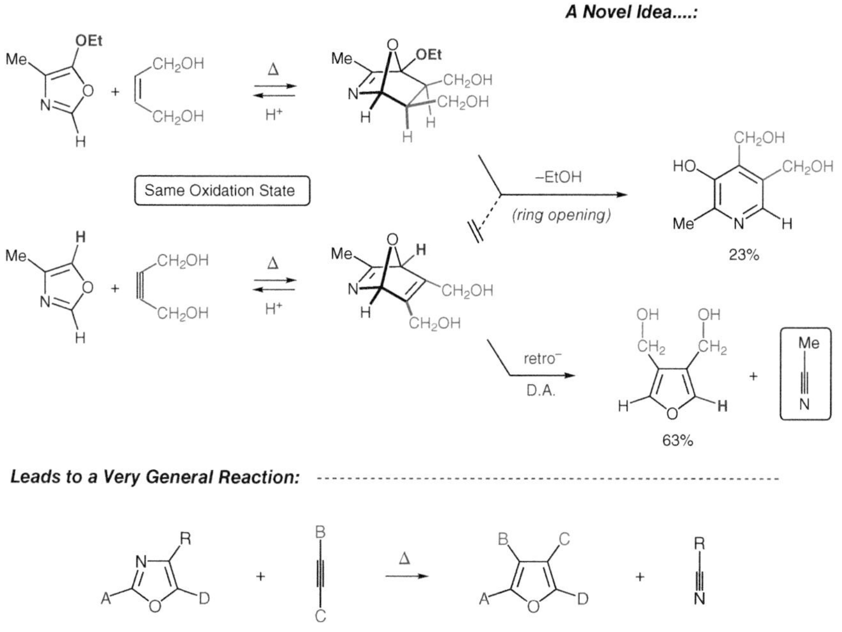

A Novel Idea....:

Leads to a Very General Reaction: -

Figure 13.10 Ring transformation of π-excessive heterocycles: Diels-Alder reactions.

choice for synthesizing highly substituted furans (bottom Figure 13.10) [26]. What follows are a few examples.

Can you think of any diene so reactive that it undergoes Diels-Alder cyclo-addition with acetylene itself? 4-Methyloxazole does [29], affording a 25% yield of furan upon heating to 190 degrees Celsius in an autoclave (top left Figure 13.11). Furthermore, reactivity increases with the addition of electron donating groups to the oxazole and electron withdrawing groups to the alkyne (i.e., normal electron demand). Take, for example, the reaction of 4-methyl-5-ethoxyoxazole with dimethyl acetylenedicarboxylate (DMAD), which gives a 53% yield of the anticipated furan simply on standing at RT in ether (top right of Figure 13.11) [30]. Even very sterically hindered oxazoles undergo facile reaction with DMAD, albeit at somewhat higher temperatures (middle row of figure). In this instance the target was the tetrasubstituted furan shown to the right, bearing an ethoxy group, two carbomethoxy groups, and a methyl ester tethered to a three-carbon chain. One would be hard put to think of another means of synthesizing this compound, which was a projected starting material for a natural product synthesis. In the present case, however, it was obtained in 69% yield simply upon heating the oxazole and alkyne components to 80 degrees Celsius [31]. Focusing now on the bottom row of Figure 13.11, let us examine the outcome of a Diels-Alder reaction involving a fused-ring oxazole, as initially

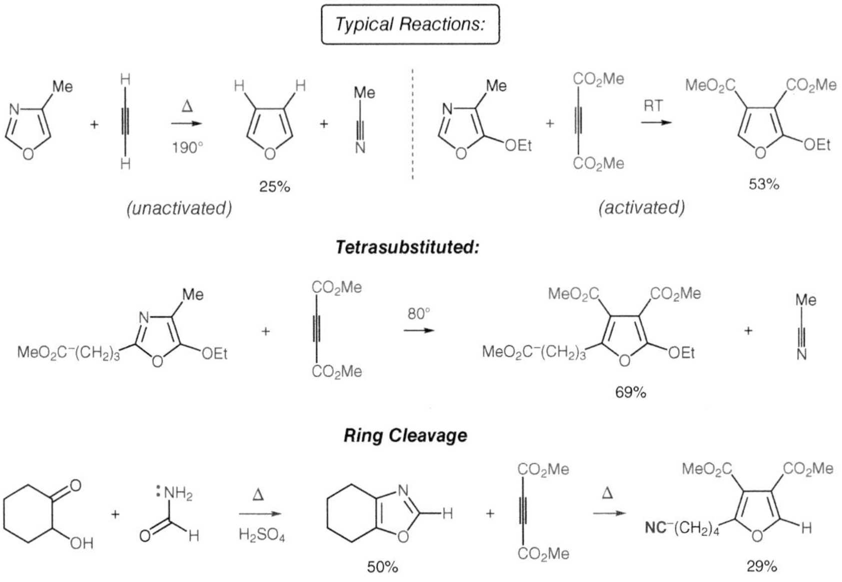

Figure 13.11 Ring transformation of π-excessive heterocycles: Diels-Alder reactions.

explored by Kondrat'eva et al. in their pioneering investigations in this area. Oxazoles of this class are readily prepared by condensation of formamide with the appropriate cyclic acyloin, as illustrated for the cyclohexane derivative shown in the center [32]. On cycloaddition with DMAD, this material reacts in the normal fashion to give an intermediate adduct, and the adduct expels a nitrile to afford a substituted furan. But to where has the nitrile gone? Nowhere! It remains attached to the four carbon tether at C-2 of the product [33].

Continuing along the same path we come to the classic studies of Wasserman et al., who investigated a similar ring cleavage but employing singlet oxygen as a dienophile (Scheme 13.5). Here the presumed adduct shown in brackets also collapsed with ejection of a tethered nitrile group, but by generating a mixed anhydride at the terminus (curly arrows). On workup these intermediates readily lost carbon monoxide, giving excellent yields of the corresponding ω-cyano carboxylic acids [34].

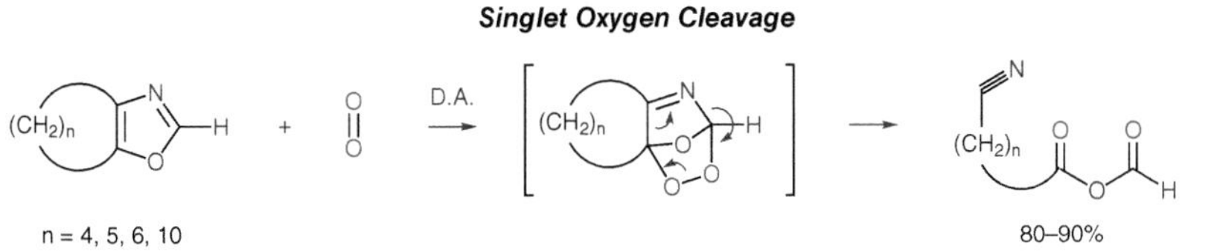

Singlet Oxygen Cleavage

n = 4, 5, 6, 10   80–90%

Scheme 13.5

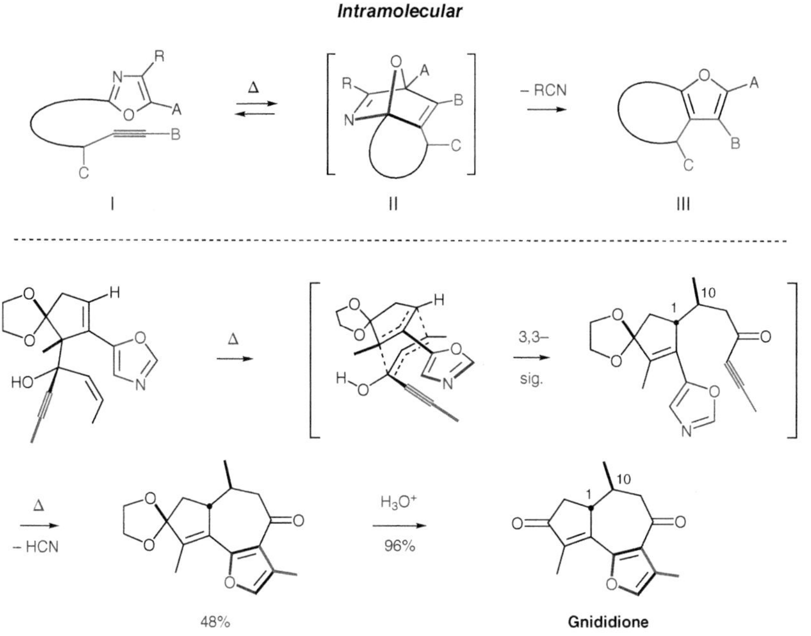

Intramolecular

Scheme 13.6 Ring transformation of π-excessive heterocycles: Diels-Alder reactions.

Finally, we close this chapter with a brief discussion of intramolecular oxazole-alkyne cycloadditions, which generate fused ring furan derivatives (top Scheme 13.6) [35]. The term "bis-heteroannulation" has been suggested for such a process, in which two ring systems are formed as part of the same reaction sequence. One of these rings is carbocyclic in nature, while the other consists of a substituted furan nucleus (hence the prefix "hetero"). Transformations of this type are of considerable synthetic utility, since the appended groups A, B, and C are transposed in an unequivocal fashion, via intermediate II, to the final annulated product III. It turns out that substitution patterns of the type found in III are quite common in nature, particularly in a large family of natural products known as furanosesquiterpenes. Several of these have been synthesized utilizing this methodology.

We will take as one example the synthesis of gnididione, a furanosesquiterpene distinguished by incorporating a guaiane skeleton (bottom Scheme 13.6) [36]. The principle challenges here were threefold, consisting of (1) generation of the perhydroazulene ring system characteristic of the guaianes; (2) regiospecific incorporation of the furan ring; and (3) control of relative stereochemistry between C1 and C10. The cis substitution pattern found in gnididione is rare among members of this class, leading to the suggestion that the alternative trans relationship might better fit the experimental data. How to differentiate these two by total synthesis? The authors chose as a key intermediate the tertiary alcohol shown at the left center of the scheme, which was properly disposed for undergoing a chemoselective oxy-Cope rearrangement. Now, follow the fate of each of the substituents through the course of this 3,3-sigmatropic rearrangement. To begin, we know that such rearrangements generally prefer a chair-like transition state, as represented by the first structure in brackets. And, since we have begun with a cis alkene, the stereochemical relationship between C1 and C10 is unequivocally set (second structure in brackets). But things do not stop here. The alkyne group has thus far been along "just for the ride," but now finds itself ideally positioned for an intramolecular Diels-Alder reaction with the nearby oxazole, and even in activated form. In this manner, all three synthetic challenges were met in a single step (48% overall yield), with acid hydrolysis providing the natural product. So, on this basis, can you suggest an experiment for preparing isognididione (i.e., having the opposite stereochemical relationship between C1 and C10)? Also, this might be a good time to return to Figure 11.8 in Chapter 11, and fill in the missing steps leading to petasalbine.

Problems for Practice [37]

References

1 Diels, O.; Alder, K. *Ann. Chem.* **1928**, *460*, 98–122.
2 Lee, M. W.; Herndon, W. C. *J. Org. Chem.* **1978**, *43*, 518.
3 Diels, O.; Alder, K. *Chem. Ber.* **1929**, *62*, 554–562.
4 Woodward, R. B.; Baer, H. *J. Am. Chem. Soc.* **1948**, *70*, 1161–1166
5 Alder, K.; Rickert, H. F. *Ann. Chem.* **1936**, *524*, 180–189.
6 (a) Politis, J. K.; Nemes, J. C.; Curtis, M. D. *J. Am. Chem. Soc.* **2001**, *123*, 2537–2547. (b) Sroji, J. Janda, M.; Stibou, I. *Collect. Czech. Chem. Commun.* **1970**, *35*, 3478–3480.
7 Wong, H. N. C. *Synthesis* **1984**, 787–790.
8 Weis, C. D. *J. Org. Chem.* **1962**, *27*, 3693–3695.
9 Webb, I. D.; Borcherdt, G. T. *J. Am. Chem. Soc.* **1951**, *73*, 752–753.
10 Laszlo, P.; Lucchetti, J. *Tetrahedron Lett.* **1984**, *25*, 4387–4388.
11 Eugster, C. H.; Bosshard, P. *Helv. Chim. Acta* **1963**, *46*, 815–851.
12 Paul. R. *Bull. Soc. Chim. Fr.* **1943**, 163–168.
13 Abarca, B.; Ballesteros, R.; Enriquez, E.; Jones, G. *Tetrahedron* **1985**, *41*, 2435–2440.
14 Longley, Jr., R. I.; Emerson, W. S. *J. Am. Chem. Soc.* **1950**, *72*, 3079–3081.
15 (a) Stork, G.; van Tamelen, E. E.; Friedman, J.; Burgstahler, A. W. *J. Am. Chem. Soc.* **1951**, *73*, 4501–4501. See also, (b) Stork, G.; van Tamelen, E. E.; Friedman, L. J.; Burgstahler, A. W. *J. Am. Chem. Soc.* **1953**, *75*, 384–392.
16 (a) Dauben, W. G.; Kessel, C. R.; Takemura, K. H. *J. Am. Chem. Soc.* **1980**, *102*, 6893–6894. For a preparative scale version, see (b) Dauben, W. G.; Gerdes, J. M.; Smith, D. B. *J. Org. Chem.* **1985**, *50*, 2576–2578.
17 Grieco, P. A.; Nunes, J. J.; Gaul, M. D. *J. Am. Chem. Soc.* **1990**, *112*, 4595–4596.
18 Dauben, W. G.; Lam, J. Y. L.; Guo, Z. R. *J. Org. Chem.* **1996**, *61*, 4816–4819.
19 Lee, C. K.; Hahn, C. S. *J. Org. Chem.* **1978**, *43*, 3727–3729.
20 Mandell, L.; Blanchard, W. A. *J. Am. Chem. Soc.* **1957**, *79*, 6198–6201.
21 (a) Gribble, G. W.; Allen, R. W.; LeHoullier, C. S.; Eaton, J. T.; Easton, Jr., N. R.; Slayton, R. I.; Sibi, M. P. *J. Org. Chem.* **1981**, *46*, 1025–1026. (b) Gribble, G. A.; Sibi, M. P.; Kumar, S.; Kelly, W. J. *Synthesis* **1983**, 502–504.
22 Clayton, S. C.; Regan, A. C. *Tetrahedron Lett.* **1993**, *34*, 7493–7496,
23 Otten, A.; Namyslo, J. C.; Stoermer, M.; Kaufmann, D. E. *Eur. J. Org. Chem.* **1998**, 1997–2001.
24 See, however, Kumamoto, K.; Fukada, I.; Kotsuki, H. *Angew. Chem. Int. Ed.* **2004**, *43*, 2015–2017.
25 González, J.; Taylor, E. C.; Houk, K. N. *J. Org. Chem.* **1992**, *57*, 3753–3755.
26 Levin, J. I.; Laakso, L. M. in *Oxazoles: Synthesis, Reactions, and Spectroscopy*, Part A, Palmer, D. C., Ed., John Wiley & Sons, Inc., Hoboken, New Jersey, **2003**.
27 (a) Harris, E. E.; Firestone, R. A.; Pfister, 3rd, K.; Boettcher, R. R.; Cross, F. J.; Currie, R. B.; Monaco, M.; Peterson, E. R.; Reuter, W. *J. Org. Chem.* **1962**, *27*,

2705–2706. See also, (b) Firestone, R. A.; Harris, E. E.; Reuter, W. *Tetrahedron* **1967**, *23*, 943–955.

28 Cohen, A.; Haworth, J. W.; Hughes, E. G. *J. Chem. Soc.* **1952**, 4374–4383.

29 König, H.; Graf, F.; Weberndörfer, V. *Ann. Chem.* **1981**, 668–682.

30 Grigg, R.; Jackson, J. L. *J. Chem. Soc. (C)* **1970**, 552–56.

31 Arrick, B. A., *Bachelor of Arts thesis*, Wesleyan University, Middletown, Connecticut, **1977**.

32 Bredereck, H.; Gompper, R. *Chem. Ber.* **1954**, *87*, 726–732.

33 Kondrat'eva, G. Y.; Medvedskaya, L. B. Ivanova, Z. N. *Bull. Acad. Sci. USSR, Div. Chem. Sci. (Engl. Transl.)* **1971**, *20*, 2148–2150.

34 Wasserman, H.; Druckrey, E. *J. Am. Chem. Soc.* **1968**, *90*, 2440–2441.

35 Jacobi, P. A. in *Advances in Heterocyclic Natural Product Synthesis*, Vol. *2*, Pearson, W. H., Ed., Jai Press, Inc., Greenwich, Connecticut, **1992**.

36 Jacobi, P. A.; Selnick, H. G. *J. Org. Chem.* **1990**, *55*, 202–209.

37 Problems for practice 4: (a) Wolthuis, E.; Jagt, D. V.; Mels, S.; De Boer, A. *J. Org. Chem.* **1965**, *30*, 190–193. (b) Wineholt, R. L.; Wyss, E.; Moore, J. A. *J. Org. Chem.* **1966**, *31*, 48–52. (c) Allen, C. F. H.; Gilbert, M. R.; Young, D. M. *J. Org. Chem.* **1937**, *2*, 227–234; 235–244. (d) Padwa, A.; Hartman, R. *Tetrahedron Lett.* **1966**, *7*, 2277–2281. (e) Bonnett, R.; Gale, I. A. D.; Stephenson, G. F. *J. Chem. Soc.* **1965**, 1518–1519.

14

Heterocycles as Synthons

"What's in a name?" goes the famous literary passage, borrowed here in reference to "synthons," not roses. Surely this has something to do with piecing things together? Actually, yes and no. It might also relate to taking things apart! The word itself can be traced to a 1967 paper by E. J. Corey [1], another of the giants of organic synthesis, and winner of the Nobel Prize in chemistry in 1990. The Nobel committee specifically cited "his development of the theory and methodology of organic synthesis," of which retrosynthetic analysis was a key component. Indeed, so useful is this type of analysis, that it is now taught as a matter of routine when introducing beginning students to the art of organic synthesis (i.e., work backward!). In this context, Corey coined the term synthon for "Structural units within a molecule which are related to possible synthetic operations (and, therefore, to the reverse operations of degradation) [1]." And the name served well for many years. Over time, though, the original meaning blurred, and in 1988 Corey noted that "the word synthon has now come to be used to mean synthetic building block rather than retrosynthetic fragmentation structures [2]." This is how we shall use it.

With this as introduction, let us enter into the realm of heterocycles as synthons, beginning with furans as 1,4-dicarbonyl equivalents (Figure 14.1). Recall that furans themselves are acid labile, but require rather forcing conditions to effect hydrolysis. Oftentimes, then, decomposition intervenes, with the result that this route to 1,4-dicarbonyl derivatives has limited practical value (for an exception, though, see Büchi's synthesis of *cis*-jasmone in Scheme 12.5). A better solution involves initial de-aromatization, which can be accomplished utilizing either bromine in methanol, or electrolytically in the same solvent. In either case the furan ring has been oxidized, affording generally high yields of 2,5-dihydro-2,5-dimethoxyfurans (top of figure). Mild acid hydrolysis then produces the unsaturated 1,4-dicarbonyl compounds shown in brackets, which are in the proper oxidation state for direct conversion to pyridazines. Alternatively, catalytic hydrogenation returns us to the furan oxidation level, and provides an excellent route to saturated 1,4-dicarbonyl derivatives (cf. also Figure 11.15).

Introductory Heterocyclic Chemistry, First Edition. Peter A. Jacobi.
© 2019 John Wiley & Sons Ltd. Published 2019 by John Wiley & Sons Ltd.

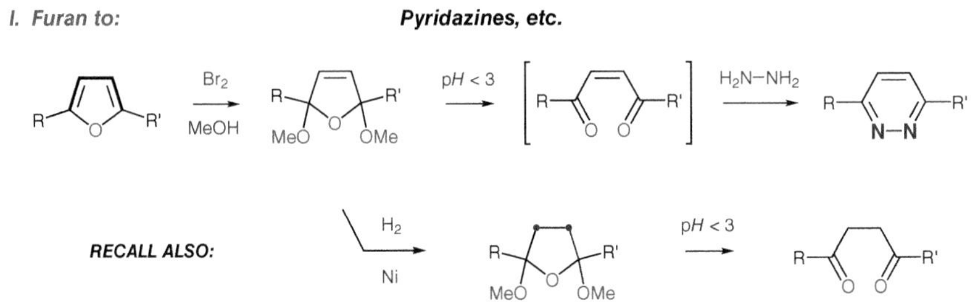

I. Furan to: **Pyridazines, etc.**

Figure 14.1 Heterocycles as synthons

So where else might this transformation find utility? As this page is being written, organic chemists are commemorating the 100th anniversary of perhaps the most groundbreaking synthesis of the early twentieth century, placing biomimetic strategy on a firm footing. The target of interest was the σ-plane symmetric alkaloid tropinone (see below), originally prepared by Richard Willstätter in some 20 odd steps beginning with cycloheptanone (1901) [3]. The overall yield for this synthesis was only ~0.75%, but it was state of the art for the era:

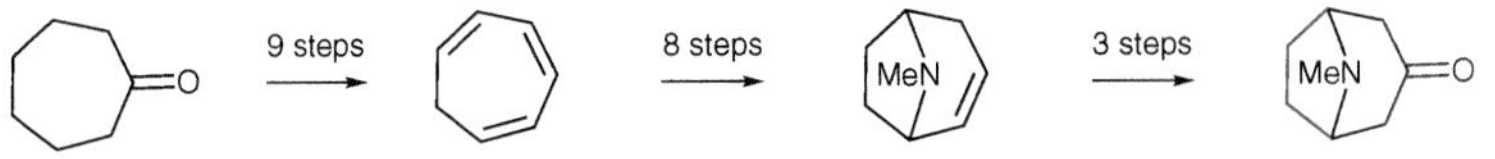

Willstätter synthesis of tropinone (0.75% overall yield)

And then along came Robert Robinson (yes the same!), who in 1917 published a single step synthesis of the same compound in 42% yield (top right Scheme 14.1) [4]. Robinson's reasoning, so eloquently espoused in the original account, was that the components in brackets, consisting of succindialdehyde (red), methylamine (black) and acetone dicarboxylic acid (blue), might undergo a double Mannich-like condensation leading directly to tropinone. And so they did, forever changing the field of natural products chemistry. However, in the experimental section Robinson alludes to one frustrating aspect. At the time there was no reliable source of pure succindialdehyde, which was typically obtained by nitrous acid decomposition of the corresponding dioxime (itself derived by ring opening of pyrrole).

Substitute furan as a synthon and the problem is well along to being solved, via the intermediacy of 2,5-dimethoxytetrahydrofuran (top left Scheme 14.1). Acid hydrolysis of this material, utilizing dilute HCl, is exceptionally clean, allowing for systematic optimization. Thus, in the "modern" era, and with careful control of pH, tropinone is obtained in upward of 90% yield [5]. The Robinson-Schöpf condensation [6], as it has come to be known, is often cited as the epitome of a double-Mannich reaction, and it has found widespread utility in synthesizing biologically active tropane alkaloids [7].

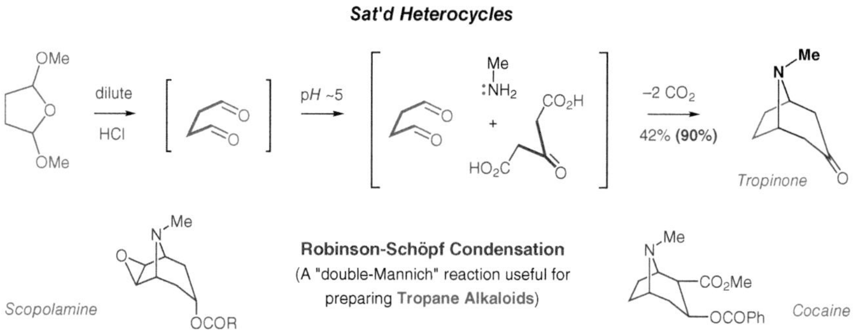

Sat'd Heterocycles

Scheme 14.1 Furan as a synthon.

Let us take a different tack now, and explore how furans might be employed in the synthesis of pyridines. In fact, this is not an entirely new area, for we have previously described how 2-acylfurans are in the proper oxidation state for direct conversion into 3-pyridinols (cf. Figure 5.2). As one example, 2-*n*-butyrylfuran affords a 74% yield of 2-*n*-propyl-3-pyridinol upon heating to 170 degrees Celsius in a sealed tube with ethanolic ammonia (top Scheme 14.2) [8]. The presumed intermediate, shown in brackets, is not isolated, but rather undergoes direct cyclization with loss of a molecule of ammonia. Although of mechanistic interest, however, such aminolysis conditions are much too harsh for synthesizing more fragile pyridine derivatives, and they even fail for the parent ring system (bottom of scheme).

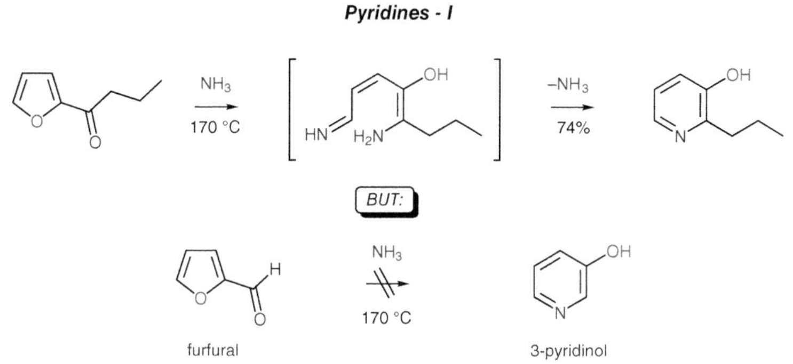

Pyridines - I

Scheme 14.2 Furan as a synthon.

As an alternative, it is better to start with the amino group already in place, which may require an extra step or two, but pays enormous dividends in overall yields. One of the early contributors to this area was the Danish chemist Clauson-Kaas, who in 1955 reported a remarkably efficient synthesis of

Pyridines - II

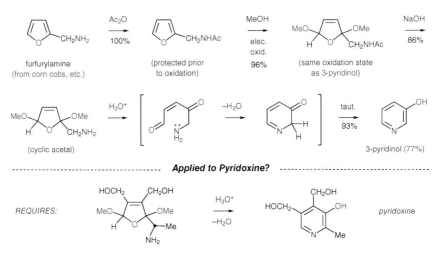

Scheme 14.3 Furan as a synthon.

3-pyridinol (top Scheme 14.3) [9]. Adding to its appeal, the starting material for this synthesis was only two steps removed from corn cobs and other agricultural waste products, via reductive amination of furfural (vide supra). So how did Clauson-Kaas convert furfurylamine to the final product, which required a total of only four steps? To begin, we should ask ourselves whether the starting material and product are in the same oxidation state, since we are always on the lookout for "thermodynamic wells." In this case, though, and after a few moments with pencil and paper, we see that the answer is no. Instead, hydrolytic ring opening of furfurylamine would provide an intermediate that, in principle, might cyclize to pyridine itself. Clearly, then, the furan ring must be oxidized, an operation that will require initial protection of the amino group. This was achieved in essentially quantitative yield by acylation with acetic anhydride. With only three steps left to go, the requisite oxidation was carried out via electrolysis in methanol (96%), and the protecting group was cleaved with NaOH (86%). We come now down to the final intermediate shown at the middle left of the scheme, leaving us with but a single transformation to achieve our goal. And what a striking transformation it was! With aromaticity no longer an issue, mild acid hydrolysis cleaved the cyclic acetal, producing a dicarbonyl derivative that was ideally constituted for intramolecular imine formation. Simple tautomerization then gave a 93% yield of 3-pyridinol, bringing the overall yield to 77% [9]. All in all, a very nice example of green chemistry.

And we can build complexity from here. Thus, by now the direction we are heading will be obvious from a perusal of the bottom of Scheme 14.3, where we take this concept one step further. Could such a strategy also be applied to the synthesis of pyridoxine? It would seem so, by incorporating the highly substituted furan derivative shown to the left. But how to introduce all of these groups? You already have the tools necessary.

A key intermediate in the Clauson-Kaas synthesis of pyridoxine was 3,4-(bis) acetoxymethyl-furan, which was readily prepared on large scale from simple starting materials (top Scheme 14.4) [10]. But rather than reproduce the literature solution, let us leave this as a small project for the reader (two methods, please). Simplifying your task, the target molecule has a plane of symmetry, eliminating any concern of regiochemical control in subsequent steps. And where does that leave us? Looking ahead to the final objective, there still remained two carbons to add to the furan skeleton, in addition to the nitrogen destined to occupy the pyridine ring. The first of these objectives was cleanly accomplished employing a Friedel-Crafts acylation, which you will recall requires special conditions for acid sensitive furan rings (cf. Figure 12.1). In the present instance the combination of acetic anhydride and boron trifluoride provided a satisfactory solution, affording a 72% yield of the 2-acyl derivative shown at the top left of the scheme. And again, of course, it does not matter which side of the furan ring undergoes substitution. Next, the missing nitrogen comes into play, and it must be added to the same position occupied by the carbonyl group. There are a number of ways to approach this, including

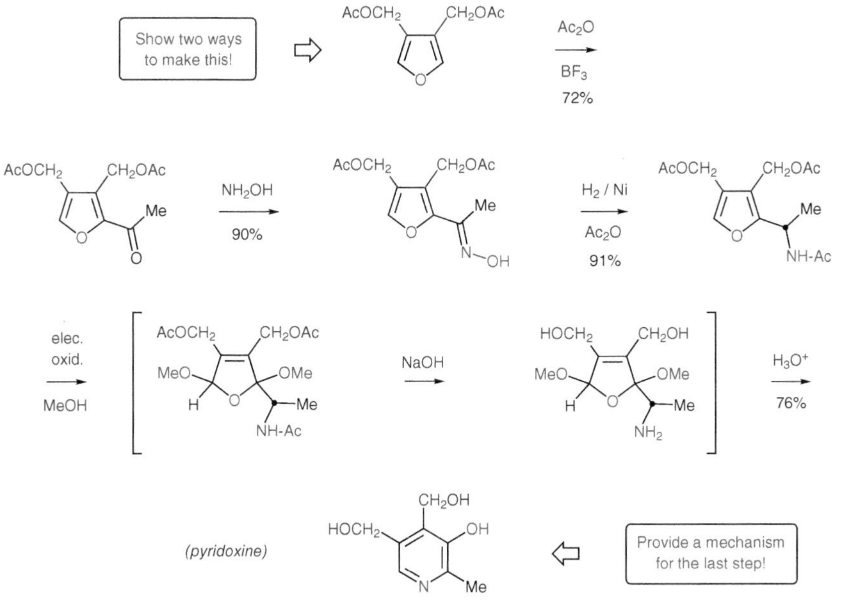

Scheme 14.4 Putting it all together.

reductive amination to arrive at the amine oxidation state directly. However, the authors elected to prepare the oxime derivative, providing a nicely crystalline intermediate for purification (90% yield). And well they did, since many of the remaining steps were carried out in situ. These included, first, hydrogenation of the oxime, employing a nickel catalyst in acetic anhydride as solvent. Thus was obtained the acylated amine derivative illustrated at the top right of the scheme, which turned out to be the last compound isolated (91% yield). From here commenced the most impressive sequence of all, in which the acetamide derivative was converted in 76% overall yield to pyridoxine. This was accomplished via (1) electrolytic oxidation; (2) base-induced cleavage of the three acetyl groups; and (3) acid catalyzed hydrolysis/rearrangement [10]. Again, it will be left to the reader to provide a mechanism for this last step. But what a remarkable synthesis, and readily adaptable to large-scale production.

Nor should we limit ourselves to synthesizing nitrogen heterocycles. To this end, let us return to Scheme 14.2 and ask the question "What happens if we substitute H_3O^+ for ammonia?" Certainly we are not going to form pyridines, but the chemistry world is full of thermodynamic wells that do not happen to be heterocycles. For example, with a little bit of practice, we can readily make the connection that 2-acetylfuran is in the same oxidation state as the tricarbonyl derivative shown in brackets, which in turn should be convertible to catechol:

That is not to say, of course, that such a transformation actually succeeds in the laboratory. In fact, the conditions required are simply too harsh, given that the furan ring is stabilized by the electron withdrawing carbonyl group. But it does start us to wondering if a favorable pathway forward might be devised. And evidently, Clauson-Kaas and his co-workers were thinking along the same lines.

Indeed, the Danish group provided the first practical solution to this transformation, taking the carbonyl group out of play in the very first step (Scheme 14.5) [11]. This was conveniently effected simply upon warming the starting material in dry methanol containing a catalytic amount of *p*-toluenesulfonic acid (TsOH), along with trimethyl orthoformate (TMOF) as water scavenger. Moreover, it proved unnecessary to isolate the ketal shown in brackets, which was directly submitted to electrolytic oxidation in methanol. In this manner was obtained a 69% yield of a dihydrofuran derivative that contains four methoxy substituents and is completely devoid of aromatic character (top right of scheme). And how can we be certain that the product is truly one oxidation state higher than our starting 2-acetylfuran? Do not be misled

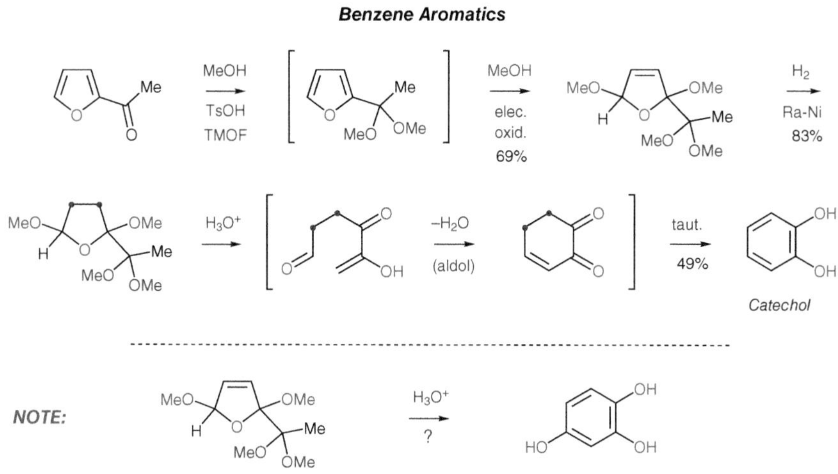

Benzene Aromatics

Scheme 14.5 Furan as a synthon.

by the fact that there are no carbonyl groups present, or that the furan ring now contains only one double bond. To satisfy yourself of oxidation level, picture in your mind's eye effecting a 1,4-elimination of MeOH, by which aromaticity would be regained. Then ask, is this material in the same oxidation state as the starting ketal? Clearly not, since it contains an extra methoxy group bonded to the furan ring, and is therefore one oxidation state higher. Spend a few moments getting comfortable with this analysis, since it is a line of reasoning that will prove useful in the future. Anyway, to get back on path the authors simply reduced the alkene double bond, providing the very labile tris-ketal derivative shown at the middle left of the scheme (83%). Thus, in two steps they have synthesized a substrate that, while in the same oxidation state as 2-acetylfuran, is far more reactive toward acid hydrolysis. This was accomplished by brief heating in 1N HCl, which undoubtedly leads to the enolized tricarbonyl derivative shown in brackets. From here, though, we are only an aldol condensation and tautomerization away from producing catechol, which was obtained in 49% yield [11]. Finally, let us carry out one additional "thought" experiment and hydrolyze the initial product of electrolytic oxidation (bottom Scheme 14.5). Curiously, this reaction pathway was apparently not pursued, for it would seem to provide an excellent route to 1,2,4-trihydroxybenzene.

Many variants of this methodology are known, a number of which produce more highly substituted benzene aromatics. One such example is outlined in Scheme 14.6, where the starting material is the commodity chemical methyl furoate, and the target is the catechol derivative shown to the middle right [12]. As with 2-acetylfuran, methyl furoate is stable to direct acid hydrolysis, bearing, as it does, a strongly electron withdrawing carbomethoxy group. However,

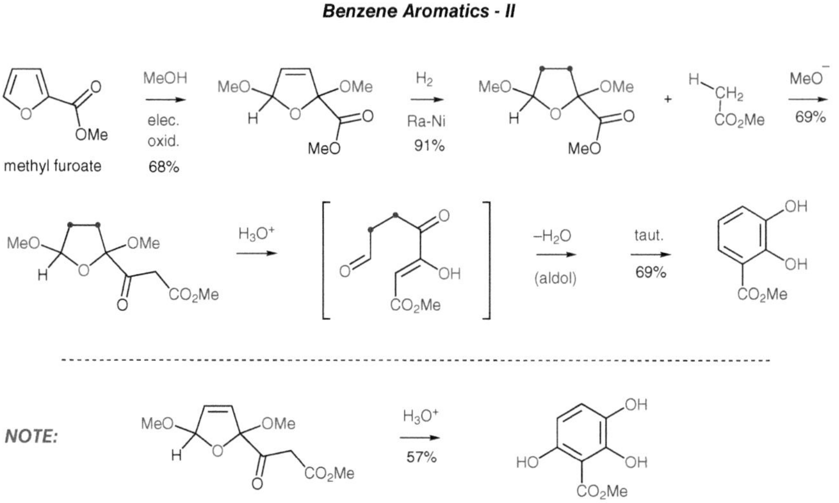

Benzene Aromatics - II

Scheme 14.6 Furan as a synthon.

by now we know that this necessitates only a minor detour, consisting of electrolytic oxidation followed by catalytic hydrogenation. In this fashion, and using the same reasoning as above, we arrive at a 2,5-dimethoxytetrahydrofuran derivative that is in the same oxidation state as the starting material, but much more readily hydrolyzed (~60% overall yield). But before hydrolysis, there is still chemistry to be accomplished, as we are lacking a carbon atom necessary to complete the benzene skeleton, as well as an additional carbomethoxy group. These pieces were installed employing a mixed Claisen condensation, whose success can be traced to the fact that the furoic acid derivative has no α-protons. Now came the pivotal transformation, as the resultant β-ketoester was subjected to acid hydrolysis, presumably affording the tricarbonyl species shown in brackets. However, neither this material, nor any other open chain products were isolated. In their stead was obtained a 69% yield of methyl 2,3-dihydroxybenzoate [12].

Closing this section, note that the reaction pathway described as a "thought" experiment at the bottom of Scheme 14.5 was actually quite successful in the present instance. Thus, acid hydrolysis of the 2,5-dimethoxydihydrofuran derivative shown at the bottom left of Scheme 14.6 gave a 57% yield of methyl 2,3,6-trihydroxybenzoate [12].

So where from here in our discussion of heterocycles as synthons? Up to now we have devoted much attention to furan and its derivatives, serving, as they do, as versatile precursors to a diverse range of ring systems. The targets themselves might be saturated heterocycles, as in the case of the Robinson-Schöpf condensation leading to tropinone (Scheme 14.1), or aromatic heterocycles, as with pyridazines (Figure 14.1) and 3-hydroxypyridines (Scheme 14.3). Or, they might not be heterocycles at all (cf. Schemes 14.5 and 14.6)! The unifying factor is that in each example the furan nucleus is serving as a 1,4-dicarbonyl "equivalent," needing only to be liberated by acid hydrolysis. And how is hydrolysis effected? Occasionally this can be carried out directly, but more often than not it is beneficial to incorporate a preliminary oxidative de-aromatization step, producing a non-aromatic cyclic ketal (cf. Figure 14.1). Hydrolysis at this stage affords an unsaturated 1,4-dicarbonyl derivative that is an oxidation state higher than the starting material. Alternatively, catalytic hydrogenation returns us to the furan oxidation level, and provides ready access to saturated 1,4-dicarbonyl derivatives. In either case, it is difficult to imagine a more efficient precursor.

But let us move away from oxidative and hydrolytic ring transformations, and focus for a bit on reductive processes. For the most part, catalytic hydrogenation of furans and pyrroles follows the expected course, affording the corresponding tetrahydro derivatives employing a variety of catalyst systems (top Figure 14.2). With furan the catalyst of choice is Raney nickel (Ra-Ni), with the caveat that hydrogenolysis and concomitant ring opening can be a

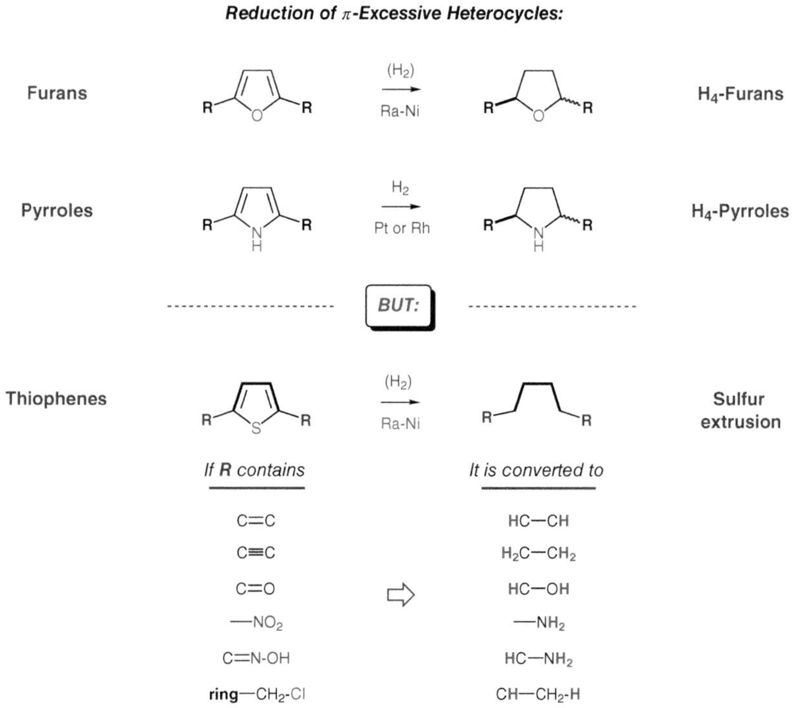
Reduction of π-Excessive Heterocycles:
Furans
(H2)
Ra-Ni
H4-Furans
Pyrroles
H2
Pt or Rh
H4-Pyrroles
BUT:
Thiophenes
(H2)
Ra-Ni
Sulfur extrusion
If R contains
It is converted to
C=C
HC—CH
C≡C
H2C—CH2
C=O
HC—OH
—NO2
—NH2
C=N-OH
HC—NH2
ring—CH2-Cl
CH—CH2-H

Figure 14.2 Heterocycles as a synthon.

competing reaction pathway. Pyrrole, on the other hand, is best hydrogenated over either Pt or Rh, and does not generally suffer hydrogenolysis [13]. And what about stereochemistry? Where this is an issue, cis- and trans-mixtures are often obtained, with the cis-isomer usually predominating. In the final analysis, though, such reductions are of limited value in the context of these ring systems serving as synthons (i.e., building blocks for other materials).

This brings us down to thiophene, where the chemistry involved is considerably more interesting (bottom Figure 14.2). As most readers will be aware, sulfur is a poison for many hydrogenation catalysts, rendering reduction of thiophene to tetrahydrothiophene impractical (at least by catalytic hydrogenation). However, the combination of H_2/Ra-Ni is an exception, effecting not only clean hydrogenation of the π-system, but also extruding sulfur (H_2 is shown in parentheses since it is typically provided by the catalyst) [14]. Consider, for example, the case of a simple 2,5-disubstituted thiophene that is subjected to exhaustive hydrogenation over Ra-Ni. The end result is that all evidence of the thiophene ring essentially disappears, producing an alkane chain in which the central four carbon atoms were once a part of the thiophene template. Of course, many unsaturated functional groups attached to the ring will also be reduced, as will benzylic-like chlorides (bottom Figure 14.2).

How does this fit in with reactions we have already discussed? On the one hand, it is the high stability of the thiophene template that renders this methodology so useful. That is, it will stand up to almost any conditions we care to throw at it, save Ra-Ni hydrogenolysis! And yet, thiophene is also perhaps the easiest of the π-excessive ring systems to functionalize—remember our analogy to playing chemical hard ball? This opens up a wealth of possibilities.

For example, suppose we wished to synthesize a long chain aldehyde. As illustrated at the top of Figure 14.3, this might be readily accomplished beginning with Vilsmeier-Haack formylation of the appropriate 2-alkylthiophene, itself prepared via alkylation of the corresponding 2-lithio derivative (cf. Figure 12.9). Acetal formation, followed by Ra-Ni hydrogenolysis, would then afford the desired aldehyde in protected form (for specific examples, see the references [14]). Or, perhaps our target is a long chain ketone? In this case we would employ a Friedel-Crafts acylation, again beginning with a suitably substituted thiophene (second equation from top). The remainder of the sequence would mirror that for the aldehyde, consisting of ketalization, Ra-Ni hydrogenolysis and acid hydrolysis. Note in particular that the choice of R and R' is extremely flexible.

Alicyclic derivatives are also well represented, in which the fused ring might be either aliphatic [15] or aromatic in character (bottom Figure 14.3) [14]. In many such cases the thiophene ring will have served as a template for appending a carbocyclic ring, which is revealed in its entirety upon hydrogenolysis. Alternatively, both rings might be constructed simultaneously utilizing

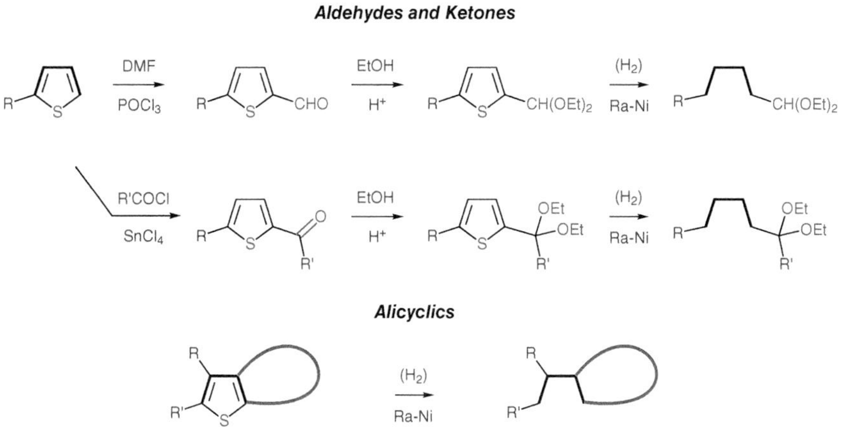

Aldehydes and Ketones

Alicyclics

Figure 14.3 Thiophene as a synthon.

thiazole-alkyne Diels-Alder chemistry, in analogous fashion to that previously described for preparing furans (cf. also Scheme 13.6):

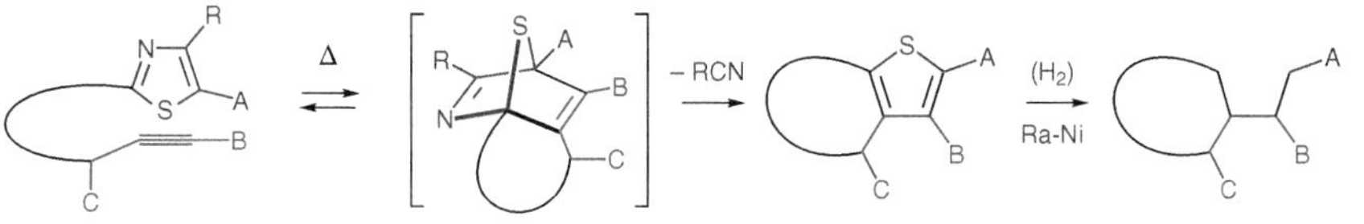

Once again, the appended substituents A, B, and C are transposed in unequivocal fashion, via intermediate II, to the fused ring thiophene III. Ra-Ni hydrogenolysis then provides the desired carbocycle IV.

An interesting example of this last approach is illustrated in Scheme 14.7, wherein the target molecule incorporates a sesquiterpene skeleton of the eremophilane class (bottom right of scheme) [16]. As with all sesquiterpenes, members of this class contain 15 carbon atoms, biosynthetically derived by combining three isoprene units. In the final product, one of these units is highlighted in red, which also draws attention to a common structural feature. Namely, one often finds an isopropyl, or isopropyl derived group attached to C7. So how is this functionality to be introduced? The authors chose to exploit thiophene chemistry, in which the penultimate intermediate was the thiazole-alkyne derivative shown at the top right of the scheme (note that the synthesis of the precursor thiazole alcohol is described in Figure 11.9). Thermolysis of this material then led directly to the anticipated thiophene ketone, which upon Ra-Ni hydrogenolysis afforded an ~74% overall yield of the desired eremophilane.

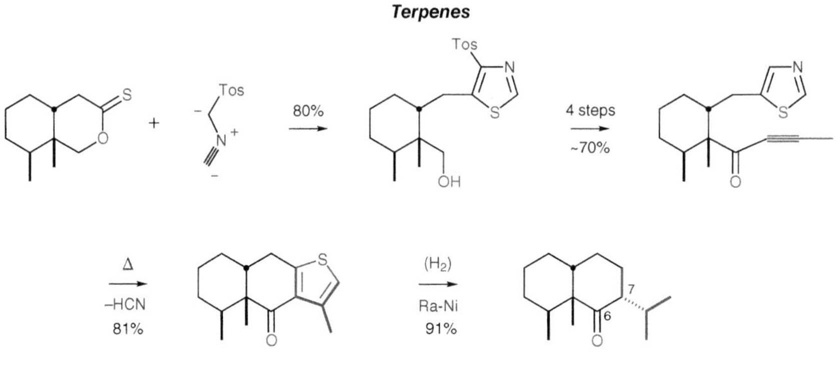

Terpenes

Scheme 14.7 Thiophene as a synthon.

Not surprisingly, thiophene is also a favorite template for synthesizing long-chain carboxylic acids, which can be either branched or linear in nature. A straightforward example of the second category is provided at the top of Scheme 14.8, where the target is pentadecanoic acid, a relatively rare 15-carbon fatty acid isolated from cow milk fat [17]. All told, the synthesis required five steps, combining three commercially available starting materials in reactions that we have already discussed. The first of these involved

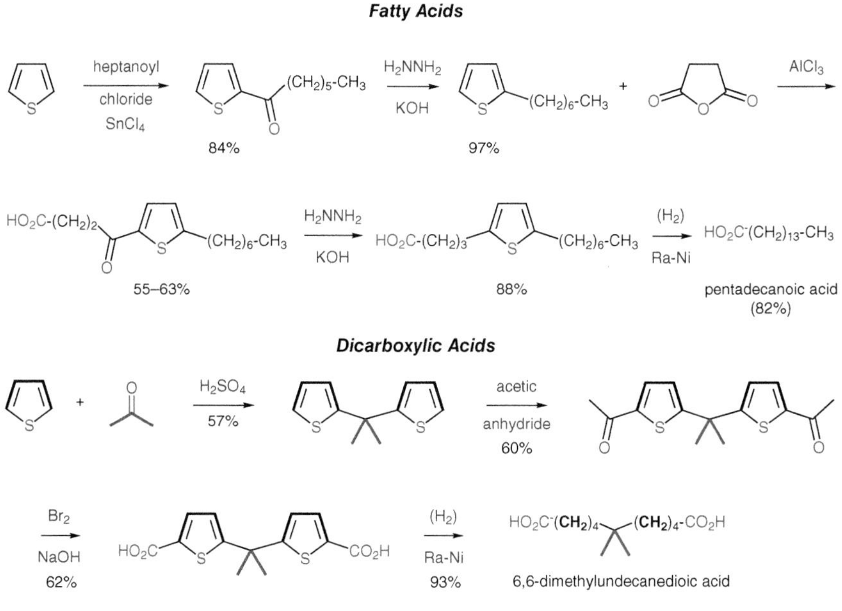

Fatty Acids

Dicarboxylic Acids

Scheme 14.8 Thiophene as a synthon.

Friedel-Crafts acylation of thiophene with *n*-heptanoyl chloride/SnCl$_4$, which afforded an 84% yield of the expected 2-thiophene ketone. Wolff-Kishner reduction then gave 2-*n*-heptylthiophene in 97% yield. With 11 of the requisite 15 carbon atoms now in place, it remained to extend the latent aliphatic chain by an additional three methylene units, capped by a carboxylic acid. This was accomplished in two steps, consisting of Friedel-Crafts acylation with succinic anhydride/AlCl$_3$, followed in turn by a second Wolff-Kishner reduction (~50% overall yield). No other regioisomers were isolated. Lastly, Ra-Ni hydrogenolysis completed the synthesis, affording an 82% yield of pentadecanoic acid.

Many similar examples can be found in the literature, but let us diverge a little from our discussion of aliphatic mono-carboxylic acids with a concise synthesis of 6,6-dimethylundecanedioic acid (bottom Scheme 14.8) [17]. Here, advantage was taken of the large scale condensation of thiophene with acetone, which upon heating in 72% H$_2$SO$_4$ gave a 57% yield of the bis-thienyl adduct shown. Mechanistically, this transformation undoubtedly proceeds via the intermediacy of a tertiary alcohol, the product of initial electrophilic aromatic substitution. However, this intermediate undergoes rapid S$_N$1 displacement by a second molecule of thiophene. Given the number of possible side reactions, the yield obtained is impressive. But perhaps most striking of all, this reaction stands as testimony to the acid stability of thiophene—both furan and pyrrole would have suffered certain decomposition. It remained now to introduce the two terminal carboxyl groups, and at this point the authors chose an interesting route. One might have thought this operation could be accomplished by direct lithiation/carboxylation, for which there was ample precedent (cf. Figure 12.9). Rather, though, the same goal was achieved by a two step procedure, consisting of bis-acylation followed by bromoform reaction. Coming down now to the bottom left of the scheme, Ra-Ni hydrogenolysis afforded the target dicarboxylic acid in 93% yield.

Finally, we close this section on thiophene as a synthon with two examples that are somewhat "out of the mold" (Scheme 14.9). The first of these might come under the heading of "you can't get too much of a good thing." That is, if a single thiophene is a good four carbon chain extender, then 2,2'-bithiophene should serve in much the same capacity as an eight carbon chain extender. Such is indeed the case, as illustrated by the transformation shown at the top of Scheme 14.9 [18]. Thus, the title compound was converted in three steps, and excellent overall yield, to the medium chain length diketone derivative shown at the top right.

And so we arrive at the last example of thiophene chemistry, in which the target compound is the 14-membered ring lactone shown at the bottom right of Scheme 14.9 [19]. Classically, such a macrocycle would be constructed by formation of the lactone bond last, in a process requiring high dilution conditions. However, in the present case this key linkage was formed early on,

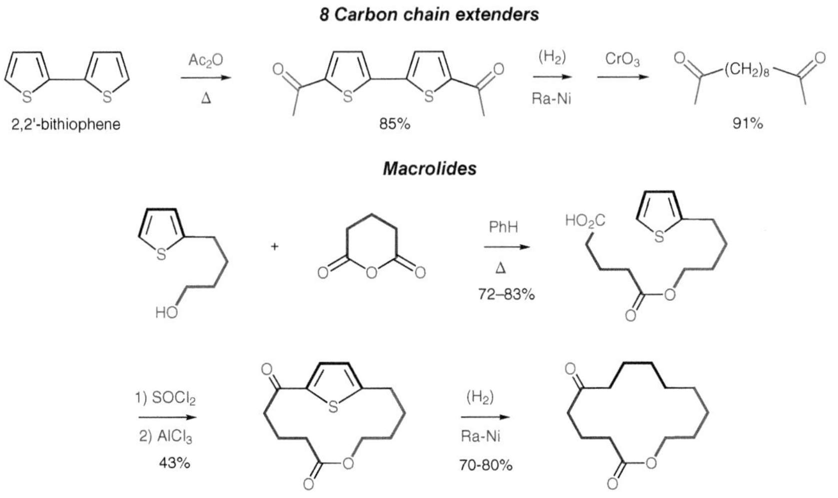

8 Carbon chain extenders

Macrolides

Scheme 14.9 Thiophene as a synthon.

beginning with the readily available thiophene alcohol shown at the left center. Note that this starting material already contains nine of the final total of 14 ring atoms, with four contributed by the thiophene template. The remaining five were now added by acylation with glutaric anhydride, producing high yields of the carboxylic acid derivative illustrated at the middle right. And how is this product related to the final macrocycle? Two points are worth emphasizing. First, although one might be skeptical at first, models clearly show that the carboxyl group and C5 of the thiophene ring are within favorable bonding distance, requiring only suitable activation for an intramolecular Friedel-Crafts acylation. Second is the natural proclivity of a 2-substituted thiophene ring to undergo electrophilic substitution at C5. In any event, acid chloride formation, followed by AlCl$_3$ catalysis, gave a 43% yield of the thiophene ketone shown at the bottom left, which on Ra-Ni hydrogenolysis afforded the desired macrolide directly (70–80% yield; interestingly, in this instance the ketone carbonyl was not reduced) [19].

Having reached this juncture, we take our leave of thiophene, but we are not excusing ourselves entirely from sulfur containing heterocycles. It remains to explore the consequences of Ra-Ni cleavage of isothiazoles, that π-excessive ring system that you will recall was first synthesized in serendipitous fashion by R. B. Woodward (cf. Figure 11.13). The reasons for Woodward's interest in this area will become clear shortly. For the moment, though, let us content ourselves with the general case of a fully substituted isothiazole reacting with Ra-Ni, for which a reasonable reaction pathway is outlined below:

Allylic Amines

Thus, we expect first that sulfur will be rapidly extruded, leading to the α,β-unsaturated imine shown in brackets. The fate of this intermediate would then depend upon the reaction conditions, with extended hydrogenation presumably leading to the fully saturated amine derivative shown to the right. Alternatively, quenching at the imine stage, followed by NaBH$_4$ reduction, should provide ready access to highly substituted allylic amines. And it was this possibility that struck a resonant chord with RBW.

Why? If you are inclined to read just one of Woodward's natural product syntheses, choose the remarkable story of the conversion of 3-methyl-4-carbomethoxyisothiazole to colchicine, transcribed from a Harvey Lecture presented in 1963 [20]:

(from figure 11.13)   Colchicine   O—Me

Time and space will limit our discussion of this feat, but we can provide at least a flavor of the author's reasoning. On the surface, colchicine is not a tremendously complex molecule, bearing, as it does, a single chiral center. Still, the first two total syntheses were only recorded in 1959 [21], nearly 140 years following its isolation (1820), and a full 10 years after its structure determination. This gives some idea of the synthetic challenges associated with the A,B,C-ring skeleton, which was unprecedented at the time. And how did Woodward view these challenges? First, he gave great consideration to the presence of the allylic amino substituent in ring B, in particular due to the high oxidation state of the adjacent tropolonoid ring C. That is, almost any de novo route to ring C would require multiple oxidations, and therefore protection of the amino group from the start. Moreover, the means of protection would have to be robust. Second was the presence of two fused 7-membered rings, which are generally more difficult to prepare than their 5- or 6-membered ring counterparts. To gain entropic advantage, Woodward proposed to construct both of these rings appended to a stable, planar platform, a scaffolding that might subsequently be removed. And what was the perfect scaffold? Why, isothiazole, of course, which even though it was an unknown ring system, met all of RBW's criteria! Surely the nitrogen in isothiazole derivatives would be only feebly basic, while aromatic stabilization would enable it to stand up to a multitude of chemical manipulations. At the proper moment, though, Ra-Ni hydrogenolysis would unveil the most sensitive portion of the colchicine molecule, consisting of C3 through C5 (see above).

It goes without saying that Woodward's choice of starting material was audacious, but how did it perform in practice? Most would say quite well, although we will only be able to highlight the key steps (Scheme 14.10). The synthesis began by taking advantage of the special properties associated with the C3 methyl group, which being "benzylic-like," was readily brominated with N-bromosuccinimide (NBS). This material was then directly converted to the corresponding Wittig reagent, which underwent smooth condensation with commercially available 3,4,5-trimethoxybenzaldehyde to install ring A (top center structure in scheme). Next, in another four steps nearly the entire carbon backbone of colchicine was put in place, consisting of (1) diimide reduction of the carbon-carbon double bond; (2) reduction of the C4 ester to the aldehyde oxidation level; (3) condensation with the Wittig reagent derived

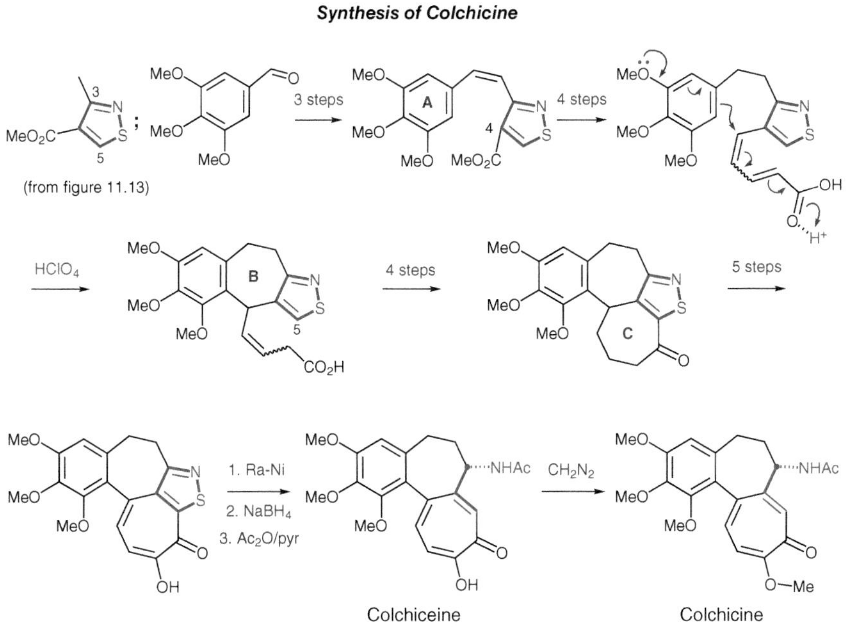

Synthesis of Colchicine

Scheme 14.10 Woodward's synthesis of colchicine.

from methyl 4-bromocrotonate (also commercially available); and (4), ester hydrolysis (top right structure in scheme). So far we are following along the classic lines of a Woodward synthesis, involving simple starting materials and reagents, leading at some point to a striking skeletal reorganization. And the present case was no exception. Thus, treatment of this last material with 70% $HClO_4$ led to clean formation of ring B, via the mechanistic pathway indicated by curly arrows. Ring C was then closed in routine fashion, initiated by double bond reduction and carboxylation at the relatively acidic C5 position.

At this stage began the considerable uphill climb in oxidation state for ring C, culminating with the aromatic tropolone derivative shown at the bottom left of the scheme. Woodward clearly delighted in describing this process, remarking at various steps on the "beautifully crystalline" nature of intermediates. However, as he also commented, the investigation had "now entered a phase which was tinged with melancholy." That is, it was time to dismantle the isothiazole core, which "had served so admirably in every anticipated capacity." And dismantle it they did, in an efficient three-step sequence consisting of Ra-Ni hydrogenolysis, $NaBH_4$ reduction, and acylation with Ac_2O in pyridine [20]. The material thus obtained was identical in all respects with an authentic sample of colchiceine (i.e., desmethylcolchicine), and was readily converted to colchicine by methylation with diazomethane.

From here it is but a short ascent up the periodic table to explore the chemistry of isoxazoles, the chemical "first-cousins" of isothiazoles. In the process, we have only substituted sulfur with oxygen, so how much could change? Quite a bit, actually. Thus, isothiazoles are excellent precursors to allylic and saturated amines, as we have just seen making use of Ra-Ni hydrogenolysis. In contrast, isoxazoles frequently play the role of masked 1,3-dicarbonyl derivatives, the reverse of one of their most common means of synthesis (top left Figure 14.4; cf. also Figure 11.12). That being the case, of what value are such transformations? This would seem much akin to synthesizing a furan from a 1,4-dicarbonyl compound, only to turn around immediately and subject it to hydrolysis. The difference of course is that there are a number of alternate strategies for preparing isoxazoles and their derivatives, including 1,3-dipolar cycloaddition of nitrile oxides with alkynes (top right of figure). This is a topic that we will explore in greater detail in Chapter 15, but for now suffice it to say that such cycloadditions represent a special case of a concerted $4\pi + 2\pi$ ring forming process.

And how do we go about unmasking an isoxazole ring to reveal its 1,3-dicarbonyl core? Not surprisingly, direct hydrolysis is usually impractical, given that we are working against aromatic stability. Instead, two indirect approaches have been taken. The first of these involves alkylation of the ring nitrogen with the powerful ethylating agent Meerwein's reagent, which serves to further polarize the π-cloud (middle Figure 14.4). The resultant salt then undergoes facile base hydrolysis, producing N-ethylhydroxylamine as a

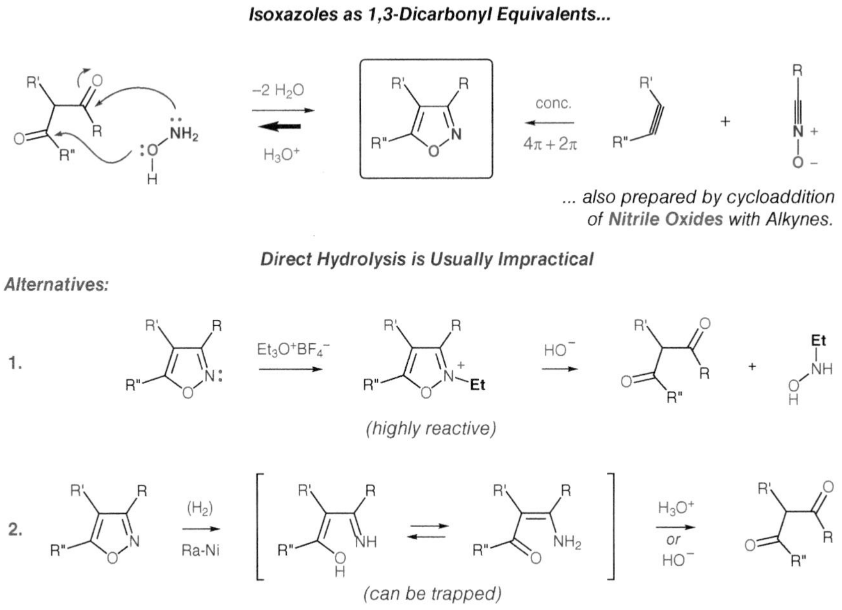

Isoxazoles as 1,3-Dicarbonyl Equivalents...

−2 H$_2$O

H$_3$O$^+$

conc.

$4\pi + 2\pi$

... also prepared by cycloaddition of Nitrile Oxides with Alkynes.

Direct Hydrolysis is Usually Impractical

Alternatives:

1. Et$_3$O$^+$BF$_4^-$ → *(highly reactive)* → HO$^-$

2. (H$_2$) Ra-Ni → *(can be trapped)* → H$_3$O$^+$ or HO$^-$

Figure 14.4 Heterocycles as synthons.

byproduct. Alternatively, the relatively weak N-O bond can be cleaved by Ra-Ni hydrogenolysis, leading to the vinylogous amide shown in brackets (bottom of figure) [22]. In certain cases, such intermediates can be trapped to give other heterocyclic ring systems, as we will see shortly. But they also suffer mild acid or base hydrolysis to 1,3-dicarbonyl derivatives.

Can you envision that 4-(chloromethyl)-3,5-dimethylisoxazole might serve as a stable surrogate for methyl vinyl ketone (MVK), especially in Robinson annulations? The eminent organic chemist Gilbert Stork did [23]. The title compound is shown in the shadow box at the top left of Scheme 14.11, and as the $ sign indicates, it is commercially available. It can also be prepared in two steps via acid catalyzed condensation of acetylacetone with hydroxylamine, followed by treatment with formaldehyde in HCl.

So how can this material function as an annulation reagent? Let us illustrate first with a simple example, in which the target is the parent $\Delta^{1,9}$-2-octalone ring system highlighted at the bottom right of the scheme [23]. To begin, the chloromethyl substituent in our starting material is highly reactive, undergoing ready substitution with the pyrrolidine enamine derived from cyclohexanone. And where from here? Note that the bold bonds in the alkylation product encompass all of the atoms found in MVK, and it remains only to free them from the isoxazole skeleton. This was initiated by reductive cleavage of the N-O bond, which was smoothly accomplished either with H$_2$/Pd/C (very slow)

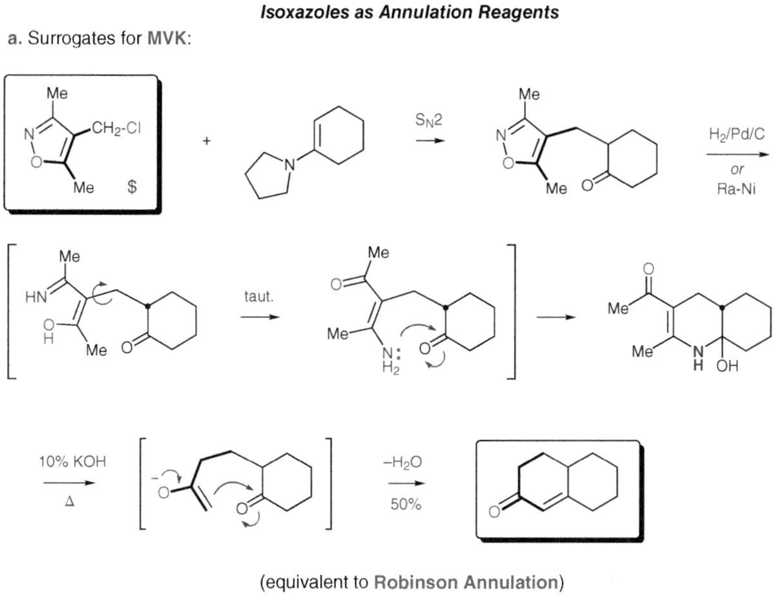

Isoxazoles as Annulation Reagents

a. Surrogates for **MVK**:

(equivalent to **Robinson Annulation**)

Scheme 14.11 Isoxazole as a synthon.

or with Ra-Ni (fast!). In either case, the resultant vinylogous amide underwent in situ addition across the cyclohexanone carbonyl group to generate the carbinolamide species shown to the middle right. Conveniently, this material precipitated directly from the reaction mixture, and it was utilized without further purification. Now came the final unveiling step, in which the crude product was heated in 10% aqueous KOH. The mechanistic pathway for the ensuing hydrolysis was elucidated in great detail, in particular with respect to the timing of β-keto cleavage. However, this is a discussion perhaps best left for another time, or for the reader to peruse in the original literature [24]. For our purposes, we need only emphasize that the penultimate intermediate in this process is the ketone enolate shown in brackets, which undergoes facile intramolecular aldol condensation. The end result was a 50% overall yield of $\Delta^{1,9}$-2-octalone, based upon the initial isoxazole alkylation product.

But what of more complex targets, such as the tetracyclic ring systems characteristic of steroids and D-homosteroids? Stork and his co-workers were not remiss in this regard, devising, for example, the "bis-annulation" reagent highlighted at the top left of Scheme 14.12 [25]. Count the carbon atoms in this reagent (bold bonds), and you will see that it is well constituted for introducing the entire A,B-ring backbone of *d,l*-D-homotestosterone in a single step. Indeed, straightforward alkylation with the C,D-ring octalone shown at the top center of the scheme afforded a 55% yield of the advanced intermediate

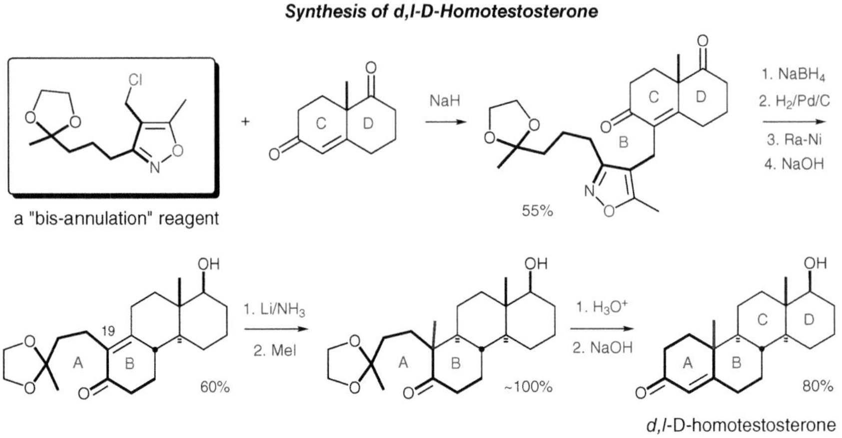

Synthesis of d,l-D-Homotestosterone

Scheme 14.12 Isoxazole as a synthon.

drawn to the right, awaiting only further bond reorganization. This was initiated by $NaBH_4$ reduction of the D-ring carbonyl group, followed by catalytic hydrogenation of the octalone double bond. With the stage thus set, the time had arrived to unmask the isoxazole carbonyl groups, by this operation laying the foundation for ring B. As in the simpler model system outlined in Scheme 14.11, this was accomplished by Ra-Ni hydrogenolysis followed by base hydrolysis, delivering ring B in 60% overall yield. It is hard to imagine a more "step-efficient" conversion, involving late stage β-keto cleavage and aldol condensation (vide supra). But let us pause here for a moment and see where we have yet to go. All of the ring A carbons are in place, as, in fact, they have been from the beginning. Lacking still is the C19 angular methyl group (steroid numbering), which at this point was introduced in essentially quantitative yield by reductive methylation. This brings us down to the bottom center of the scheme, where we have only to free the protected ketone derivative and carry out an intramolecular aldol condensation. As indicated, this was effected in 80% yield by mild acid hydrolysis followed by base-induced ring closure.

To this point we have had little to say about the isoxazolium salts formed upon N-ethylation of isoxazoles, aside from the fact that they undergo hydrolytic ring opening in aqueous alkali (cf. Figure 14.4). Generally, this reaction is carried out at room temperature, and if properly substituted such species can also participate in annulation reactions. However, at elevated temperatures an entirely new reaction pathway becomes operative. Take, for example, the case of the isoxazolium salt shown at the top left of Scheme 14.13, which with hot 1N NaOH affords a 47% yield of the phenol derivative at the bottom left [26]. As to a mechanism, it is clear that the C5-methyl group comes into play, given the regiochemical features of the final product. And what special properties

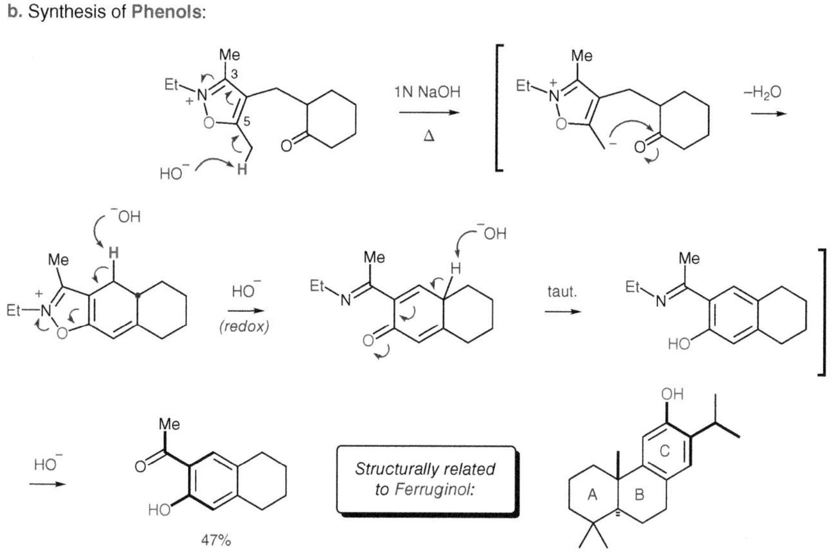

b. Synthesis of Phenols:

Scheme 14.13 Isoxazole as a synthon.

does this group possess? Most notably, it is relatively acidic, by virtue of its position in conjugation with the positively charged iminium bond (of course, the same argument holds for the C3-methyl group, but it is more sterically hindered). In any event, proton abstraction leads to a stabilized carbanion, which is ideally positioned to undergo an intramolecular aldol-like condensation with the cyclohexanone carbonyl group. This takes us down to the second structure in the brackets, and here the reaction might have stopped, were it not for two factors. First is the weakness of the N-O bond, which being highly polarized is primed for reductive cleavage. And second is our old friend, the drive to aromaticity. Both of these objectives can be met by a two step sequence triggered by an internal redox reaction, the nature of which is indicated by the curly arrows. In essence, this reaction corresponds to a 1,4-elimination in which the "leaving group" is the neutral imine derived by N-O bond scission. Tautomerization then affords the corresponding phenol, which is followed by imine hydrolysis to the methyl ketone. As the authors note, the product of this sequence is structurally related to the B,C-ring core of ferruginol, a naturally occurring diterpenoid phenol.

What better test for this new methodology than a total synthesis of ferruginol [26]? The starting materials were the readily available β-keto ester shown at the top left of Scheme 14.14, together with the (by now) familiar chloromethyl isoxazole derivative placed to its immediate right (cf. also Scheme 14.11). By the simple process of NaH-mediated alkylation nearly all pieces of the ferruginol skeleton were put into place, requiring only N-ethylation to activate

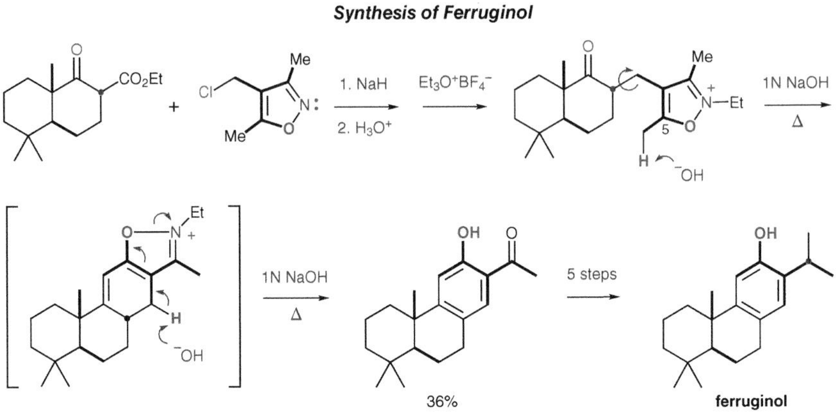

Synthesis of Ferruginol

Scheme 14.14 Isoxazole as a synthon.

the isoxazole C5 methyl group for proton abstraction. This was cleanly accomplished employing Meerwein's reagent, providing an intermediate isoxazolium salt that was only a single step removed from a known precursor to ferruginol (top right in scheme).

And now came the moment of truth. Without purification, the isoxazolium salt was heated in 1N NaOH, initiating a reaction sequence analogous to that outlined in Scheme 14.13. Thus, aldol-like condensation between the C5 methyl group and the decalone carbonyl provides the transient intermediate shown in brackets, which upon internal redox reaction and imine hydrolysis leads to the final product. On first impression, the overall yield of 36% seems relatively modest, and one wonders if a better result might be obtained with a non-nucleophilic base. However, bearing in mind the complexity of this transformation, and the known pathway forward from product to ferruginol (5 steps) [27], the experimental outcome takes on a more favorable light.

Return now briefly to Scheme 14.11, our jumping off point into the topic of isoxazoles as annulation reagents. Therein was described a simple model system in which reductive cleavage of the N-O bond generates a vinylogous amide intermediate, this last species being captured by the nearby carbonyl group to afford a cyclic carbinolamide (middle right of scheme). Next, alkaline hydrolysis produced a ketone enolate, which underwent rapid intramolecular aldol condensation to give $\Delta^{1,9}$-2-octalone in 50% overall yield. For convenience, the first two steps of this sequence have been reproduced in Scheme 14.15, with the cyclic carbinolamide shown to the lower left. But here we can ask another question: Suppose one omits the base hydrolysis step and simply allows the carbinolamide derivative to stand in solution exposed to air? The not-so-surprising result is that increasing amounts of the fused ring pyridine derivative illustrated to the far right are formed. And from this observation it was but

c. Synthesis of **Pyridines**:

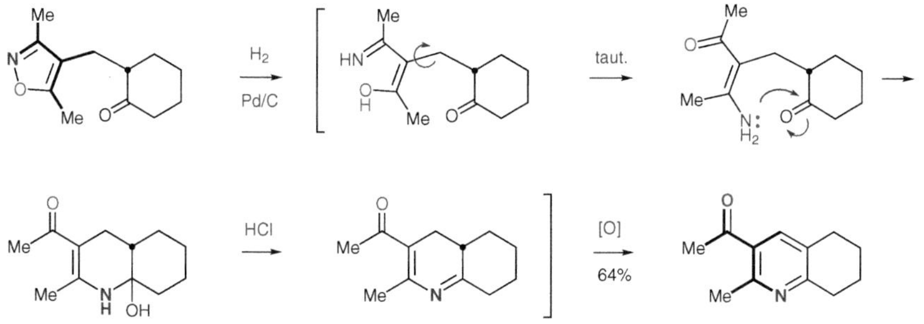

Scheme 14.15 Isoxazole as a synthon.

a small step to optimize this transformation, involving initial acid catalyzed dehydration to afford the dihydropyridine shown as the last structure in brackets, followed by in situ oxidation [28]. In the example at hand nitrous acid was employed as the oxidant (64% overall yield), but in other cases air sufficed.

For clarity, let us summarize what we have learned thus far about isoxazoles as annulation reagents, focusing on three main classes of reactions (Figure 14.5). Starting off to the left (*path a*), the N-O bond of isoxazoles is readily cleaved by hydrogenolysis, somewhat slowly employing Pd/C as catalyst, but rapidly with Ra-Ni. The resultant cyclic carbinolamides undergo facile base hydrolysis, ultimately producing a ketone enolate that is ideally positioned for intramolecular aldol condensation. In this manner the isoxazole ring is functioning as a methyl vinyl ketone surrogate, and the net transformation is equivalent to a Robinson annulation (see, for example, Schemes 14.11 and 14.12). Turning now to the center column in Figure 14.5 (*path b*), N-ethylation utilizing Meerwein's reagent serves two purposes, that of reducing the

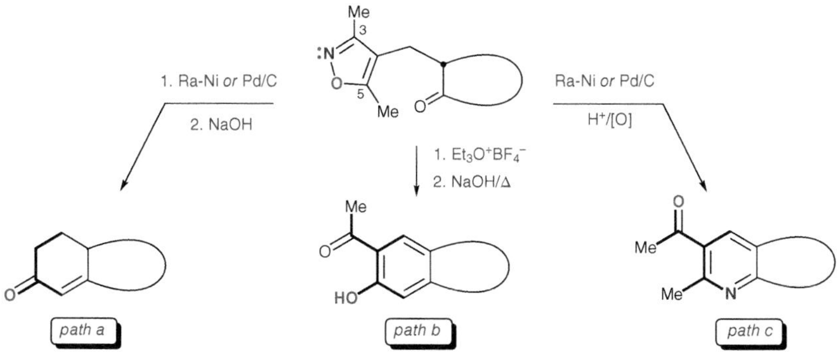

SUMMARY - *Isoxazoles as Annulation Reagents*

Figure 14.5 Isoxazole as a synthon.

aromatic stability of the isoxazole ring, and of significantly increasing the acidity of the C3- and C5-methyl groups. Note that at room temperature the principle reaction on treatment with dilute alkali involves ring hydrolysis, producing 1,3-dicarbonyl derivatives that might also undergo annulation. However, with more concentrated NaOH, and reflux temperatures, proton abstraction predominates, favoring the C5-methyl group for steric reasons. The stabilized carbanion thus produced adds to the nearby carbonyl, providing an aldol-like condensation product that aromatizes to the corresponding phenol. This occurs via an internal redox reaction, followed by tautomerization and imine hydrolysis (cf. Schemes 14.13 and 14.14). Lastly, in perhaps the simplest transformation of all (*path c*), hydrogenolysis of the isoxazole N-O bond is again effected utilizing either Pd/C or Ra-Ni catalysis, to produce a vinylogous amide that undergoes ring closure to a cyclic carbinolamide (cf. Scheme 14.15). So far we are following along the same route as *path a*, where base hydrolysis led to a Robinson annulation-like outcome. However, under the conditions of *path c*, acid catalyzed dehydration yields a dihydropyridine, which is readily oxidized to afford fused-ring pyridines.

Does this conclude our discussion on isoxazole chemistry? Not quite, and then only for the time being. For one, it remains to comment on a special property of C3-unsubstituted isoxazoles, which undergo irreversible ring scission upon treatment with moderately strong bases (top Scheme 14.16) [29]. In essence this transformation can be viewed as an E2-elimination, facilitated by the relatively high acidity of H3, as well as the weak nature of the N-O bond (cf. curly arrows). In the reaction shown, the "leaving group" is the otherwise difficultly accessible anion of cyanoacetone, which exhibits typical nucleophilic properties. For example, generation in the presence of mesityl oxide (red) affords a 75% yield of the cyano-substituted cyclohexenone derivative illustrated to the right, via initial conjugate addition followed by intramolecular aldol condensation [29].

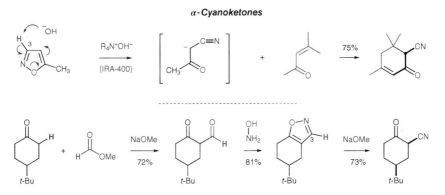

Scheme 14.16 Isoxazole as a synthon.

And where else might this ring scission find applicability? To answer, let us briefly take off our heterocyclic chemist's cap, and consider this transformation from a different perspective. Suppose you wished to introduce a nitrile group to the α-position of a ketone, as illustrated for the case of 4-*t*-butylcyclohexanone at the bottom of Scheme 14.16. The method of choice turns out to be a three-step sequence consisting of formylation, isoxazole formation, and ring-cleavage employing NaOMe [30].

With this we are nearing the end of our initial foray into "Heterocycles as Synthons," although the subject matter is so broad that it will resurface from time to time in remaining chapters. Along the way we have learned a considerable deal about the synthetic utility of various π-excessive heterocycles, including, most recently, isothiazoles and isoxazoles. Let us tie things together with some general observations regarding pyrazoles, the last of our triad of common 1,2-azoles:

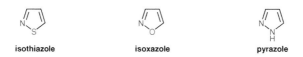

| isothiazole | isoxazole | pyrazole |

Much of pyrazole chemistry mirrors that of isoxazoles, which is not surprising given their close structural resemblance. As with isoxazoles, pyrazoles are frequently prepared from 1,3-dicarbonyl compounds, substituting a hydrazine derivative for hydroxylamine (top left, Figure 14.6; cf. also Figure 11.12). Also in common with isoxazoles, pyrazoles can be considered 1,3-dicarbonyl equivalents by the reverse of this process. However, this would be an illogical cycle, were it not for the fact that alternative syntheses exist. The most prominent of these involves 1,3-dipolar cycloaddition of alkynes with nitrile imines, a species which is isoelectronic with nitrile oxides (top right of figure). The products in this case are 1-substituted pyrazoles, and yields are generally high (vide infra).

Alternatively, if one wished to synthesize a (1H)-pyrazole, recourse would be taken to 1,3-dipolar cycloaddition of an alkyne with a mono-substituted diazoalkane, shown in one of its resonance representations in the center row of Figure 14.6. As indicated, the initial product undergoes rapid tautomerization to achieve aromaticity. Of course, both diazoalkanes and nitrile imines are highly reactive, and the latter in particular must be generated in situ. One means of accomplishing this is outlined at the bottom of Figure 14.6, starting with simple hydrazide derivatives of the type shown to the left. Treatment of these materials with PCl_5 affords the corresponding hydrazonoyl chlorides, which on 1,3-elimination give the dipolar species.

Finally, we close this chapter exploring some interesting chemistry of 3-pyrazolidinones and 5-pyrazolones, compounds related to pyrazoles but lacking an aromatic sextet. A representative example of the first ring system is

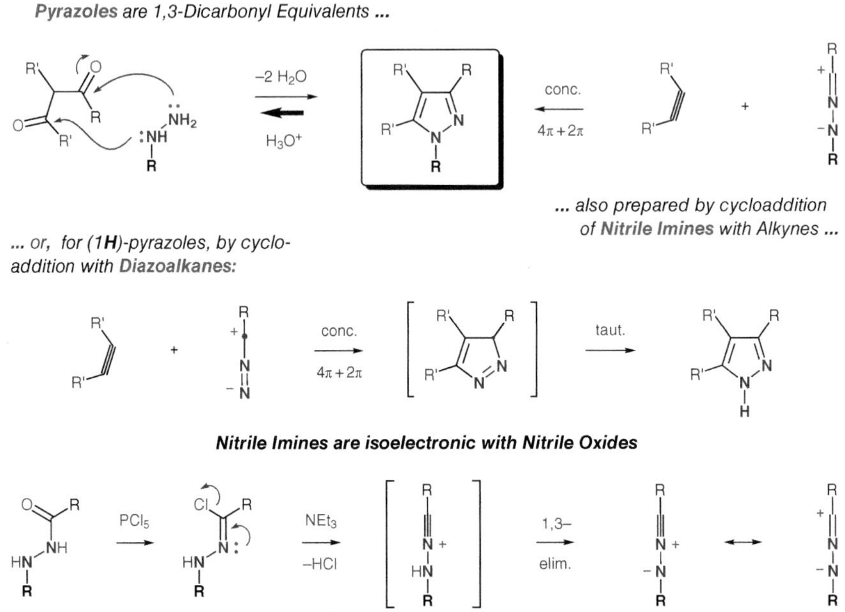

Pyrazoles are 1,3-Dicarbonyl Equivalents ...

−2 H$_2$O

H$_3$O$^+$

conc.

4π + 2π

... also prepared by cycloaddition of Nitrile Imines with Alkynes ...

*... or, for (1**H**)-pyrazoles, by cyclo-addition with Diazoalkanes:*

conc.

4π + 2π

taut.

Nitrile Imines are isoelectronic with Nitrile Oxides

PCl$_5$

NEt$_3$

−HCl

1,3−

elim.

Figure 14.6 Heterocycles as synthons.

shown in the shadow box at the top of Scheme 14.17, along with two possible means of synthesis. In both of these, the hydrazide bond would be generated by acylation of a hydrazine with an appropriate carboxylic acid derivative, while the aliphatic C-N bond looks to be the outcome of conjugate addition. But in what order these steps? To the top left is given the option of initial conjugate addition of hydrazine to an appropriate unsaturated ester, which would generate an intermediate alkyl hydrazine well positioned for intramolecular acylation (curly arrows). According to "Baldwin's Rules," this pathway corresponds to a 5-*exo-trig* ring closure, and it is geometrically favorable (i.e., proper orbital overlap is feasible) [31]. Alternatively, at the top right of the scheme, a pre-formed hydrazide is subjected to intramolecular conjugate addition. By Baldwin's conventions, this ring closure corresponds to an unfavorable 5-*endo-trig* process, in which orbital overlap would introduce a significant degree of strain. This is easily seen with models. In any event, the experimental facts are that α,β-unsaturated hydrazides of the type shown at the top right are thermally stable to ring closure up to 200 degrees Celsius [32].

And what does all of this have to do with "Heterocycles as Synthons?" We are not yet through with our discussion on 3-pyrazolidinones, which undergo a fascinating degradation on treatment with common oxidants such as mercuric oxide (HgO) or thallium trinitrate (TTN) [33]. Under these conditions the hydrazide N-N bond is rapidly transformed to the azene oxidation level

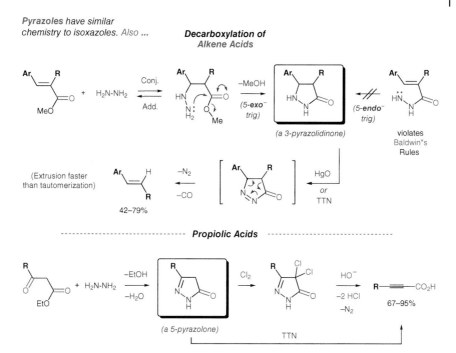

Scheme 14.17 Pyrazoles as a synthon.

(center Scheme 14.17), providing an intermediate that undergoes spontaneous extrusion of both nitrogen and carbon monoxide. The final products are alkenes of the general structure shown to the left center, formed in moderate to very good yield. Now, let us take off our heterocyclic chemist's cap once again, and examine the overall conversion we have just described. We began with an α,β-unsaturated ester derivative, and we have effected a net decarboxylation to give specifically a trans-disubstituted alkene. Can you think of any other means of accomplishing such a transformation? There are very few.

Now, as to 5-pyrazolones, these are formed in near quantitative yield by condensation of β-keto esters with hydrazine (bottom left Scheme 14.17). In this case the mechanistic pathway is quite clear, involving hydrazone formation followed by intramolecular N-acylation. And of what potential synthetic utility are these materials? Let us suppose, for example, that we wished to convert our starting β-keto ester to a propiolic acid derivative, a transformation that on paper would involve enolization and dehydration. Of course, this "paper" transformation is not practical, but we are on the right track by preliminary conversion to a 5-pyrazolone. Two paths forward are now possible. In the first of these, chlorination yields the corresponding 4,4′-geminal dichloride reproduced below, which is properly constituted for ultimate conversion to an alkyne acid. This process is initiated by 1,4-elimination of HCl to give an intermediate azene, followed by base-induced fragmentation [34]:

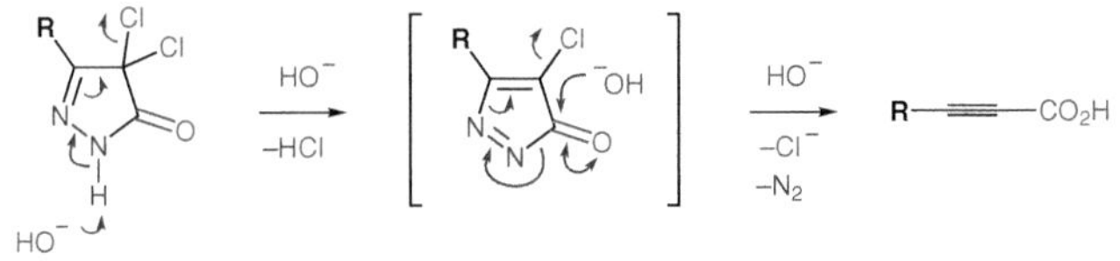

Alternatively, Taylor et al. demonstrated that 5-pyrazolones can be converted directly to the corresponding propiolic acid derivatives by reaction with thallium trinitrate (TTN; bottom Scheme 14.17). Yields are generally good to excellent [35].

So, what *is* in a name? We began this chapter with the oft-quoted question posed by Juliet to Romeo, admittedly a literary stretch in the context of heterocyclic chemistry. However, it did serve to introduce the concept of "synthon" as a synthetic building block. Have we made a case that heterocyclic ring systems are among the most versatile synthons in organic chemistry? Perhaps not yet, for we have barely scratched the surface of what is an immense field. Indeed, entire texts have been devoted to this area [36], and we will see additional examples in Chapter 15.

Problems for Practice [37]

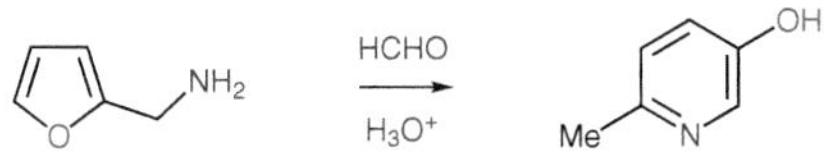

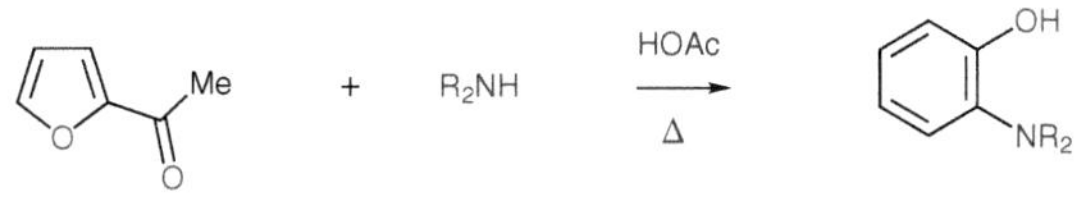

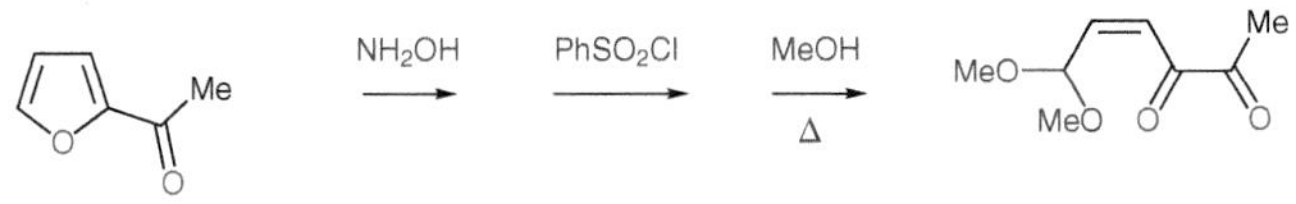

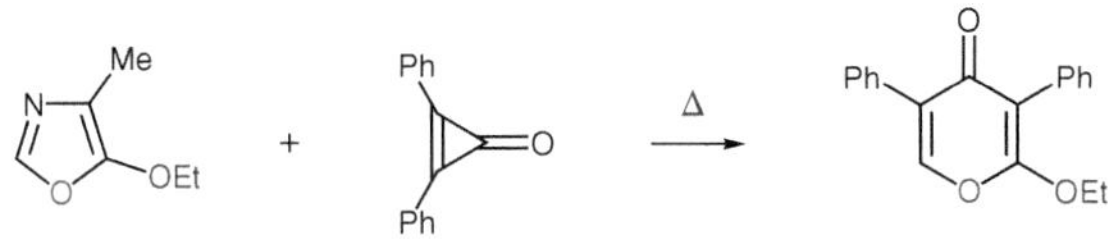

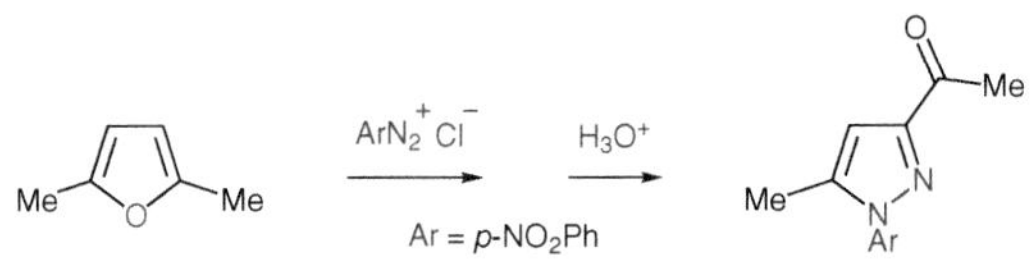

References

1 Corey, E. J. *Pure Appl. Chem.* **1967**, *14*, 30–37.
2 Corey, E. *J. Chem. Soc. Rev.* **1988**, *17*, 111–133, reference 15.
3 (a) Willstätter, R. *Ann. Chem.* **1901**, *317*, 204–265. (b) Willstätter, R. *Ibid.*
   1901, *317*, 307–374. (c) Willstätter, R. *Chem. Ber.* **1901**, *34*, 3163–3165. See
   also, (d) Willstätter, R. *Ann. Chem.* **1903**, *326*, 1–22.
4 Robinson, R. *J. Chem. Soc., Trans.* **1917**, *111*, 762–768.
5 Burks, Jr., J. E.; Espinosa, L.; LaBell, E. S.; McGill, J. M.; Ritter, A. R.;
   Speakman, J. L.; Williams, M. A.; Bradley, D. A.; Haehl, M. G.; Schmid, C. R.
   Org. Proc. Res. Dev. **1997**, *1*, 198–210.
6 Schöpf, C.; Lehman, G. *Ann. Chem.* **1935**, *518*, 1–37.
7 See, for example, Nocquet, P.-A.; Opatz, T. *Eur. J. Org. Chem.* **2016**, 1156–1164.
8 Gruber, W. *Can. J. Chem.* **1953**, *31*, 564–568.
9 Clauson-Kaas, N.; Elming, N.; Tyle, Z. *Acta Chem. Scand.* **1955**, *9*, 1–8.
10 Elming, N.; Clauson-Kaas, N. *Acta Chem. Scand.* **1955**, *9*, 23–26.
11 Nielsen, J. T.; Elming, N.; Clauson-Kaas, N. *Acta Chem. Scand.* **1955**, *9*, 9–13.
12 Clauson-Kaas, N.; Nedenskov, P. *Acta Chem. Scand.* **1955**, *9*, 27–29.
13 Hext, N. M.; Hansen, J.; Blake, A. J.; Hibbs, D. E.; Hursthouse, M. B.; Shishkin,
   O. V.; Mascal, M. *J. Org. Chem.* **1998**, *63*, 6016–6020.
14 (a) Hauptmann, H.; Walter, W. F. *Chem. Rev.* **1962**, *62*, 347–404. (b) Belen'kii,
   L. I.; Gol'dfarb, Y. L. in *Thiophene and its Derivatives*, Gronowitz, S., Ed.;
   Vol. *44*, Part 1 in *The Chemistry of Heterocyclic Compounds*, Weisberger, A.;
   Taylor, E. C., Eds., John Wiley & Sons, Inc., New York, New York, **1985**,
   pp. 457–569.
15 Sullivan, D.; Pettit, R. *Tetrahedron Lett.* **1963**, *4*, 401–403
16 Jacobi, P. A.; Egbertson, M.; Frechette, R. F.; Miao, C. K.; Weiss, K. T.
   Tetrahedron **1988**, *44*, 3327–3338.
17 Badger, G. M.; Rodda, H. J.; Sasse, W. H. F. *J. Chem. Soc.* **1954**, 4162–4168.
18 Wynberg, H.; Logothetis, A. *J. Am. Chem. Soc.* **1956**, *78*, 1958–1961.
19 Taits, S. Z.; Alashev, F. D.; Gol'dfarb, Y. L. *Russian Chemical Bulletin* **1968**, *17*,
   550–554.
20 Woodward, R. B. in *Harvey Lectures Series 59 (1963–64),* Academic Press,
   Inc., New York, New York, **1965**, pp. 31–47.
21 (a) Schreiber, J.; Leimgruber, W.; Pesaro, M.; Schudel, P.; Eschenmoser, A.
   Angew. Chem. **1959**, *71*, 637–640. (b) Van Tamelen, E. E.; Spencer, T. A.; Allen,
   D. S.; Orvis, R. L. *J. Am. Chem. Soc.* **1959**, *81*, 6341–6342.
22 Casnati, G.; Quilico, A.; Ricca, A. Finzi, P. V. *Tetrahedron Lett.* **1966**, *7*,
   233–238.
23 Stork, G.; Danishefsky, S.; Ohashi, M. *J. Am. Chem. Soc.* **1967**, *89*, 5459–5460.
24 Stork, G.; McMurry, J. E. *J. Am. Chem. Soc.* **1967**, *89*, 5463–5464.
25 Stork, G.; McMurry, J. E. *J. Am. Chem. Soc.* **1967**, *89*, 5464–5465.
26 Ohashi, M.; Maruishi, T.; Kakisawa, H. *Tetrahedron Lett.* **1968**, *9*, 719–722.

27 King, F. E.; King, T. J.; Topliss, J. G. *J. Chem. Soc.* **1957**, 573–577.

28 Stork, G.; Ohashi, M.; Kamachi, H.; Kakisawa, H. *J. Org. Chem.* **1971**, *36*, 2784–2786.

29 Eugster, C. H.; Leichner, L.; Jenny, E. *Helv. Chim. Acta* **1963**, *46*, 543–571.

30 Kuehne, M. E. *J. Org. Chem.* **1970**, *35*, 171–175.

31 Gilmore, K.; Alabugin, I. V. *Chem. Rev.* **2011**, *111*, 6513–6556.

32 Baldwin, J. E.; Cutting, J.; Dupont, W.; Kruse, L.; Silberman, L.; Thomas, R. C. *J. Chem. Soc., Chem. Commun.* **1976**, 736–738.

33 Kent, R. H.; Anselme, J.-P. *Can. J. Chem.* **1968**, *46*, 2322–2324.

34 Carpino, L. A.; Terry, P. H.; Thatte, S. D. *J. Org. Chem.* **1966**, *31*, 2867–2873.

35 Taylor, E. C.; Robey, R. L.; McKillop, A. *Angew. Chem. Int. Ed.* **1972**, *11*, 48.

36 Meyers, A. I. *Heterocycles in Organic Synthesis;* Vol. 3 in *General Heterocyclic Chemistry Series*, Taylor, E. C.; Weissberger, A., Eds., John Wiley & Sons, Inc., New York, New York, **1974**.

37 Problems for practice 5: (a) Clauson-Kass, N.; Meister, M. *Acta Chem. Scand.* **1967**, *21*, 1104. (b) Birkhofer, L.; Daum, G. *Chem. Ber.* **1962**, *95*, 183–190. (c) Vargha, L.; Ramonczai, J.; Bite, P. *J. Am. Chem. Soc.* **1948**, *70*, 371–374; Varga, L.; Gönczy, F. *J. Am. Chem. Soc.* **1950**, *72*, 2738–2740. (d) Grigg, R.; Jackson, J. L. *J. Chem. Soc. C*, **1970**, 552–556. (e) Eastman, R. H.; Detert, F. L. *J. Am. Chem. Soc.* **1948**, *70*, 962–964.

15

1,3-Dipolar Cycloadditions—An Overview

Close your eyes and take a mental snapshot of the first image that comes to mind when you think of synthesizing a 6-membered ring. Most likely you are visualizing a Diels-Alder reaction, the classic example of a concerted $4\pi + 2\pi$ cycloaddition:

In such transformations the 4π component is a conjugated diene, and there is wide latitude over the choice of atom constituents a-d. Likewise, the dienophile might in principle be any π-unsaturated system, so long as the transition state has a total of 6π electrons distributed over six atoms.

But is this the only $4\pi + 2\pi$ cycloaddition that is symmetry allowed? Not by any means. In fact, we have already seen a number of examples wherein the 4π component is isoelectronic with allyl anion, leading to five membered ring formation (vide supra). In all cases that we shall consider the 3-atom fragments are dipolar in nature, rendering them electronically neutral. Also, they contain at least one heteroatom. The overall process is known as a 1,3-dipolar cycloaddition, and in its many variations it is equally versatile as a Diels-Alder reaction [1]. Much of the pioneering work in this area was carried out by Professor Rolf Huisgen and co-workers, beginning in the late 1950s.

High oxidation state 1,3-dipoles: There are two main classes of 1,3-dipolar cycloadditions, both of which are *syn*-stereospecific and in large part regioselective. The first of these involves so-called high oxidation state dipoles, which incorporate an additional π-bond orthogonal to an allyl anion type molecular orbital (top Figure 15.1). Thus, in one of their resonance structures they can be represented as having a triple bond, and they are linear in geometry. Note, though, that in their 1,3-dipolar canonical form the sextet atom is joined to the central core by a double bond (structures in brackets).

Introductory Heterocyclic Chemistry, First Edition. Peter A. Jacobi.
© 2019 John Wiley & Sons Ltd. Published 2019 by John Wiley & Sons Ltd.

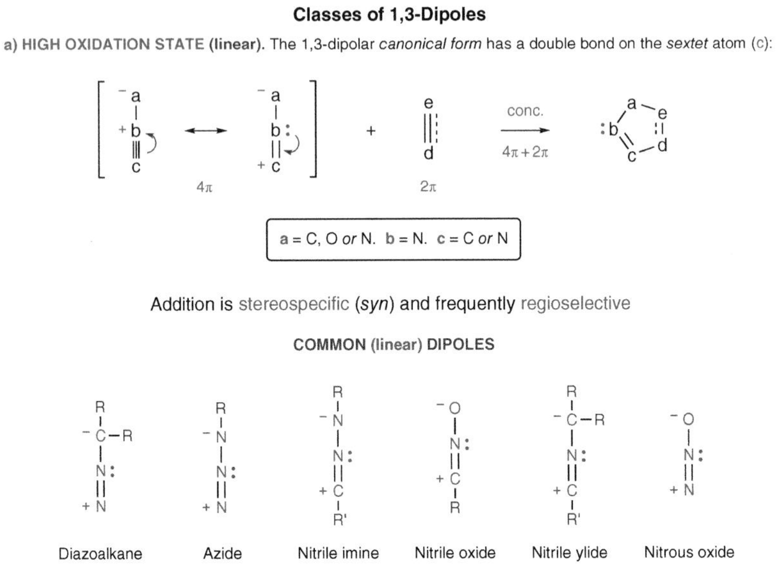

Classes of 1,3-Dipoles

a) HIGH OXIDATION STATE (linear). The 1,3-dipolar *canonical form* has a double bond on the *sextet* atom (c):

a = C, O *or* N. b = N. c = C *or* N

Addition is stereospecific (*syn*) and frequently regioselective

COMMON (linear) DIPOLES

| Diazoalkane | Azide | Nitrile imine | Nitrile oxide | Nitrile ylide | Nitrous oxide |

Figure 15.1 Synthesis and reactions of 1,3-dipoles.

Considering only second-row elements, atom "b" in such dipoles is always nitrogen, while there is more flexibility at the two termini. For example, "a" can be either C, O or N, while "c" is composed of C or N. With these restrictions, we can draw six structures for high oxidation state 1,3-dipoles, consisting of diazoalkanes, azides, nitrile imines, nitrile oxides, nitrile ylides, and finally nitrous oxide (bottom Figure 15.1).

Diazoalkanes [2]: Many 1,3-dipoles are thermally unstable, and must be generated in situ in the presence of a suitable dipolarophile, or stored at low temperature in inert solvents. Such is the case with low molecular weight diazoalkanes lacking a conjugating group at the carbon terminus, the parent member of which is diazomethane. Readers are perhaps already familiar with this species as a powerful methylating agent, particularly useful in the small scale conversion of carboxylic acids to methyl esters. However, it is also reactive as a 1,3-dipole. As to its preparation, one widely employed method involves aqueous base decomposition of commercially available N-nitroso derivatives such as Diazald[TM]:

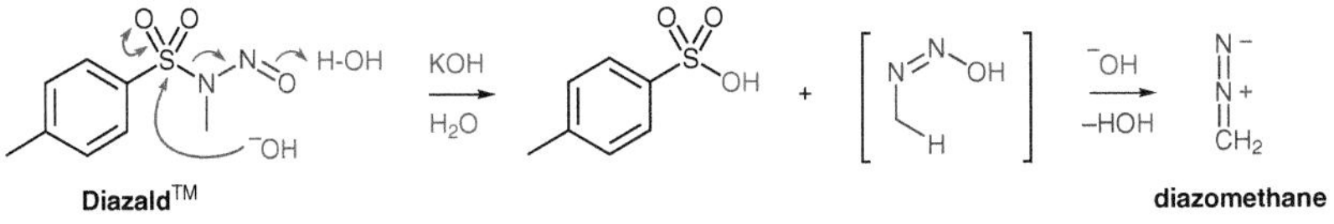

And what of more structurally diverse diazoalkanes? For alkyl- and aryl-substituted derivatives the most time-tested synthesis involves oxidation of

the corresponding hydrazones, themselves derived by condensation of an appropriate carbonyl precursor with hydrazine (top Figure 15.2). Historically the oxidant of choice has been HgO, although MnO_2 and $Pb(OAc)_4$ are also effective. On the other hand, diazoalkanes bearing strongly electron withdrawing groups are frequently best prepared by diazo transfer (middle Figure 15.2; Z=carboalkoxy, ketone, etc.). In this procedure, a stabilized carbanion undergoes nucleophilic addition to the terminal nitrogen of a sulfonyl azide, generating a transient intermediate of the type shown in brackets. The curly arrows then tell the rest of the story, involving intramolecular proton transfer, followed by ejection of a sulfonamide as leaving group. Finally, in a route more-or-less specific to α-aminoesters and analogs, N-nitrosation affords stabilized diazoalkanes of the general structure illustrated at the bottom right of Figure 15.2. We can assume that this reaction follows the usual mechanism for diazotization of an amino group, followed by loss of a proton from the resultant diazonium salt. Note that an alternative pathway involving loss of N_2 would in this case lead to a very unstable carbocation.

Let us look at some representative examples of diazoalkane cycloadditions. As with the dienophile in Diels-Alder reactions, the dipolarophile can in principle be any multiple bonded species, ranging in composition from electron rich to electron deficient. Also in common with Diels-Alder reactions, solvent effects are minimal, to be expected for a concerted reaction.

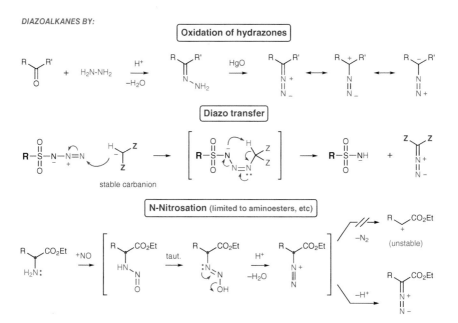

Figure 15.2 Synthesis and reactions of 1,3-dipoles.

The products, in turn, might be in the proper oxidation state for subsequent aromatization to afford 1*H*-pyrazoles. Or, they might undergo thermal or photochemical elimination of N_2 to generate cyclopropanes.

For the case of simple diazoalkanes (i.e., incorporating no conjugating substituents), cycloaddition is HOMO(dipole)–LUMO(dipolarophile) controlled, meaning that electron withdrawing groups on the dipolarophile will accelerate the reaction. This phenomenon again finds its direct analogy in "normal electron demand" Diels-Alder reactions, wherein the HOMO–LUMO gap is diminished by electron withdrawing substituents on the dienophile. Dimethyl acetylenedicarboxylate is thus both an excellent dienophile as well as dipolarophile, undergoing clean cycloaddition with a wide range of diazoalkanes (top Figure 15.3) [3]. The initially formed 3*H*-pyrazoles shown in brackets are typically not isolated, but rather undergo rapid tautomerization to afford their aromatic 1*H*-isomers.

Conversely, for diazoalkanes containing strongly electron withdrawing groups, reactivity tends to be guided more by HOMO(dipolarophile)–LUMO(dipole) interactions, much as in the case with "inverse electron demand" Diels-Alder reactions. With such species, the HOMO–LUMO gap closes with electron donating groups on the dipolarophile, with a corresponding increase in rate. A particularly nice example of a reaction of this

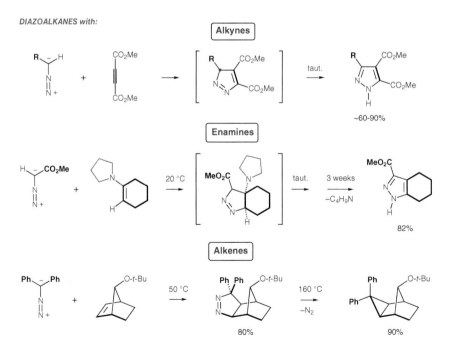

Figure 15.3 Synthesis and reactions of 1,3-dipoles.

class involves the cycloaddition of methyl diazoacetate with the pyrrolidine enamine derived from cyclohexanone (middle Figure 15.3) [4]. As indicated, these two substrates readily combine at 20 degrees Celsius, presumably to afford the primary adduct shown in brackets (regiochemistry was inferred from parallel experiments employing dimethyl diazomalonate). Once again, however, the drive toward aromaticity is irresistible, which is attained on prolonged standing via tautomerization followed by elimination of pyrrolidine (82% overall yield). In accord with the HOMO–LUMO analysis given above, diazomethane itself is unreactive toward enamines.

Finally, strained alkenes are also reactive partners, as exemplified by the stereoselective reaction of diphenyldiazomethane with *anti-7-tert-*butoxynorbornene (bottom Figure 15.3) [5]. At 50 degrees Celsius the *exo–* adduct is isolated in 80% yield, with aromatization blocked by the geminal diphenyl groups. In its stead, higher temperatures lead to smooth loss of N_2, providing a 90% yield of the derived cyclopropane.

Azides [6]: Azides are among the most widely distributed of high oxidation state 1,3-dipoles, and they have a rich and lengthy history. Indeed, the cycloaddition of dimethyl acetylenedicarboxylate and phenyl azide to give a 1,2,3-triazole was first reported by Arthur Michael in 1893 (the same Michael of Michael addition fame) [7]:

A word of caution, though—as with diazoalkanes, low molecular weight azides are notoriously shock and heat sensitive, and must be handled with great care. For example, sodium azide was the explosive of choice in older automotive air bags and airplane evacuation slides, and even phenyl azide will detonate if distilled at atmospheric pressure.

Still, there are many examples of these species that are isolable and have reasonable shelf lives, and they can often be prepared from simple starting materials. In principle, there exist five general routes to organic azides [6a], including diazo transfer to amines, diazotization of hydrazines, cleavage of triazines, and azide rearrangements. However, by far the most common procedures involve insertion of an intact N_3 group via a substitution reaction. Thus, the classic route to alkyl azides involves S_N2 displacement of a primary alkyl halide or sulfonate ester with sodium azide (top Figure 15.4). Yields are generally good to excellent, and there is wide latitude in the choice of R. For aromatic azides, advantage is often taken of the reaction of alkali azides with the appropriate diazonium salts, a transformation that on the surface appears reminiscent of a Sandmeyer reaction (bottom Figure 15.4). This is not the case, though, as labeling studies have shown

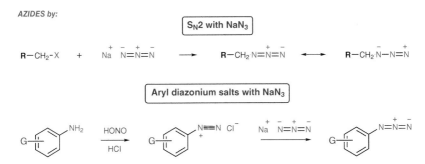

AZIDES by:

S_N2 with NaN_3

$$R-CH_2-X \quad + \quad Na \overset{+}{N}=\overset{-}{N}=\overset{+}{N} \longrightarrow R-CH_2 \overset{+}{N}=\overset{-}{N}=\overset{+}{N} \longleftrightarrow R-CH_2 \overset{-}{N}-\overset{+}{N}\equiv N$$

Aryl diazonium salts with NaN_3

Figure 15.4 Synthesis and reactions of 1,3-dipoles.

that initial attack by azide occurs on the diazonium group, producing an intermediate pentazole that subsequently ejects N_2. Successful conversions have been observed at temperatures as low as -80 degrees Celsius, and again, yields tend to be very high [6a].

Organic azides undergo concerted 1,3-dipolar cycloaddition with a wide range of alkene and alkyne dipolarophiles, the reaction being completely stereospecific. That is, *cis*–olefins afford the corresponding *cis*-substituted Δ^2-triazolines (4,5-dihydrotriazoles), while the opposite is true for *trans*–olefins:

When A≠B mixtures of regioisomers may be obtained, as is also the case with mono-substituted olefins. The composition of such mixtures can usually be rationalized by consideration of whether the reaction is HOMO(dipole) or HOMO(dipolarophile)–controlled, which in turn depends upon the nature of the substituents on both components. Recall that, in general, electron donating groups raise the energy level of molecular orbitals, while electron withdrawing groups have a lowering effect. Such effects are important, since the smaller the HOMO-LUMO gap, the faster the reaction. Regiochemistry is then determined by the best matching of orbital coefficients in the transition state leading to cycloaddition.

As with diazoalkanes, reaction is very slow with non-polarized olefins, but rates increase dramatically with angle strain. In fact, the rate of azide cycloaddition was once used as a probe of angle strain [8]. The classic example is provided by the cycloaddition of variously substituted phenyl azides with norbornene, which require only moderate warming (top Figure 15.5) [9]. Upon photolysis, the resultant *exo*-triazolines then afford near quantitative yields of the corresponding aziridines.

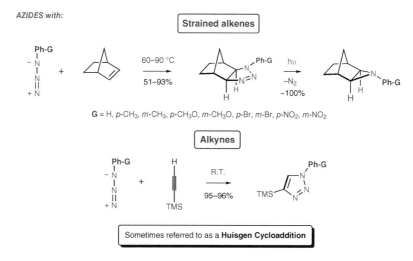

Figure 15.5 Synthesis and reactions of 1,3-dipoles.

Steric effects can also be important, in particular when the dipolarophile is an unsymmetrical or terminal alkyne. Non-catalyzed azide cycloadditions of this class tend to produce regioisomeric mixtures, reflecting near equal HOMO(dipole)/HOMO(dipolarophile)–control. Trimethylsilylacetylene, however, is an exception, undergoing clean reaction with a range of phenyl azide derivatives to give exclusively 1,4-disubstituted triazoles (bottom Figure 15.5) [10]. Clearly the TMS and phenyl groups experience severe steric crowding in the pathway leading to the 1,5-isomers.

Purely thermal azide–alkyne dipolar cycloadditions are sometimes referred to as "Huisgen cycloadditions," in honor of the first investigator to probe the scope of this concerted reaction. But we cannot let this section pass without some mention of Cu(I)-catalyzed azide-alkyne cycloadditions, a class of reactions extensively investigated by Professor K. Barry Sharpless, and which has come to be almost synonymous with "click chemistry [11]." As defined by Sharpless, click reactions *"must be modular, wide in scope, give very high yields, generate only inoffensive byproducts that can be removed by nonchromatographic methods, and be stereospecific (but not necessarily enantioselective)."* They should also take place under *simple reaction conditions*, employing either *no solvent or a solvent that is benign*, and with *simple product isolation.*

According to these stringent criteria, Huisgen cycloadditions do not qualify as click reactions, since they often require elevated temperatures and afford product mixtures. For example, the thermal cycloaddition between phenyl propargyl ether and benzyl azide required 18 hours at 92 degrees Celsius (neat), and gave a 1.6:1 mixture of 1,4- and 1,5-regioisomers (left Scheme 15.1). In contrast, the Cu(I)-catalyzed version of this same

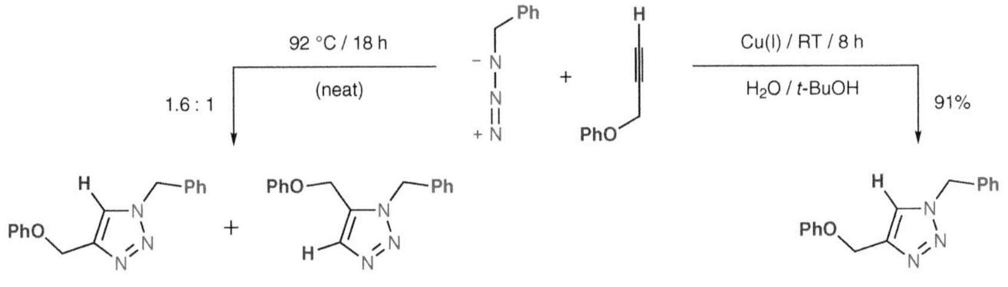

Scheme 15.1

reaction was complete in 8 hours at room temperature (H$_2$O/t-BuOH as solvent), and afforded a 91% yield of only the 1,4-isomer [12]. Why the great difference? At the heart of the answer is the fact that the two reactions follow entirely different mechanisms, with the Cu(I)-catalyzed version being stepwise in nature, initiated by copper acetylide formation (note that internal alkynes show no rate acceleration). The details, however, are beyond the scope of this text, and the interested reader is referred to the original literature [12]. In any event, due to its broad applicability, and exceptionally mild conditions, the "Sharpless cycloaddition" has many important applications in bioconjugate chemistry [13].

Nitrile imines: In comparison to diazoalkanes and azides, nitrile imines are relative newcomers to the playing field of 1,3-dipolar cycloaddition chemistry. The first reliable syntheses of members of this class were only reported in 1959 [14], initially involving thermolysis of 2,5-disubstituted tetrazoles, but more conveniently by 1,3-elimination of HCl from hydrazonoyl chlorides (top Figure 15.6). This last approach has the advantage of proceeding at room temperature or below. Even in these studies, though, the existence of such dipoles could only be inferred by capture in nucleophilic solvents, or trapping with suitable dipolarophiles. Left to their own they undergo facile dimerization, and to this day they have escaped isolation (except in argon matrices at cryogenic temperatures) [15].

Diphenyl nitrile imine is undoubtedly the most thoroughly investigated of these species, and it is highlighted in the shadow box in the middle of Figure 15.6. When generated at 20 degrees Celsius in the presence of norbornene, it undergoes rapid cycloaddition, affording an 85% yield of the *exo*–pyrazoline adduct shown to the right center of the figure [16]. Alternatively, reaction with alkyne dienophiles provides a general route to the corresponding pyrazoles, often with high regioselectivity. For example, terminal alkynes give nearly exclusively 1,3,5-trisubstituted pyrazoles (left center of figure) [16]. Even modest dipolarophiles undergo smooth cycloaddition, reflecting the fact that nitrile imines are far more reactive than diazoalkanes and azides. Thus, diphenyl nitrile imine affords a 75–77% yield

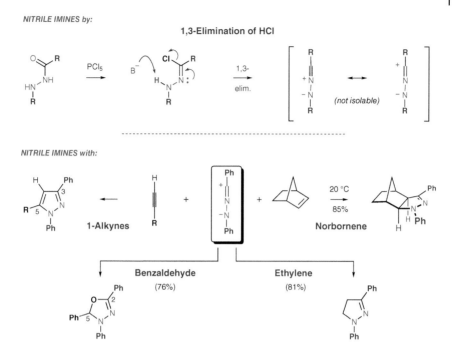

NITRILE IMINES by:

1,3-Elimination of HCl

NITRILE IMINES with:

1-Alkynes

Norbornene

Benzaldehyde
(76%)

Ethylene
(81%)

Figure 15.6 Synthesis and reactions of 1,3-dipoles.

of 2,4,5-triphenyl-1,3,4-oxadiazoline upon reaction with the normally reluctant partner benzaldehyde (bottom left of Figure 15.6) [8]. And perhaps most remarkable of all, this same dipole produces an 81% yield of 1,3-diphenyl-Δ^2-pyrazoline simply on shaking in benzene under an atmosphere of ethylene (bottom right of figure) [8].

Nitrile oxides [17]: Nitrile oxides share much in common with their chemical first cousins, nitrile imines, being among the most reactive of high oxidation state 1,3-dipoles. As such, they are generally prepared and reacted in situ, in that manner minimizing competing dimerization. Interestingly, though, sterically hindered representatives can occasionally be isolated. One means of preparation involves NEt$_3$-induced 1,3-elimination of HCl from hydroxamoyl chlorides, themselves derived either by chlorination of the corresponding aldoximes, or, as shown, on treatment of hydroxamic acids with PCl$_5$ (top Figure 15.7). A complementary route is by dehydration of primary aliphatic nitro compounds, typically employing the reagent combination of phenylisocyanate/NEt$_3$ (middle Figure 15.7). The isocyanate functions by converting one of the nitro oxygen atoms into a carbamic acid leaving group, ideally positioned for intramolecular proton abstraction and loss of the elements of aniline and carbon dioxide (cf. curly arrows). This approach is limited only by the availability of the requisite nitro starting materials.

Much of the early cycloaddition chemistry of nitrile oxides was explored with benzonitrile oxide, shown in the shadow box at the bottom of Figure 15.7. When reacted at 20 degrees Celsius with norbornene it affords a quantitative yield of the anticipated *exo*–isoxazoline, and with phenyla-cetylene it produces exclusively 3,5-diphenylisoxazole (97% yield) [8]. The regiochemical outcome of this last reaction appears to be quite general for mono-substituted alkynes not bearing strongly electron withdrawing groups (vide infra). Similarly, mono-substituted alkenes tend to favor 5-substituted Δ^2-isoxazolines, regardless of the nature of the nitrile oxide.

This all-too-brief summary does not end the story, however, as the derived isoxazolines and isoxazoles have a rich chemistry in their own right. Isoxazolines, for example, are excellent precursors to γ-amino alco-hols and β-hydroxycarbonyls (aldols), by virtue of the ease of reductive cleavage of their weak N–O bond (Figure 15.8) [18]. For the case of γ-amino alcohols, the reagent of choice is LiAlH$_4$, which effects both N–O bond cleavage as well as imine reduction (right center of figure). Alternatively, the N–O bond can be cleaved selectively employing Ra-Ni

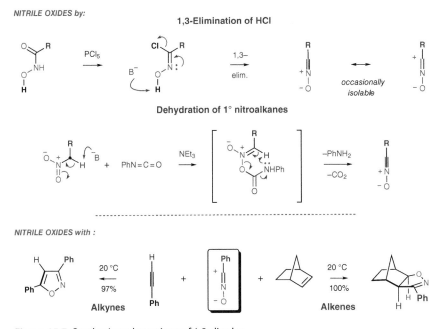

Figure 15.7 Synthesis and reactions of 1,3-dipoles.

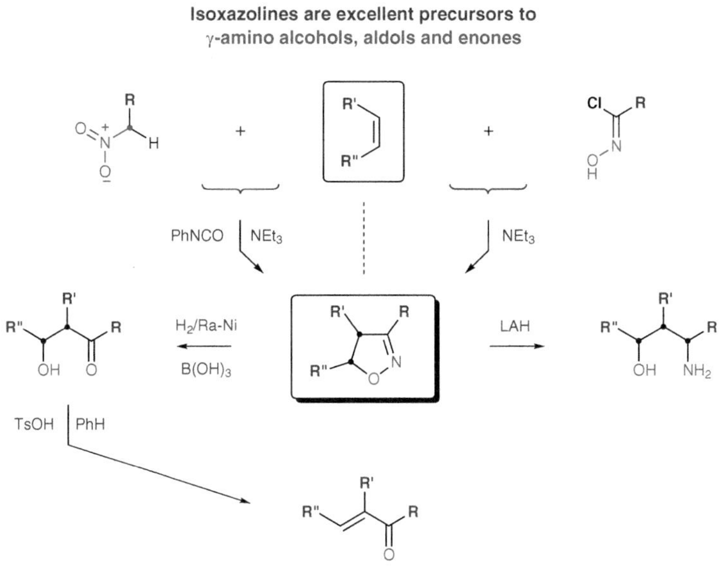

Figure 15.8 Nitrile oxides.

hydrogenolysis (left center of figure) [18,19]. Under neutral conditions, it is possible to isolate and characterize the intermediate β-hydroxyimines [19b]. However, more typically hydrogenolysis is carried out in the presence of boric acid [B(OH)$_3$], which accomplishes imine hydrolysis without deleterious epimerization or β-elimination to form enones. If desired, though, enone formation is cleanly brought about utilizing stronger acids (bottom of figure) [20].

Finally, let us close our discussion of nitrile oxides with an exquisite example of the concept of latent functionality (Scheme 15.2) [21]. The target molecule in this instance was the nickel(II) octamethyl-precorphin complex shown at the bottom right of the scheme, which would serve as a model system for a proposed total synthesis of vitamin B$_{12}$. The lead investigator on this 1975 publication was a 34-year-old Professor Robert V. Stevens, one of the most creative synthetic organic chemists of his generation. And the time frame was not long after the successful synthesis of cobyric acid (and thence vitamin B$_{12}$) by an international team led by R. B. Woodward and Albert Eschenmoser [22].

How did Stevens choose his target? Certainly it lacks the stereochemical complexity of vitamin B$_{12}$ (cf. Figure 1.3), and it would seem only a distant relative to the B$_{12}$ corrin skeleton. In this context, however, it is important to note that Eschenmoser had two years earlier solved the problem of precorphin to corphin cyclization [23], and these latter macrocycles were viewed as potential precursors to corrins. Also, the issue of the natural substitution pattern could be addressed later (after all, these *were* model studies). Rather, what Stevens identified as a pressing challenge in B_{12} synthesis was the extended vinylogous amidine chromophore found in the parent ring system. This was an area that was ripe for exploration, and his approach was entirely novel.

Focus now on the top left of Scheme 15.2, where the synthesis begins with a straightforward nitrile oxide cycloaddition [21a]. The components of this cycloaddition were a terminal alkyne and a primary aliphatic nitro compound, which upon treatment with PhNCO/NEt$_3$ afforded an 89% yield of the expected 3,5-disubstituted isoxazole with excellent regiochemical control. The heavy blue bond serves to tag that part of the skeleton contributed by the terminal alkyne, while the red atoms derive from the nitrile oxide.

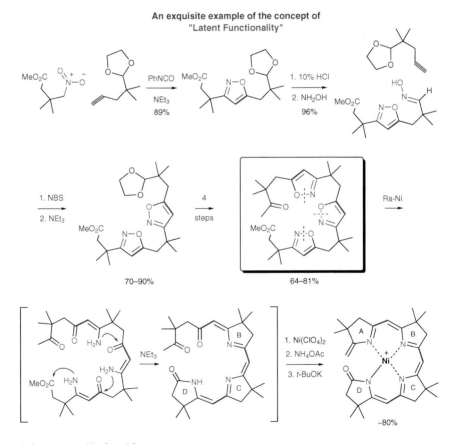

Scheme 15.2 Nitrile oxides.

This convention will be followed throughout the scheme. Next, acetal hydrolysis, followed by oxime formation set the stage for a second nitrile oxide cycloaddition, initiated by bromination of the oxime with N-bromosuccinimide (NBS), followed by dropwise addition of NEt_3. When carried out in the presence of the same alkyne dipolarophile as employed in step one, the result was a 70–90% overall yield of the *bis*–isoxazole pictured immediately to the left of the shadow box. By this stage it is apparent that we are rapidly assembling the carbon backbone of the target precorphin, taking advantage of an iterative strategy. In fact, it remained only for one additional iteration to complete the process, by the now familiar sequence of four steps (acetal hydrolysis, oxime formation, bromination and dipolar cycloaddition; can you identify the dipolarophile?). To convince yourself, count the number of carbon atoms in the *tris*–isoxazole shown in the shadow box, and compare that number with a likewise accounting of the desired precorphin. Leaving out the methoxy group of the methyl ester, you will see that the numbers are the same (28). In other words, we are ready to begin the unmasking process leading to the vinylogous amidine chromophore.

What a satisfying experience it must have been carrying out the first step in this sequence, involving simple Raney nickel hydrogenolysis. Thus was produced a near quantitative yield of the *tris*–enamide derivative shown as the first structure in brackets. Moreover, exposure of this material to even a trace of NEt_3 resulted in rapid bond reorganization, affording in one stroke rings B, C and D of the final target (cf. curly arrows). Spend a few moments working through the mechanism for this transformation, and you will per-haps agree that it deserves the adjective "exquisite." Indeed, the remaining steps seem almost routine, consisting of complexation with $Ni(ClO_4)_2$, and insertion of the final nitrogen atom employing ammonium acetate (NH_4OAc) followed by *t*-BuOK.

And where from here? As a postscript, the Stevens group came very close to achieving the synthesis of vitamin B_{12} employing this methodology [24a]. All of the appropriate substituents for rings A-D were incorporated into a *tris*–isoxazole intermediate, and there was strong precedent for the "end game" (vide supra). Unfortunately, though, the story ends on a sad note, with the sudden passing of Professor Stevens just shy of his forty-third birthday, and within sight of his goal. The "dream" would have to await another generation.

As for the remaining two high oxidation state 1,3-dipoles (cf. Figure 15.1), nitrile ylides and nitrous oxide have far less synthetic utility. In fact, there is only a smattering of reports alluding to N_2O as a viable 1,3-dipole in cycload-dition chemistry, and even in those cases the primary adducts were not iso-lated [25]. Most studies have been computational in nature [26]. So at this juncture let us shift our attention to low oxidation state 1,3-dipoles, a number of which have also been employed in complex natural product synthesis.

Low oxidation state 1,3-dipoles: What sets this class apart? As the name implies, such dipoles have a lower level of unsaturation as compared to their high oxidation state relatives, which are more propargyl-like in character. That is, they contain an additional π-bond orthogonal to an allyl anion type molecular orbital, and as a consequence are linear (cf. Figure 15.1). Low oxidation state 1,3-dipoles, in contrast, are bent (top Figure 15.9). Reflecting this difference, in their 1,3-dipolar canonical form they have only a single bond to the sextet atom (structures in brackets). Thus with alkenes they afford saturated 5-membered rings, in stereospecific and frequently regioselective fashion.

Again considering only second-row elements, "a" and "c" can be either C, N, or O, while "b" is limited to N or O. Based upon these criteria, there exist a total of twelve possible structures for 1,3-dipoles of this class. Five of the most common of these are ozone, nitrones, azomethine ylides, carbonyl ylides, and azomethine imines (bottom Figure 15.9). **Ozone**, of course, is the familiar reagent utilized for oxidative cleavage of alkenes, the first step of which is believed to be 1,3-dipolar cycloaddition to produce a so-called molozonide (Scheme 15.3) [27]. Rearrangement then affords an ozonide, which upon reductive cleavage generates two carbonyl components.

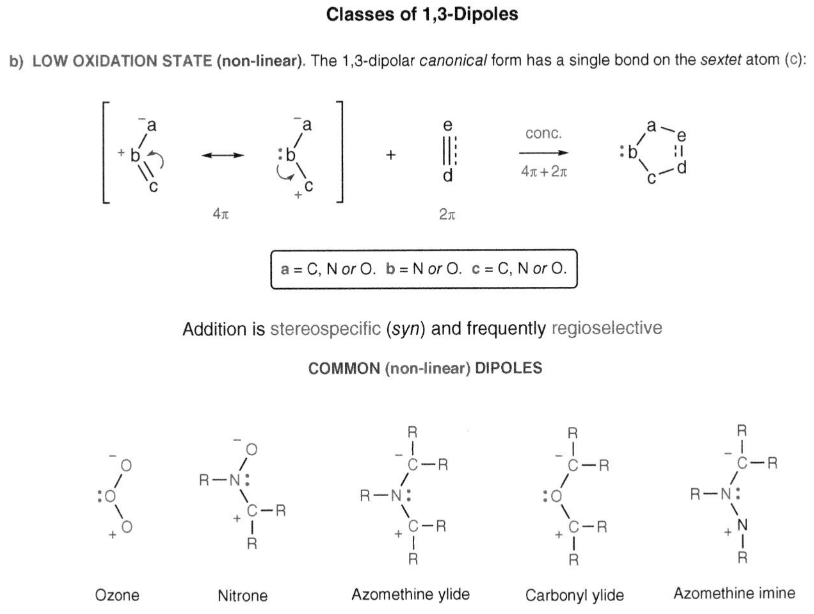

Classes of 1,3-Dipoles

b) **LOW OXIDATION STATE (non-linear)**. The 1,3-dipolar *canonical* form has a single bond on the *sextet* atom (c):

a = C, N *or* O. b = N *or* O. c = C, N *or* O.

Addition is stereospecific (*syn*) and frequently regioselective

COMMON (non-linear) DIPOLES

| Ozone | Nitrone | Azomethine ylide | Carbonyl ylide | Azomethine imine |

Figure 15.9 Synthesis and reactions of 1,3-dipoles.

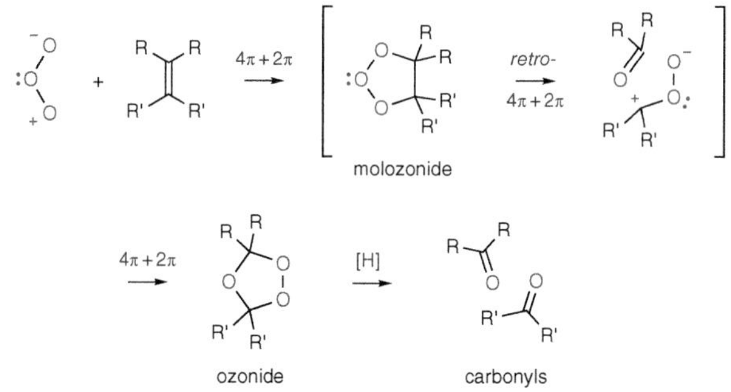

Scheme 15.3

Nitrones [28]: Nitrones rival diazoalkanes and azides in their historical significance, being first identified in 1890. Their name stems from a contraction of the term "nitrogen-ketone," emphasizing the fact that they undergo facile nucleophilic addition, much as an iminium or carbonyl functionality. It was only in the 1960s that attention was drawn to their power as 4π-components in dipolar cycloaddition reactions, again largely due to the systematic studies of Huisgen and co-workers. As we shall see, they are among the most useful of all 1,3-dipoles, generating as many as three new stereocenters in their reaction with alkene dipolarophiles. First, though, a few words about their synthesis.

The majority of nitrones are prepared by either of two methods, consisting of (a) oxidation of N,N-disubstituted hydroxylamines (top Figure 15.10), or (b) condensation of N-monosubstituted hydroxylamines with aldehydes or ketones (middle Figure 15.10) [28a]. The oxidant of choice in the first method has historically been yellow mercuric oxide, and this approach is particularly useful for cyclic nitrones. Note, however, that mixtures are possible with unsymmetrical hydroxylamines, and the starting materials themselves may be difficultly accessible. In contrast, regiochemical control is not an issue in the condensation of N-alkyl- or N-arylhydroxylamines with aldehydes or ketones. The initial step in this case involves hemiaminal formation, which is followed by 1,3-elimination of water. By whichever method, it is often of benefit to generate the desired nitrone in situ, as the more reactive members of this class undergo dimerization or trimerization upon attempted isolation.

A wide range of multiple bonded species serve as dipolarophiles, including alkenes, alkynes, nitriles, thiocarbonyls, and isocyanates. Similarly, enormous diversity is possible in the nitrone component. As pertains to reactivity, a good place to start is with some simple examples of *inter*molecular cycloadditions, employing cyclic nitrones of the general structure shown in the shadow box at the bottom of Figure 15.10 (n=1,2; R=vinyl, alkyl) [29].

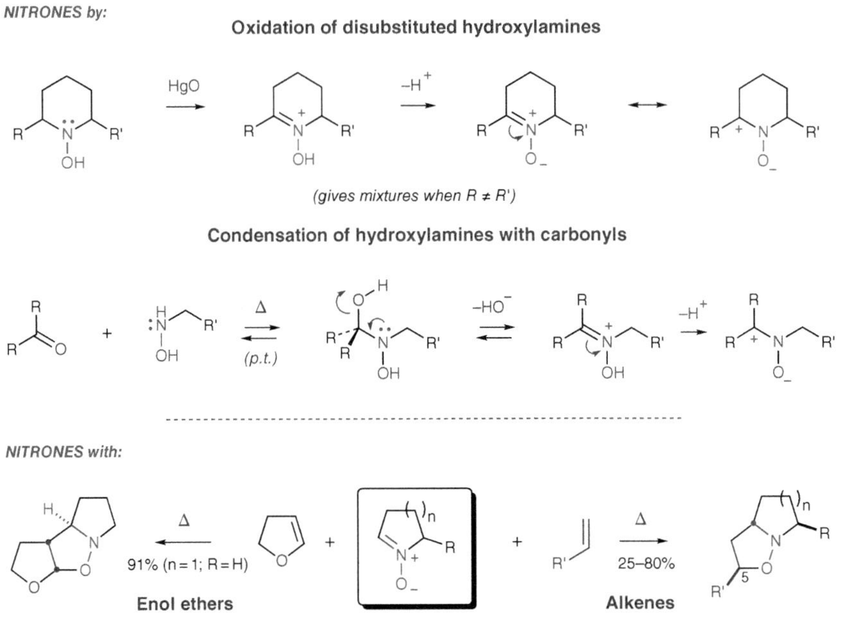

NITRONES by:

Oxidation of disubstituted hydroxylamines

(gives mixtures when R ≠ R')

Condensation of hydroxylamines with carbonyls

NITRONES with:

| Enol ethers | | Alkenes |

Figure 15.10 Synthesis and reactions of 1,3-dipoles.

Why cyclic nitrones? Mainly because they are configurationally stable, eliminating any complications due to *E,Z*-isomerization. In any event, the dipoles illustrated undergo regioselective cycloaddition with a variety of substituted alkenes, affording fused-ring isoxazolidines of the type shown to the bottom right of the figure (R'=Ph, *n*-Bu, CO$_2$Et) [29a,b]. Yields range from modest to very good and only that isomer corresponding to a 5-substituted isoxazolidine is obtained. Moreover, facial selectivity is also high, with R and R' orientated *syn* in the final products. Finally, the parent nitrone of this series (n=1, R=H) gives a 91% yield of the cycloadduct illustrated to the bottom left upon reaction with 2,3-dihydrofuran [29c]. Once again, stereo- and regiochemical control are excellent, with only 3% of epimeric material being isolated.

However, it is in *intra*molecular examples that the true versatility of nitrone cycloadditions shines through. Representatives of this class of reaction first appeared in 1959 [30a], and it was quickly realized that such transformations provided a rapid means of building complexity (Figure 15.11). LeBel et al. were early contributors to this area, demonstrating in 1962 that 4-cycloheptenecarboxaldehyde condensed with N-methylhydroxylamine to produce the (presumed) nitrone shown in brackets (equation 1) [30b]. Without isolation, this last material underwent intramolecular cycloaddition, leading in 60% yield to the tricyclic isoxazolidine shown to the top

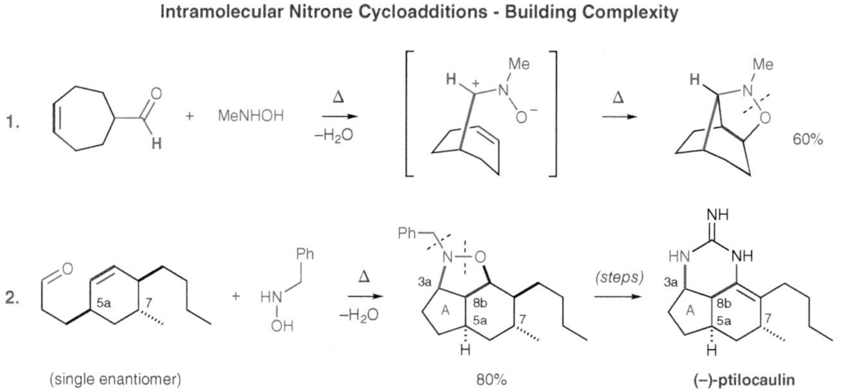

Intramolecular Nitrone Cycloadditions - Building Complexity

Figure 15.11 Synthesis and reactions of 1,3-dipoles.

right. Hydrogenolysis then afforded the corresponding amino alcohol (dashed line). In the years that followed there appeared an ever increasing number of elegant natural product syntheses building upon this methodology, which have been the subject of several excellent reviews [28]. We shall limit ourselves to a brief discussion of but two.

Fast forward now to March 1978, and to a location just off the coast of Burn Cay, Honduras. A scuba expedition is searching for bioactive compounds in the relatively shallow waters. From one species of sponge the team members isolated and purified a compound that had potent activity against both gram-positive and gram-negative bacteria, as well as high levels of cytotoxicity. This they named ptilocaulin, from *Ptilocaulis aff. P. spiculfer*, and following another three years they had secured its structure (although not the absolute stereochemistry) [31]. What an attractive synthetic target, including a relatively rare guanidine ring as well as four chiral centers in the hydrocarbon core (bottom right of Figure 15.11)! This did not escape the attention of Professor William Roush and his co-worker Alan Walts, who set out to effect an enantioselective total synthesis [32].

A key intermediate in the Roush synthesis of (-)-ptilocaulin was the enantiomerically pure cyclohexene-aldehyde shown at the start of equation 2, itself derived from naturally occurring (+)-pulegone. The details for preparing this material we need not go into, but it is worth spending a few moments summarizing its attributes. First to note is the fact that nearly all carbon atoms of the ptilocaulin skeleton are in place, save only that found in the guanidine functionality. This includes the cyclohexene ring, the C7 methyl group with its adjacent *n*-butyl chain, and a C5a aldehyde of proper length for constructing ring A. Also not to be lost, the C5a and C7 substituents have the proper relative, and (perhaps) absolute stereochemistry (this last was as yet undetermined). And now came the "tour de force," involving brief heating with N-benzylhydroxylamine to generate a transient nitrone. The product of this reaction, obtained in 80% yield, was the tricyclic isoxazolidine illustrated in the center of equation 2. In a single transformation, the authors had put into place ring A, as well as both remaining chiral centers found in ptilocaulin (C3a and C8b). Of the remaining steps, the dashed lines tell most of the story, involving first, reductive cleavage of the N–O bond, oxidation of the resultant alcohol to the corresponding ketone, and N-debenzylation. Condensation with a guanidine equivalent then gave (-)-ptilocaulin, which was identical to the naturally occurring material in all aspects save one: The absolute configuration turned out to be opposite to that synthesized. This is not unusual in endeavors of this type, where, lacking firm biosynthetic precedent, there is always a chance of choosing wrong. However, it does not diminish at all from the accomplishment.

Finally, we close this section on nitrones with a very creative synthesis of (±)-biotin (vitamin H), the so-called "beauty vitamin" of purported benefit to "Haar und Haut" (Scheme 15.4; cf. also Figure 1.3) [33]. Who would have thought that cycloheptene might serve as a viable precursor to this material, containing a fused ring tetrahydrothiophene/imidazolidinone skeleton, appended with a 5-carbon carboxylic acid side chain? But imagination came to the fore, and Confalone, Lollar et al. laid out a convincing strategy (the red stars, by the way, allow us to trace the fate of the two vinylic carbons in our starting material). Thus, it was but four short steps from cycloheptene to the mercaptoaldehyde shown as the second structure in the scheme, which, if one were bookkeeping, contains all but the urea carbon found in the final target. And what of the chiral centers? These were all introduced in a single step by the simple expedient of warming with benzylhydroxylamine. The intermediate nitrone then underwent completely stereoselective cycloaddition to afford the advanced precursor shown as the third structure in the scheme (66% yield). Once again, the dashed lines give an inkling of the remaining steps, since the way forward will certainly involve N-debenzylation

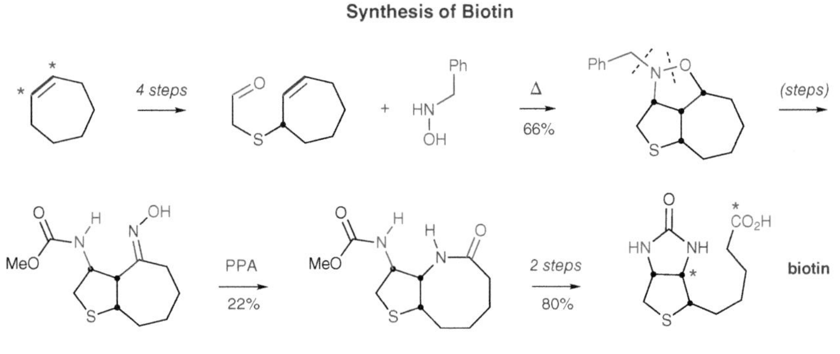

Synthesis of Biotin

Scheme 15.4 Synthesis and reactions of 1,3-dipoles.

and N–O bond cleavage. More than that, though, it will be necessary to effect an oxidative cleavage of the 7-membered ring, with insertion of the final nitrogen. This the authors accomplished by Beckmann rearrangement of the anti-oxime illustrated at the bottom left, which although modest in yield, produced an 8-membered ring lactam only two steps removed from biotin. These involved base hydrolysis, followed by cyclic urea formation with phosgene (80% yield).

Azomethine Ylides [34]: As with nitrones, azomethine ylides have found considerable utility in natural product synthesis, providing, as they do, a very general route to highly substituted pyrrolidines and Δ^3-pyrrolines. Take, for example, the "stabilized" azomethine ylide shown in the shadow box at the top of Figure 15.12. With N-phenylmaleimide this material gave an 86% yield of the cycloadduct illustrated to the top right, with exclusive *endo*-selectivity [35]. Alternatively, capture with dimethyl acetylenedicarboxylate produced a 53% yield of the Δ^3-pyrroline at the top left, again with excellent stereoselectivity [36]. In both cases, the geometry of the dipole was conserved in the final products. Finally, as a peek ahead, certain azomethine ylide cycloadditions have been used in probing one of the most storied theories in organic chemistry, the Woodward–Hoffmann rules. More about this shortly. For

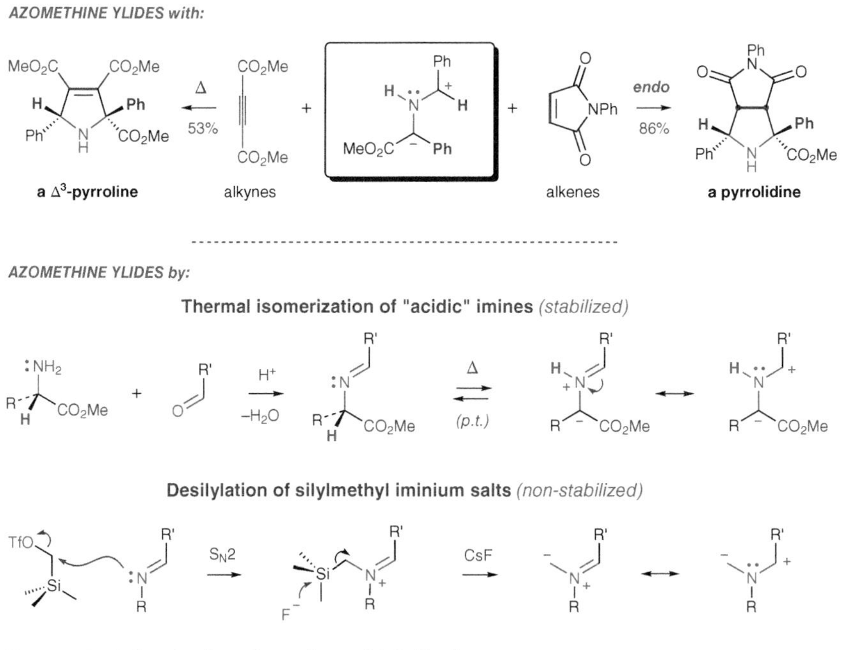

Figure 15.12 Synthesis and reactions of 1,3-dipoles.

now, though, let us back up a moment and gain some insight into their preparation.

From a conceptual standpoint, the methodology of Grigg et al. is undoubtedly the most straightforward for preparing stabilized azomethine ylides, via thermal isomerization of "acidic" imines (middle Figure 15.12) [37]. Typically, the starting materials in this approach are amino acid esters, which undergo ready imine formation with a wide range of aldehydes. Upon thermolysis, the hydrogen shown in bold red undergoes a 1,2-prototropic shift to the free electron pair on the adjacent nitrogen, generating an azomethine ylide that is typically captured in situ. Importantly, isomerization is also catalyzed by both Brønsted and Lewis acids, allowing in many cases for cycloaddition to occur at room temperature. The main limitation to this approach is that it is specific for azomethine ylides possessing powerful electron withdrawing groups on that carbon bearing a negative charge. For non-stabilized, N-substituted ylides, recourse is often taken to desilylation of silylmethyl iminium salts, themselves prepared via alkylation of aldimines with trimethylsilylmethyl triflate (bottom Figure 15.12) [38]. The example shown highlights the chemistry of Vedejs et al., in which CsF is used for cleaving the carbon–silicon bond. Major contributions to this area have also been made by Padwa [39] and Livinghouse [40].

Yet to describe is a third major method for generating azomethine ylides, involving ring opening of aziridines. In principle, such a process might be either thermal or photochemical in nature, and it could lead to either of two possible geometries (we will assume that A is more sterically demanding):

Illustrated to the right is that dipole arising from rotation of the substituents A,B in the same direction, defined as conrotatory (red arrows). To the left is shown the consequences of a disrotatory ring opening, wherein the substituents rotate in opposite directions (blue arrows). The question to be answered was which mode of ring opening would be favored? And would it be the same under both thermal and photochemical conditions? Of historical import, in 1965, Woodward and Hoffmann predicted that the isoelectronic ring system cyclopropyl anion would undergo concerted thermal ring opening exclusively in a conrotatory sense. The opposite was forecast for photochemical ring opening [41]. But at the time, experimental difficulties precluded a direct test.

Enter once again Professor Huisgen, who in 1967 devised an ingenious experiment to settle the issue for aziridines, and by analogy, for cyclopropyl anion [42]. Let us follow his reasoning. Highlighted in shadow boxes in Figure 15.13 are two isomeric aziridines, differing only in the stereochemical relationship of the two carbomethoxy substituents. Thus, in the top isomer these groups are *trans*, while in the bottom they are *cis*. Shown to the left of the *trans*-isomer is the azomethine ylide that would result from a disrotatory ring opening, while to the right is the most stable product of conrotatory opening. But how to distinguish these two ylides, since they would clearly not be isolable? The trick was to carry out the reaction in the presence of a sufficiently reactive dipolarophile, such that cycloaddition would occur faster than C–N bond rotation. The perfect combination was found employing dimethyl acetylenedicarboxylate as the 2π-component, at 100 degrees Celsius. The experimental outcome was dramatic, in that only the *cis*-Δ^3-pyrroline was obtained [42]. To quote Woodward, "The net inversion of stereochemistry observed in the thermal reaction would be extremely puzzling were it not the obvious consequence of a conrotatory opening, followed by a [4 + 2] cycloaddition [43]." Moreover, although not shown, photolysis at 10–15 degrees Celsius produced only that product

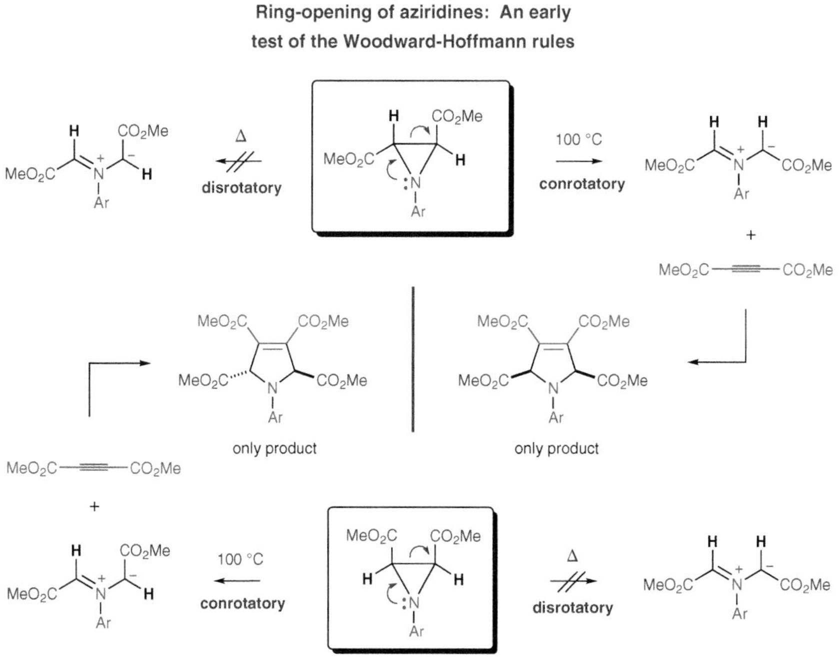

Figure 15.13 Synthesis and reactions of 1,3-dipoles.

corresponding to disrotatory ring opening! And what of the *cis*-substituted aziridine shown in the bottom shadow box? The results were equally unequivocal, in that thermal ring opening occurred with complete inversion of stereochemistry [42].

Carbonyl Ylides [44]: Not surprisingly, aziridine ring openings find their counterpart in epoxide chemistry, as evidenced in particular with oxirane substrates bearing strongly electron withdrawing and/or conjugating groups. Perhaps the best known example of a reaction of this class involves tetracyanoethylene oxide (TCEO), which on warming is in equilibrium with the corresponding carbonyl ylide shown in brackets at the top of Figure 15.14 [45]. Although there is no stereochemical "tag" to this conversion, we can predict with a high degree of confidence that ring opening is conrotatory (i.e., allowed by Woodward and Hoffmann), an assertion that is supported by subsequent studies with non-symmetric epoxides [46]. And what a reactive dipolar species it is! Simply heating TCEO in benzene (130–150 degrees Celsius) affords a 31% yield of the *mono*-adduct shown to the middle right, whose structure was proven by chemical correlation and degradation studies. Furthermore, with ethylene there is obtained an 87% yield of 2,2,5,5-tetracyanotetrahydrofuran, and with acetylene a 71% yield of the dihydrofuran derivative illustrated at the bottom right [45].

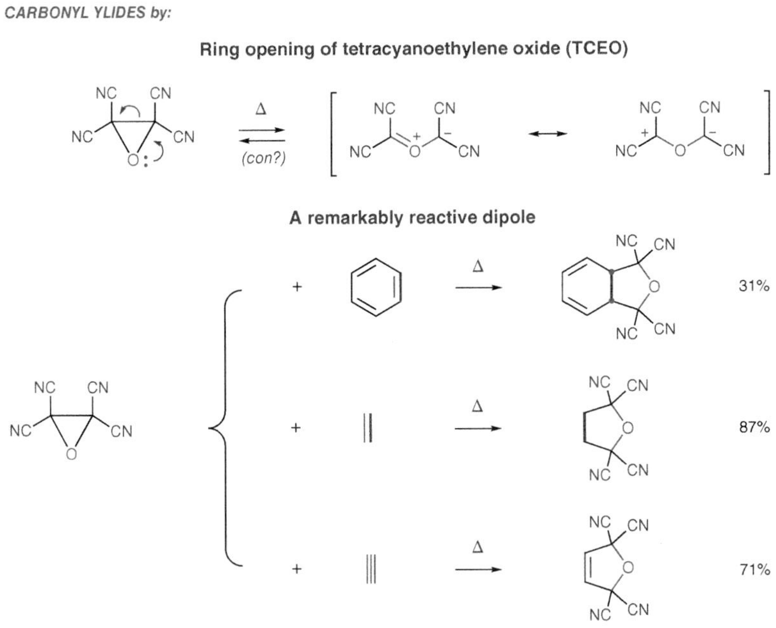

Figure 15.14 Synthesis and reactions of 1,3-dipoles.

These are reactivity patterns that are unmatched by other 1,3-dipoles we are surveying.

Note that in principle the same carbonyl ylide derived by ring opening of an epoxide might be generated by orbital overlap of a singlet carbene with an electron pair on an appropriate carbonyl derivative (curly arrow):

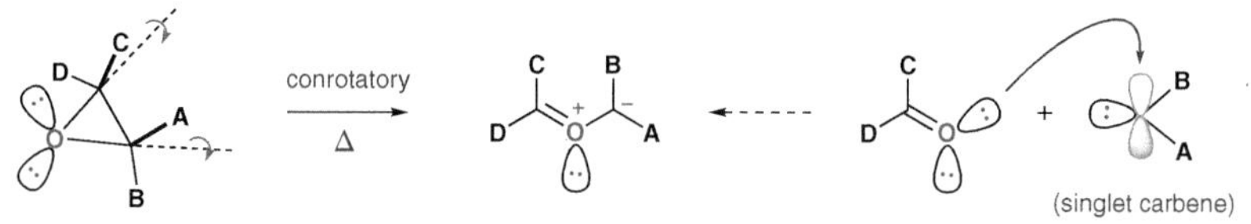

However, this approach is rarely practical for preparative scale work, mainly due to the highly reactive nature of free carbenes, with their attendant myriad of potential side reactions. Much more satisfactory results are obtained by capture of *stabilized* metallocarbenoids with carbonyl compounds (top Figure 15.15) [44]. In such cases at least one of the groups A,B must be strongly electron withdrawing (typically a carbonyl group), while there is a wide latitude of choices for C and D. The carbenoids themselves

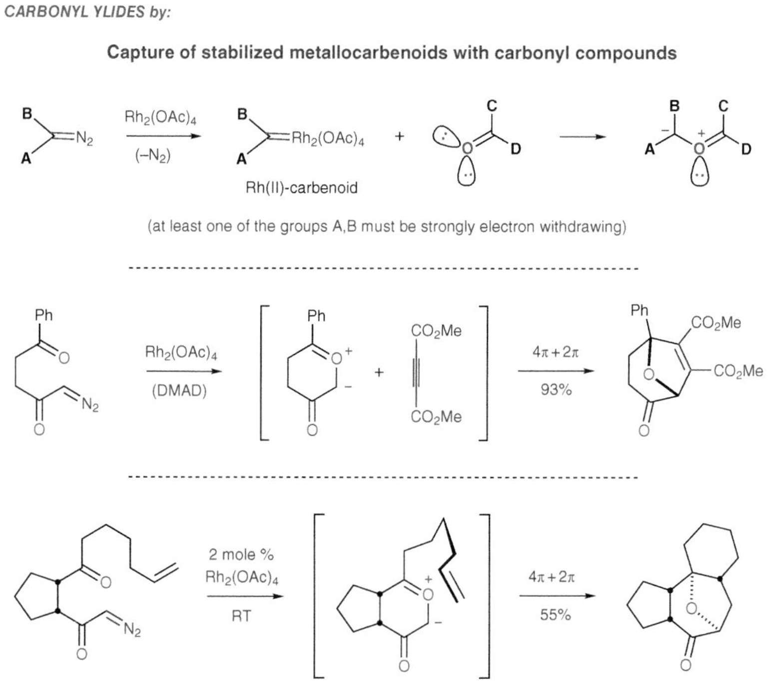

CARBONYL YLIDES by:

Capture of stabilized metallocarbenoids with carbonyl compounds

Figure 15.15 Synthesis and reactions of 1,3-dipoles.

are readily prepared by metal catalyzed decomposition of stabilized diazo derivatives, most commonly employing a rhodium(II) salt [47]. As expected, these last species are highly electrophilic on carbon, and with carbonyl compounds undergo facile conversion to carbonyl ylides.

Cyclic carbonyl ylides are also conveniently accessed utilizing this methodology, involving internal capture of a stabilized metallocarbenoid. The example shown in the center of Figure 15.15 comes from the group of Padwa et al., who pioneered much of the methodology in this area [48]. In this case the dipolarophile was dimethyl acetylenedicarboxylate (DMAD), and the yield of cycloadduct was 93%. The example shown at the bottom of the figure is by McMills, Wright et al., and was part of model studies for a projected synthesis of phorbol [49]. It nicely illustrates how all facets of carbonyl ylide formation and capture might be intramolecular in nature.

Azomethine Imines [50]: Last but by no means least in our consideration of low oxidation state 1,3-dipoles, azomethine imines are comparable to nitrones in their ability to generate complexity. One of the oldest means for preparing these species involves deprotonation of π-deficient N-amino salts (top Figure 15.16), themselves typically derived by N-amination of the parent heterocyclic ring systems with hydroxylamine-O-sulfonic acid or its

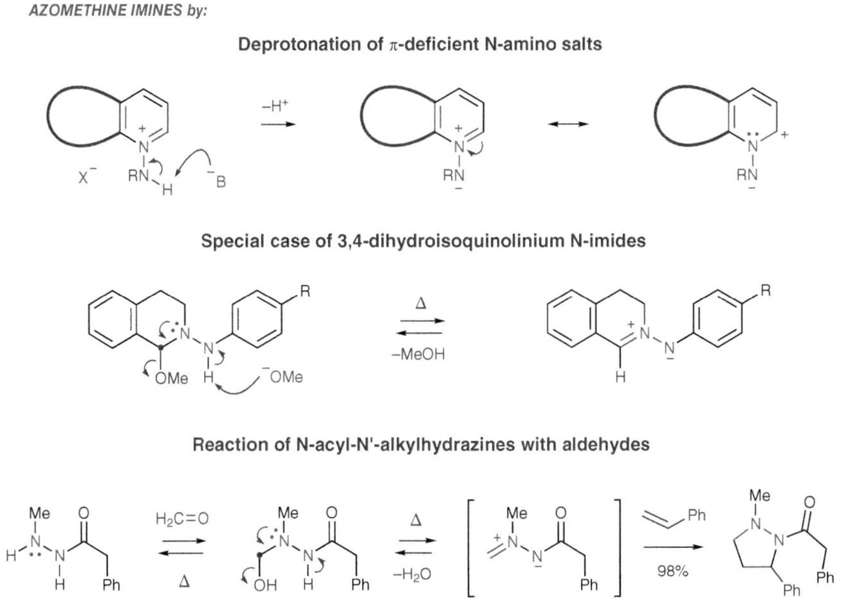

Figure 15.16 Synthesis and reactions of 1,3-dipoles.

equivalent (R=H). In such examples, the formal carbon-nitrogen double bond of the dipole occupies part of a heteroaromatic ring system, most commonly pyridine, quinoline, or isoquinoline. Occasionally azomethine imines of this class can be isolated and purified, in particular if the imide nitrogen bears a stabilizing substituent (i.e., R=acyl or sulfonyl). More often, though, the reactive dipole is generated in situ in the presence of a suitable dipolarophile.

Mention must also be made of the special case of 3,4-dihydroisoquinolinium N-imides, formed in reversible fashion by moderate warming (50–80 degrees Celsius) of hemiaminal derivatives of the type shown to the left center of Figure 15.16. This is a colorful transformation to observe, as the normally pale yellow solution of hemiaminal transforms to a bright orange or deep red indicative of the dipole, depending upon the nature of R. On cooling the original hue is restored. Azomethine imines of this type react so readily with a wide variety of alkenes that they have been suggested as analytical reagents for the identification of liquid olefins [8].

This brings us down to by far the most general means of synthesizing azomethine imines, involving reaction of N-acyl-N'-alkylhydrazines with aldehydes. The specific example shown at the bottom of Figure 15.16 was reported in 1970 by Doctor Wolfgang Oppolzer, then of Sandoz AG, who took a lead role in developing this chemistry [51]. The key feature involves condensation of N-methyl-N'-phenacetylhydrazine with paraformaldehyde in the presence of styrene. Undoubtedly this reaction involves the transient formation of the 1,3-dipole shown in brackets, initiated by hemiaminal formation, which is followed by 1,3-elimination of water. To this end, the experimental conditions could not be more straightforward. The reactants were simply mixed in toluene and heated at reflux for three hours under a nitrogen atmosphere. A Soxhlet extractor filled with molecular sieves sufficed to remove water as formed, and the crude product was crystallized directly from ether. The result was a 98% yield of the adduct illustrated to the bottom right, obtained as a single regioisomer.

From here it was but a short step for Oppolzer to extend this methodology to intramolecular cycloadditions, publishing in 1972 the example shown at the top of Scheme 15.5 [52a]. Interestingly, the geometry of the starred (*) stereocenter in the product was not assigned, although studies to be described below would strongly suggest the β-configuration for the phenyl group. But putting aside stereochemistry for the moment, of what further use could such adducts serve? To answer that question, let us perform a "thought" experiment in which we carry out three reductive processes, consisting of (1) cleavage of the labile N–N bond; (2) N-debenzylation; and (3) reduction of the lactam carbonyl group (cf. dashed lines). The product of this experiment would be the diamino derivative illustrated at the bottom left of the scheme, which moreover, would seem an ideal candidate for ring

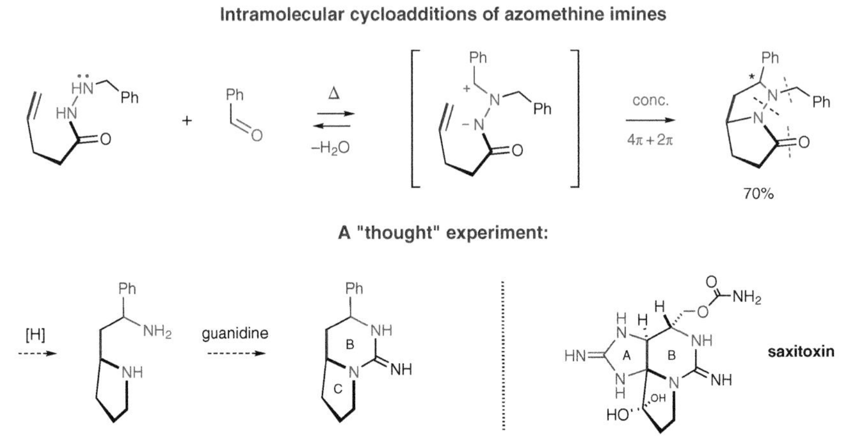

Intramolecular cycloadditions of azomethine imines

70%

A "thought" experiment:

saxitoxin

Scheme 15.5 Synthesis and reactions of 1,3-dipoles.

closure employing guanidine. And what have we accomplished? In this fashion we have constructed a fused-ring heterocycle that contains many of the structural features found in rings B and C of saxitoxin, the paralytic agent of the Alaskan butter clam *Saxidomus giganteus*.

A little bit of background is in order. Saxitoxin has also been isolated from clams and mussels collected off both the east and west coasts of the mainland United States, which were feeding on certain marine dinoflagellates [53]. Perhaps you have heard of the so-called "red tide," when it is unsafe to harvest shellfish? The red color is produced by high concentrations of such organisms, which contain saxitoxin, and which multiply rapidly (or bloom) under favorable conditions of temperature and light. The toxin is stored for lengthy periods in both mussels and clams. It has been estimated that a single dose of 0.2–1.0 mg would prove fatal in humans, thereby ranking it about 55 times more poisonous than strychnine, and 1100 times more toxic than sodium cyanide [54]. Symptoms of poisoning in humans begin with a numbness in the lips, tongue and fingertips within a few minutes after ingestion. This is rapidly followed by weakness in the legs, arms and neck, progressing to a general muscular incoordination. Ultimately, death occurs from respiratory paralysis.

By this point, many readers are probably questioning the wisdom of synthesizing such a material, but it does have beneficial uses as well. Saxitoxin acts by selectively blocking the entrance to sodium channels in neuron membranes, thereby preventing the transient Na^+ ion conductance increases associated with action potentials. The blockage of the sodium channel, although very strong, is also totally reversible. As such, saxitoxin is well suited as a probe of normal and afflicted tissue, and it is an excellent tool for the study of synaptic and neuromuscular transmissions [55]. Medical researchers have used this agent for the labeling, characterization, and isolation of sodium channel components, which has opened new avenues in the study of various nerve disorders. The first synthesis of saxitoxin was accomplished in elegant fashion by Professor Yoshito Kishi and co-workers, of Harvard University (1977) [56]. Some years later (1984), Jacobi et al. reported a second synthesis of saxitoxin, making use of azomethine imine chemistry [57]. This route is outlined in Scheme 15.6, and in its initial phase closely follows the precedent established by Oppolzer (cf. Scheme 15.5).

Note the similarity between the Jacobi group's starting material for saxitoxin (compound **A**, top left of Scheme 15.6), and Oppolzer's alkene-tethered hydrazide (top left, Scheme 15.5). In **A**, the alkene is replaced with a 2-imidazolone as dipolarophile, and there is an appended dithiane ring. Also, methyl glyoxylate hemimethylacetal (MGA) substitutes for benzaldehyde as the active carbonyl component. Brief heating of **A** with MGA, in the presence of BF$_3$ etherate as catalyst, afforded a 65–75% yield of the adduct shown to the top right, as a single epimer at the starred (*) position. Of course, this was the wrong relative configuration for what would eventually become C5 in saxitoxin, but this was easily rectified by epimerization. Moreover, in three additional steps the investigators arrived at the thiocarbamate illustrated to the left center of the scheme, in which every carbon

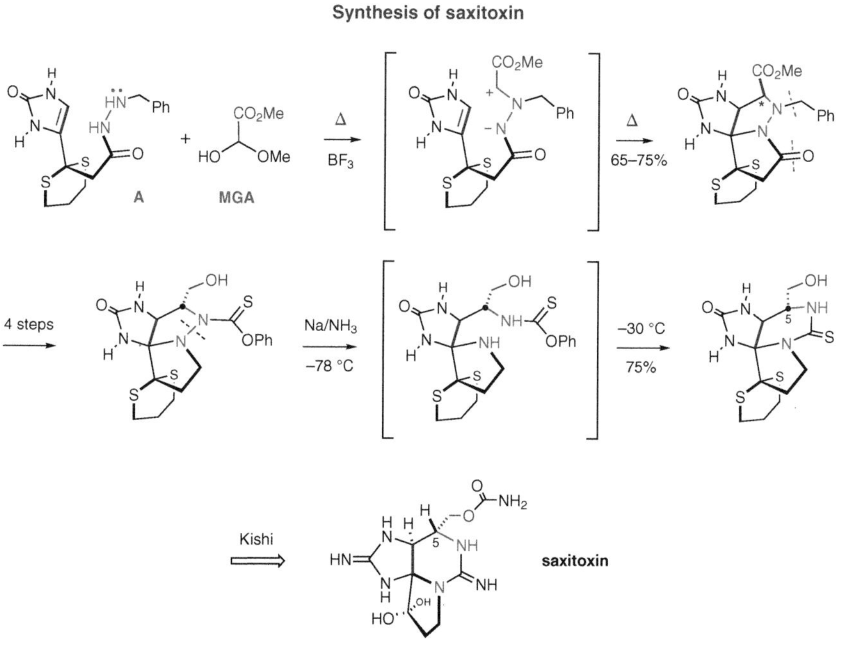

Synthesis of saxitoxin

Scheme 15.6 Synthesis and reactions of 1,3-dipoles.

atom, as well as all stereocenters of saxitoxin were in place. It remained mainly to effect a skeletal reorganization, which was initiated by Na/NH$_3$ reduction of the N–N bond at -78 degrees Celsius. At this temperature the intermediacy of the diamino derivative shown in brackets could be detected by TLC. However, upon slight warming this material underwent smooth intramolecular cyclization, affording a 75% overall yield of the key tricyclic thiourea shown to the middle right. To complete the synthesis required converting the urea and thiourea groups to guanidines, carbamylation of the C5 hydroxymethyl group, and hydrolysis of the dithiane protecting group. These steps followed readily from the precedent of Kishi et al.

References

1 For excellent reviews in this area, see(a) *1,3-Dipolar Cycloaddition Chemistry*, Vol. *1*, Padwa, A., Ed.; Vol. 6 in *General Heterocyclic Chemistry Series*, Taylor, E. C.; Weissberger, A., Eds., John Wiley & Sons, Inc., New York, New York, **1984**. (b) *1,3-Dipolar Cycloaddition Chemistry*, Vol. *2*, Padwa, A., Ed.; Vol. 6 in *General Heterocyclic Chemistry Series*, Taylor, E. C.; Weissberger, A., Eds., John Wiley & Sons, Inc., New York, New York, **1984**. (c) *Synthetic Applications of 1,3-Dipolar Cycloaddition Chemistry Toward Heterocycles and Natural Products*, Padwa, A.; Pearson, W. H., Eds.; Vol. *59* in *The Chemistry of Heterocyclic Compounds*, Taylor, E. C.; Wipf, P., Eds., John Wiley & Sons, Inc., New York, New York, **2002**. (d) Carruthers, W. *Cycloaddition Reactions in Organic Synthesis*; Vol. 8 in *Tetrahedron Organic Chemistry Series*, Baldwin, J. E.; Magnus, P. D., Eds., Pergamon Press, Oxford, UK, **1990**, pp. 269–331.

2 For a compendium of preparative methods, see(a) Regits, M.; Maas, G. *Diazo Compounds: Properties and Synthesis*, Academic Press, Inc., Orlando, Florida, **1986**. For a general survey of reactivity, see(b) Regitz, M.; Heydt, H. in Chapter 4 of reference 1a.

3 cf. Scheme 127 in reference 2b.

4 Huisgen, R.; Reissig, H.-U. *Angew. Chem. Int. Ed. Engl.* **1979**, *18*, 330–331.

5 Wilt, J. W.; Malloy, T. P.; Mookerjee, P. K.; Sullivan, D. R. *J. Org. Chem.* **1974**, *39*, 1327–1336.

6 For a comprehensive review on the synthesis and reactivity of organic azides, see Bräse, S.; Gil, C.; Knepper, K.; Zimmermann, V. *Angew. Chem. Int. Ed.* **2005**, *44*, 5188–5240. See also Chapter 5 in reference 1a.

7 Michael, A. *J. Prakt. Chem.* **1893**, *48*, 94–95.

8 Huisgen, R. *Angew. Chem. Int. Ed.* **1963**, *2*, 565–598.

9 Scheiner, P.; Schomaker, J. H.; Deming, S.; Libbey, W. J.; Nowack, G. P. *J. Am. Chem. Soc.* **1965**, *87*, 306–311.

10 Zanirato, P. *J. Chem. Soc., Perkin Trans. 1* **1991**, 2789–2796.

11 Kolb, H. C.; Finn, M. G.; Sharpless, K. B. *Angew. Chem. Int. Ed.* **2001**, *40*, 2004–2021.

12 Rostovtsev, V. V.; Green, L. G.; Fokin, V. V.; Sharpless, K. B. *Angew. Chem. Int. Ed.* **2002**, *41*, 2596–2599.

13 See, for example, Amblard, F.; Cho, J. H.; Schinazi, R. F. *Chem. Rev.* **2009**, *109*, 4207–4220.

14 Huisgen, R.; Seidel, M.; Saur, J.; McFarland, J. W.; Wallbillich, G. *J. Org. Chem.* **1959**, *24*, 892–893.

15 Bégué, D.; Qiao, G. G.; Wentrup, C. *J. Am. Chem. Soc.* **2012**, *134*, 5339–5350.

16 Huisgen, R.; Seidel, M.; Wallbillich, G.; Knupfer, H. *Tetrahedron* **1962**, *17*, 3–29.

17 The synthesis and reactivity of nitrile oxides have been extensively reviewed in Chapter 3 of reference 1a, and Chapter 6 of reference 1c. See also reference 1d, pp. 285–297.

18 Kozikowski, A. P. *Acc. Chem. Res.* **1984**, *17*, 410–416.

19 (a) Curran, D. P. *J. Am. Chem. Soc.* **1983**, *105*, 5826–5833. (b) Curran, D. P.; Fenk, C. J. *Tetrahedron Lett.* **1986**, *27*, 4865–4868.

20 Kozikowski, A.P.; Li, C.-S. *J. Org. Chem.* **1987**, *52*, 3541–3552.

21 (a) Stevens, R. V.; Christensen, C. G.; Cory, R. M.; Thorsett, E. *J. Am. Chem. Soc.* **1975**, *97*, 5940–5942. See also,(b) Stevens, R. V.; DuPree, L. E.; Edmonson, W. L.; Magid, L. L.; Wentland, M. P. *J. Am. Chem. Soc.* **1971**, *93*, 6637–6643.

22 Woodward, R. B. *Pure Appl. Chem.* **1973**, *33*, 145–177.

23 Müller, P. M.; Farooq, S.; Hardegger, B.; Salmond, W. S.; Eschenmoser, A. *Angew. Chem. Int. Ed.* **1973**, *12*, 914–916.

24 (a) Stevens, R. V.; Beaulieu, N.; Chan, W. H.; Daniewski, A. R.; Takeda, T.; Waldner, A.; Williard, P. G.; Zutter, U. *J. Am. Chem. Soc.* **1986**, *108*, 1039–1049. For a fundamentally different approach, see(b) Wang, H.; Tassa, C.; Jacobi, P. A. *Org. Lett.* **2008**, *10*, 2837–2840.

25 See, for example, Banert, K.; Plefka, O. *Angew. Chem. Int. Ed.* **2011**, *50*, 6171–6174.

26 (a) Nguyen, L. T.; De Proft, F.; Chandra, A. K.; Uchimaru, T.; Nguyen, M. T.; Geerings, P. *J. Org. Chem.* **2001**, *66*, 6096–6103. (b) Houk, K. N. *J. Am. Chem. Soc.* **1972**, *94*, 8953–8955.

27 For a review of ozone cycloaddition chemistry, see Chapter 11 of reference 1b.

28 The chemistry of nitrones has been extensively reviewed. See, for example,(a) Confalone, P. N.; Huie, E. M. *Org. React.* **1988**, *36*, 1–173. (b) Brandi, A.; Cardona, F.; Cicchi, S.; Cordero, F. M.; Goti, A. *Org. React.* **2017**, *94*, 1–529. See also Chapter 9 in reference 1b, Chapter 1 in reference 1c, and pages 298–313 in reference 1d.

29 (a) Ali, S. A.; Wazeer, M. I. M. *Tetrahedron* **1993**, *49*, 4339–4354, (b) Lathbury, D.; Gallagher, T. *Tetrahedron Lett.* **1985**, *26*, 6249–6252. (c) Iwashita, T.; Kusumi, T.; Kakisawa, H. *Chem. Lett.* **1979**, 1337–1340.

30 (a) LeBel, N. A.; Whang, J. J. *J. Am. Chem. Soc.* **1959**, *81*, 6334–6335. (b) LeBel, N. A.; Slusarczuk, G. M. J.; Spurlock, L. A. *J. Am. Chem. Soc.* **1962**, *84*, 4360–4361.

31 Harbour, G. C.; Tymiak, A. A.; Rinehart, Jr., K. L.; Shaw, P. D.; Hughes, Jr., R. G.; Mizsak, S. A.; Coats, J. H.; Zurenko, G. E.; Li, L. H.; Kuentzel, S. L. *J. Am. Chem. Soc.* **1981**, *103*, 5604–5606.

32 Roush, W. R.; Walts, A. E. *J. Am. Chem. Soc.* **1984**, *106*, 721–723.

33 (a) Confalone, P. N.; Lollar, E. D.; Pizzolato, G.; Uskokovic, M. R. *J. Am. Chem. Soc.* **1978**, *100*, 6291–6292. (b) Confalone, P. N.; Pizzolato, G.; Confalone, D. L.; Uskokovic, M. R. *J. Am. Chem. Soc.* **1980**, *102*, 1954–1960.

34 For an overview of azomethine ylide chemistry, see Chapter 6 in reference 1a, Chapter 3 in reference 1c, and pages 272–284 in reference 1d.

35 Grigg, R.; Gunaratne, H. Q. N.; Kemp, J. *J. Chem. Soc., Perkin Trans. 1* **1984**, 41–46.

36 Grigg, R.; Kemp, J.; Sheldrick, G.; Trotter, J. *J. Chem. Soc., Chem. Commun.* **1978**, 109–111.

37 Grigg, R. *Chem. Soc. Rev.* **1987**, *16*, 89–121.

38 Vedejs, E.; Martinez, G. R. *J. Am. Chem. Soc.* **1979**, *101*, 6452–6454.

39 Padwa, A.; Haffmanns, G.; Tomas, M. *J. Org. Chem.* **1984**, *49*, 3314–3322.

40 (a) Smith, R.; Livinghouse, T. *J. Org. Chem.* **1983**, *48*, 1554–1555. (b) Smith, R.; Livinghouse, T. *Tetrahedron* **1985**, *41*, 3559–3568.

41 Woodward, R. B.; Hoffmann, R. *J. Am. Chem. Soc.* **1965**, *87*, 395–397.

42 Huisgen, R.; Scheer, W.; Huber, H. *J. Am. Chem. Soc.* **1967**, *89*, 1753–1755.

43 Woodward, R. B.; Hoffmann, R., *The Conservation of Orbital Symmetry*, Verlag Chemie, GmbH, Weinheim, Germany, **1970**, p.58.

44 For an overview of carbonyl ylide chemistry, see Chapter 4 in reference 1c.

45 (a) Linn, W. J.; Webster, O. W.; Benson, R. E. *J. Am. Chem. Soc.* **1963**, *85*, 2032–2033. (b) Linn, W. J.; Benson, R. E. *J. Am. Chem. Soc.* **1965**, *87*, 3657–3665. (c) Linn, W. J. *J. Am. Chem. Soc.* **1965**, *87*, 3665–3672.

46 (a) Huisgen, R. *Angew. Chem. Int. Ed. Engl.* **1977**, *16*, 572–585. (b) Brokatzky-Geiger, J.; Eberbach, W. *Heterocycles* **1981**, *16*, 1907–1912. (c) Wong, J. P. K.; Fahmi, A. A.; Griffin, G. W.; Bhacca, N.S. *Tetrahedron* **1981**, *37*, 3345–3355. For a theoretical study, see (d) Houk, K. N.; Rondan, N. G.; Santiago, C.; Gallo, C. J.; Gandour, R. W.; Griffin, G. W. *J. Am. Chem. Soc.* **1980**, *102*, 1504–1512.

47 Wong, F. M.; Wang, J.; Hengge, A.C.; Wu, W. *Org. Lett.* **2007**, *9*, 1663–1665.

48 Padwa, A.; Fryxell, G. E.; Zhi, L. *J. Am. Chem. Soc.* **1990**, *112*, 3100–3109.

49 McMills, M. C.; Zhuang, L.; Wright, D. L.; Watt, W. *Tetrahedron Lett.* **1994**, *35*, 8311–8314.

50 For an overview of azomethine imine chemistry, see Chapter 7 in reference 1a and Chapter 12 of reference 1b. See also reference 8.

51 Oppolzer, W. *Tetrahedron Lett.* **1970**, *11*, 2199–2204.

52 (a) Oppolzer, W. *Tetrahedron Lett.* **1972**, *13*, 1707–1710. For a review including this methodology, see(b) Oppolzer, W. *Angew. Chem. Int. Ed. Engl.* **1977**, 16, 10–23.

53 (a) Schantz, E. J.; Mold, J. D.; Stanger, D. W.; Shavel, J.; Riel, F. J. Bowden, J. P.; Lynch, J. M.; Wyler, R. S.; Riegel, B.; Sommer, H. *J. Am. Chem. Soc.* **1957**, *79*, 5230–5235. (b) Schantz, E. J. *Pure Appl. Chem.* **1980**, *52*, 183–188.

54 Mosher, H. S.; Fuhrman, F. A.; Buchwald, H. D.; Fischer, H. G. *Science* **1964**, *144*, 1100–1110.

55 Ritchie, J. M.; Rogart, R. B. *Proc. Natl. Acad. Sci. USA* **1977**, *74*, 211–215.

56 (a) Tanino, H.; Nakata, T.; Kaneko, T.; Kishi, Y. *J. Am. Chem. Soc.* **1977**, 99, 2818–2819. See also,(b) Hannick, S. N.; Kishi, Y. *J. Org. Chem.* **1983**, *48*, 3833–3835.

57 Jacobi, P. A.; Martinelli, M. J.; Polanc, S. *J. Am. Chem. Soc.* **1984**, *106*, 5594–5598.

16

Back to Basics

So where from here? Many paths forward might be taken, engaging in an almost limitless number of additional, and fascinating topics. But time and space have their constraints. Rather, in our next-to-last chapter let us go "back to basics." The author confesses to a slight play on words here, since "basics" could connote a discussion of more introductory level, or it could simply refer to a class of compounds of certain pK_a. In this case it applies to both.

Suppose you were tasked with preparing a large "library" of nitrogen heterocycles, perhaps for the purpose of biological screening. What would you choose as starting materials? There is no question you would include in your "tool kit" a wide assortment of amines and carbonyl derivatives, since imine formation and aldol condensations are fundamental to heterocyclic synthesis (cf. Figure 5.1). However, you would also do well to lay in an ample supply of nitriles and amidines, whose chemistry we explore more fully below.

To begin, nitriles are only feebly basic and even less so nucleophilic, much preferring to react as electrophiles by virtue of their highly polarized carbon–nitrogen triple bond (top Figure 16.1) [1]. In fact, one synthesis of amidines involves nucleophilic addition of amines to nitriles, a reaction that when intermolecular requires forcing conditions (not so, of course, with the intramolecular variant shown in the center of Figure 16.1). In any event, the product amidines are very different from nitriles in their chemical properties. Not only are they potent nucleophiles but they are amongst the strongest of neutral organic bases [2]. Why should this be the case? It is all a matter of orbitals. Note at the bottom of Figure 16.1 that a typical amidine has two non-bonding electron pairs, one of which occupies an sp^2-orbital orthogonal to the π-system of the carbon-nitrogen double bond, while the other is in a p-orbital in conjugation with the imine. Where would an electrophile (including H^+) prefer to attack? If at the p-orbital the reaction pathway leads off to the left, affording a cation that is strongly inductively destabilized. No resonance delocalization is possible. On the other hand, attack at the sp^2-orbital presents an entirely different situation, wherein a developing positive charge can be stabilized by

Introductory Heterocyclic Chemistry, First Edition. Peter A. Jacobi.
© 2019 John Wiley & Sons Ltd. Published 2019 by John Wiley & Sons Ltd.

Back to Basics...

Some Properties of Nitriles and Amidines

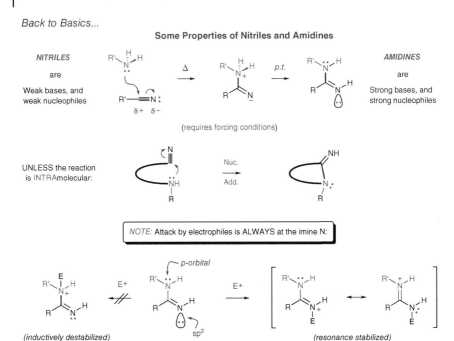

Figure 16.1 Back to basics.

resonance (structures in brackets). So powerful is this effect that certain amidines have found use in organic synthesis as so-called "super bases."

As to their preparation, we have already seen that direct intermolecular addition of amines across the carbon-nitrogen triple bond of nitriles has limitations, due to the generally forcing conditions required. Much more satisfactory is a two step sequence initiated by a Pinner reaction [3], for which a typical procedure involves treatment of a nitrile with ethanolic HCl at low temperatures (top Figure 16.2). The product under these conditions is an imidate ester hydrochloride, derived by ethanolysis of the initially formed imidoyl chloride shown in brackets. Not surprisingly, such "Pinner salts" are very reactive toward nucleophilic substitution, and with primary amines afford generally good–excellent yields of *mono*-substituted amidines. And what of more highly substituted amidines? One route to these materials involves aminolysis of orthoesters, in which case the initial product is an N-substituted imidate ester, derived via the mechanism outlined in brackets in the middle of Figure 16.2 [4]. Subsequent nucleophilic displacement with a second equivalent of amine then affords a disubstituted amidine.

Amidines themselves are also subject to nucleophilic displacement, behaving in much the same fashion as an acyl derivative. A straightforward example involves acid-catalyzed hydrolysis to afford the corresponding amide, but this is

Nitriles and Amidines...

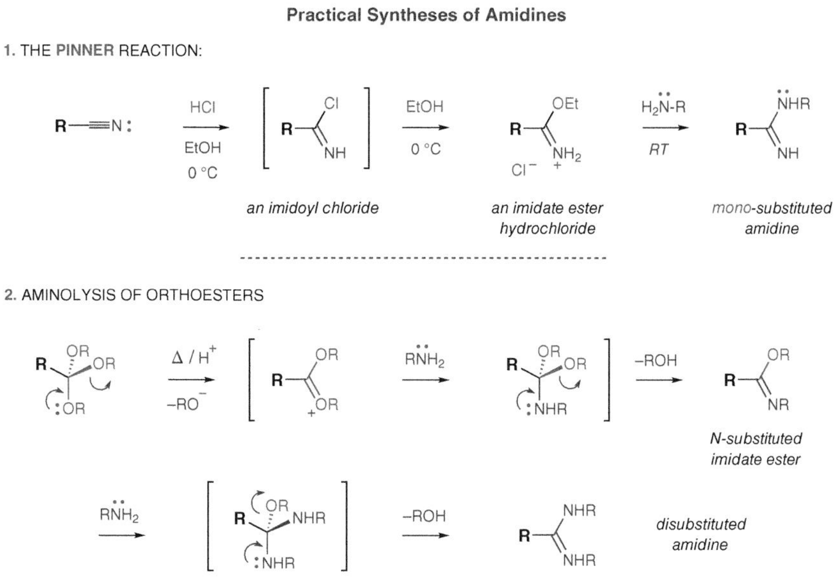

Practical Syntheses of Amidines

1. THE PINNER REACTION:

2. AMINOLYSIS OF ORTHOESTERS

Figure 16.2 Nitriles and amidines.

a transformation of limited practical value (equation 1, Figure 16.3). Indeed, much effort has been devoted to the reverse of this process, converting amides to amidines. Of far greater utility is the reaction shown in equation 2, known as a transamination, and not to be confused with the enzyme catalyzed process of the same name involved in amino acid biosynthesis (cf. Chapter 1). In amidine chemistry, transamination simply refers to the fact that one amine component has been exchanged with another, in the present case illustrated by the conversion of a *mono*-substituted amidine to an unsymmetrical disubstituted amidine. Exchanges of this type are of great importance in nitrogen heterocyclic chemistry, providing access to a variety of ring skeletons. Also worthy of attention is the reaction of amidines with active methylene compounds, in some ways reminiscent of a mixed Claisen condensation, but with loss of an amine instead of an alcohol (equation 3). Malononitrile (pK_a 11) is a particularly effective nucleophile in such condensations, affording with formamidine, for example, high yields of aminomethylenemalononitrile (AMM) (Y,Z=CN; R=H) [5].

For their part, nitriles are moderately good electrophiles, undergoing intermolecular nucleophilic addition with such species as lithium amides and alkoxides (in principle reversible), as well as alkyl lithiums and hydroxide (irreversible; cf. equation 1, Figure 16.4). And, as we have already noted (middle Figure 16.1), *intra*molecular amidine formation can be especially favorable. Under basic conditions, aliphatic nitriles undergo ready deprotonation at the

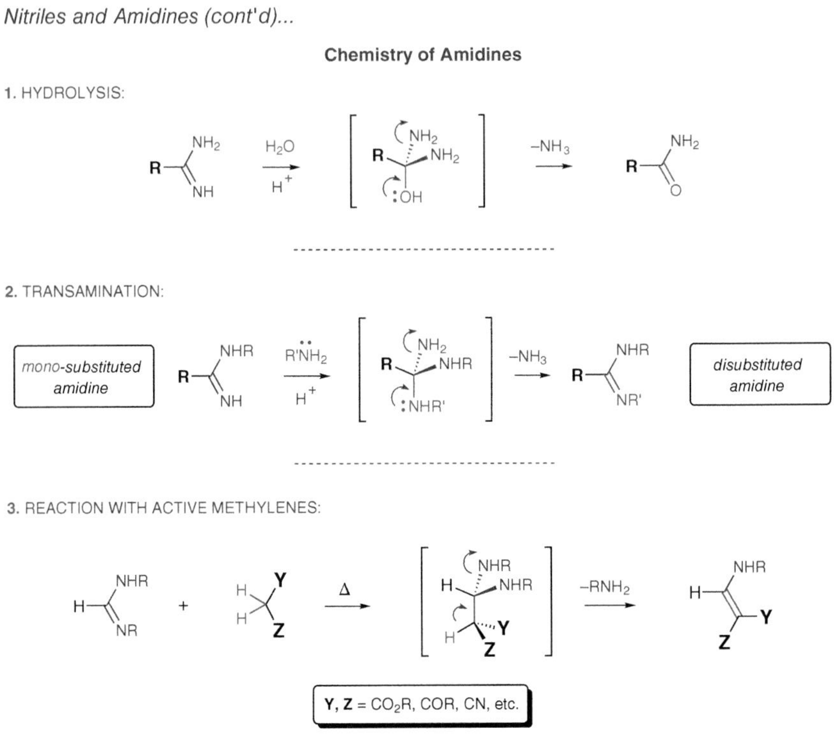

Nitriles and Amidines (cont'd)...

Chemistry of Amidines

1. HYDROLYSIS:

---

2. TRANSAMINATION:

mono-substituted amidine

disubstituted amidine

---

3. REACTION WITH ACTIVE METHYLENES:

Y, Z = CO$_2$R, COR, CN, etc.

Figure 16.3 Nitriles and amidines.

α-position, being in the same pK_a range (~25) as aliphatic esters (equation 2). The resultant resonance stabilized carbanions can then be captured with a variety of electrophiles, including, of course, amidines and other nitriles. An interesting variant of this last category is the intramolecular condensation of tethered dinitriles to give cyclic enaminonitriles, known as a Thorpe-Ziegler reaction (left equation 3)[6,7]. Ring structures as large as 33 carbon atoms have been prepared utilizing this methodology, and it is frequently superior in yields to the closely related Dieckmann cyclization [7]. Finally, to top off our "tool kit," we need only emphasize that any condensation leading to an aromatic ring is almost certain to be energetically downhill, falling into that thermodynamic well we have so often spoken of. Included in this class is the condensation of cyclic enaminonitriles with nitriles, which provides a versatile route to fused-ring pyrimidines (right equation 3) [8]. With this last example we are now fairly well versed in the chemistry of amidines and nitriles, enough so to examine some common threads in mechanistic analysis and synthesis.

An important consideration is that some ring forming reactions may be simpler than they appear, a theme which is nicely illustrated by the trimerization

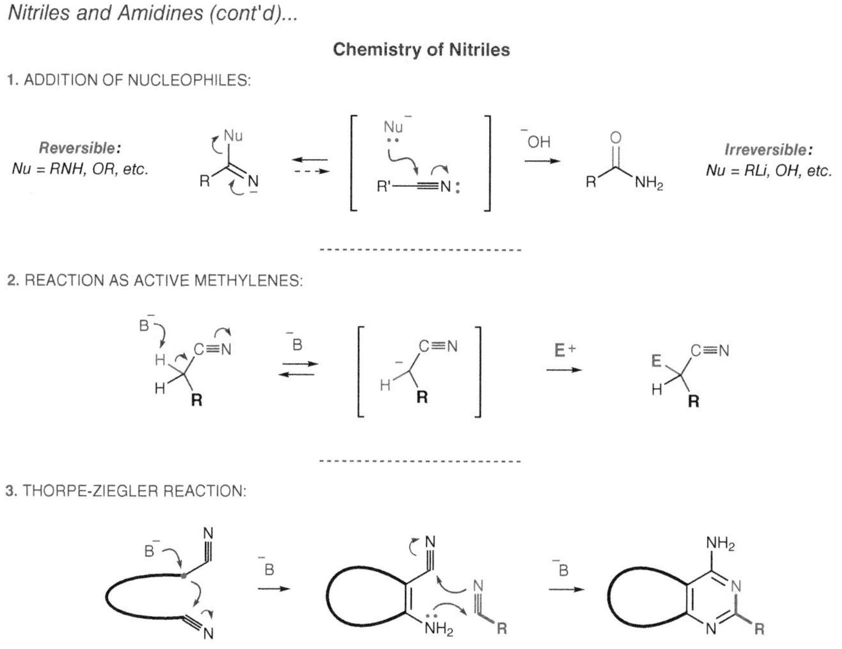
Nitriles and Amidines (cont'd)...
Chemistry of Nitriles
1. ADDITION OF NUCLEOPHILES:
Reversible:
Nu = RNH, OR, etc.
Irreversible:
Nu = RLi, OH, etc.
2. REACTION AS ACTIVE METHYLENES:
3. THORPE-ZIEGLER REACTION:

Figure 16.4 Nitriles and amidines.

of aliphatic nitriles to give 4-aminopyrimidines (Scheme 16.1) [9]. The experimental conditions for this transformation have been well worked out, requiring only catalytic quantities of potassium *tert*-butoxide and brief microwave heating. Also, yields are high and the synthesis can be effected on multi-gram

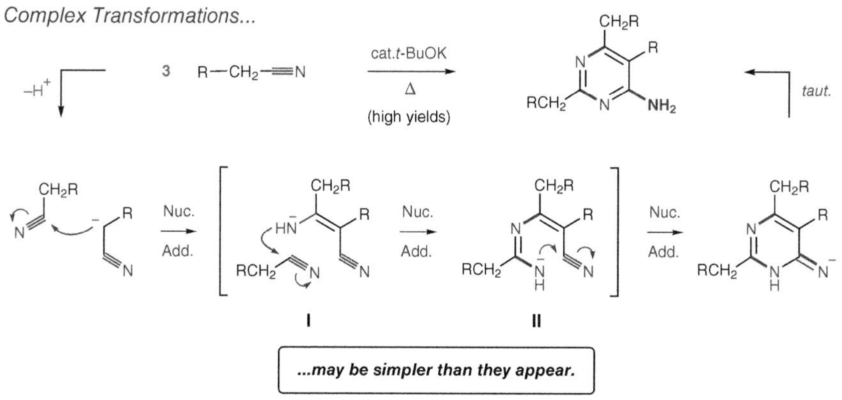
Complex Transformations...
cat.t-BuOK
(high yields)
taut.
Nuc.
Add.
...may be simpler than they appear.

Scheme 16.1 Complex transformations.

scales. But suppose you had been the first to observe this reaction, which is actually quite a bit older than the optimized conditions described above. Would this result have surprised you? And more to the point, could you propose a reasonable pathway leading from starting material to product? The answer to the first question is perhaps a qualified yes, since this was not an obvious route to pyrimidines at the time. However, with the benefit of hindsight, the mechanism seems clear-cut.

Why so? As a start, the base-catalyzed self condensation of aliphatic nitriles to afford enaminonitriles was already a well known process, producing the first intermediate shown in brackets. In fact, we have just finished examining the intramolecular version of this transformation, in the form of the Thorpe-Ziegler reaction (cf. equation 3, Figure 16.4). It is always reassuring to have such strong precedent for the first step in a mechanism; but of equal importance is the end game, and that last crucial step leading to ring closure. Let us look ahead, then, to the final product, focusing on the amino group highlighted in bold, and its position adjacent to a ring nitrogen. In a retrosynthetic sense, such a substitution pattern is a clear marker for an intramolecular addition of an amine across a carbon–nitrogen triple bond, drawing us back to the second intermediate in brackets. We are now left only to "connect the dots" in order to convert intermediate **I** to **II**, and once again we have sound precedent. Bridging the gap is the third equivalent of nitrile, which undergoes nucleophilic addition by the enaminonitrile (curly arrows).

Having gained some experience, let us try our hand at a mechanism slightly wider in scope, and see how it relates to retrosynthetic analysis. What is the first feature that catches your eye on inspecting the reaction at the top of Scheme 16.2 [10]? The starting materials consist of formamidine, shown in red, and malononitrile (blue), heated together in a 3:1 ratio. And the product is the pyrimido[4,5-*d*]pyrimidine highlighted in the shadow box. You are probably thinking, there is that amino group in bold again, right? And adjacent to a ring nitrogen! So with little further ado we could write the last step in a synthetic sequence leading to this material, involving intramolecular nucleophilic addition of an amidine across a nitrile (last structure in brackets). But how did we come to this point, which seems so far removed from our starting materials? One possibility is that our amidine intermediate is the result of a transamination reaction between formamidine and 4-amino-5-cyanopyrimidine, the compound drawn just to the left. This makes good chemical sense, and as reassurance, the proposed pyrimidine is in fact isolable when the reaction is conducted with only two equivalents of formamidine. So let us continue working our way backward and, need we say it, there is that tell-tale pattern again, consisting of an amino group adjacent to a ring nitrogen. The next disconnection thus leads to the first structure in brackets, and from there by transamination back to aminomethylenemalononitrile (AMM). Lastly, if AMM looks familiar

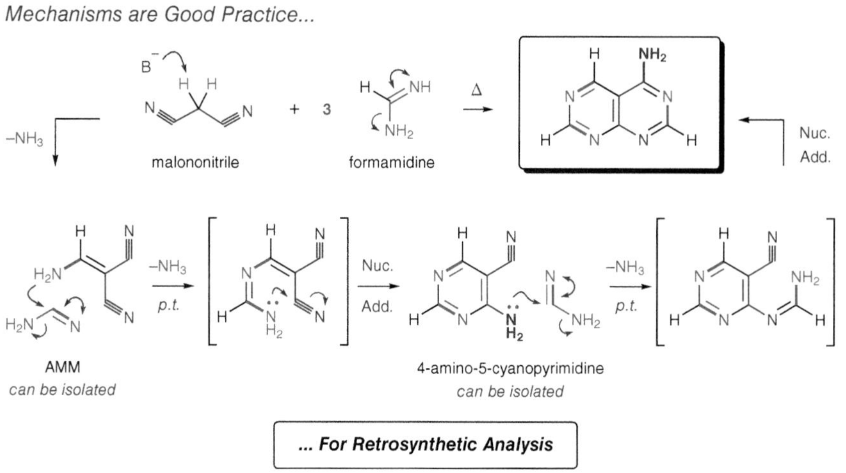

Mechanisms are Good Practice...

Scheme 16.2 Mechanisms are good practice.

to you there is good reason, since we have earlier described its preparation by condensation of malononitrile with formamidine (equation 3, Figure 16.3). In the forward direction, then, this mechanism is quite repetitive, consisting of nucleophilic substitution, transamination, nucleophilic addition, transamination, and a last nucleophilic addition. And how does the synthesis work? Remarkably well for a "one pot" conversion, averaging ~75% yield per step [10]. Amidines and nitriles are indeed versatile building blocks.

References

1 *The Chemistry of the Cyano Group*, Rappoport, Z., Ed., John Wiley & Sons, Ltd., London, UK, **1970**.

2 (a) *The Chemistry of Amidines and Imidates*, Vol. *1*, Patai, S., Ed., John Wiley & Sons, Ltd., London, UK, . (b) *The Chemistry of Amidines and Imidates*, Vol. 2, Patai, S.; Rappoport, Z., Eds., John Wiley & Sons, Ltd., Chichester, UK, 1991.

3 Roger, R.; Neilson, D. G. *Chem. Rev.* **1961**, *61*, 179–211.

4 Taylor, E. C.; Ehrhart, W. A. *J. Org. Chem.* **1963**, *28*, 1108–1112.

5 Jaffe, G. M.; Rehl, W. R., *Synthesis of Aminomethylene Malononitrile*, US Pat. 3660463 A, **1972**.

6 (a) Fleming, F. F.; Shook, B. C. *Tetrahedron* **2002**, *58*, 1–23. (b) Schaefer, J. P.; Bloomfield, J. J. *Organic Reactions* **1967**, *15*, 1–203.

7 Taylor, E. C.; McKillop, A., *The Chemistry of Cyclic Enaminonitriles and o-Aminonitriles*, Vol. *7* in *Advances in Organic Chemistry: Methods and Results*, Taylor, E. C., Ed., John Wiley & Sons, Inc., New York, New York, **1970**.

8 Chercheja, S.; Simpson, J. K.; Lam, H. W. *Tetrahedron* **2011**, *67*, 3839–3845.

9 Baxendale, I. R.; Ley, S. V. *J. Comb. Chem.* **2005**, *7*, 483–489.

10 Taylor, E. C.; Ehrhart, W. A. *J. Am. Chem. Soc.* **1960**, *82*, 3138–3141.

17

A Brief Synopsis

With our discussion now drawing to a close, let us spend a few moments surveying the terrain we have covered. We began with a thumbnail sketch of some biologically important heterocycles of nature—far from an extensive list, to be sure, but enough to perhaps whet our appetites for further exploration. Think DNA and RNA, vitamins and tetrapyrroles, antibiotics and enzymes (*Chapter 1*). Early on, we also learned something of the difference in chemical reactivity between π-deficient and π-excessive heteroaromatic ring systems, the former being much akin to nitrobenzene in their behavior toward electrophiles, while the latter mimic to a certain extent anisole (*Chapter 2*). And what was the take home message from *Chapter 3*? Partly that heterocycles should in principle be easier to synthesize than their carbocyclic analogs, making good use of the free electron pair(s) on the heteroatoms. But beware of unusual molecular rearrangements! To rationalize these, it was necessary to up our level of understanding of some of the physical properties of these species (*Chapter 4*). In the aggregate, then, these four chapters served as a foundation for the chemistry to come.

The next five chapters dealt exclusively with π-deficient heterocycles, with a strong emphasis on synthetic methodology. We started off with so-called "de Novo" syntheses, in which most (if not all) of the substituents to be found in the final product were present in the starting materials (*Chapter 5*). An ideal synthesis utilizing this approach is one in which the target molecule is formed directly in the proper oxidation state, dropping into a "thermodynamic well." Moving on from here, *Chapter 6* initiated our coverage of introduction of new substituents, with the advice of first taking a "cleansing rinse." That is, it was necessary to put aside much of our pre-conceived notions about how to functionalize aromatic rings. Absent powerful electron donating groups, electrophilic aromatic substitution reactions are simply not viable. Rather, nucleophilic substitutions rule the day, but not without complications of their own (remember ANRORC?). In any event, for sheer versatility it is hard to overstate the usefulness of heterocyclic N-oxides,

Introductory Heterocyclic Chemistry, First Edition. Peter A. Jacobi.
© 2019 John Wiley & Sons Ltd. Published 2019 by John Wiley & Sons Ltd.

which facilitate *both* electrophilic substitution and nucleophilic addition (*Chapter 7*). We spent a fair amount of time examining how this could be possible, and introduced in preliminary fashion the concepts of strategy and tactics in heterocycle synthesis. Many of these same reactions are also applicable to quinolines and isoquinolines, which, however, merited their own treatment in *Chapter 8*. Nor could we end this section without touching on the topic of manipulation of existing substituents (*Chapter 9*). Recall that π-deficient heterocycles are strongly electron withdrawing, rendering alkyl groups in the α- or γ-ring positions moderately acidic. As such they undergo much of the chemistry typical of active methylene compounds, including base-catalyzed condensations and substitutions of all sorts. By the same token, vinyl substituents in these positions undergo facile conjugate addition, due to electron delocalization into the ring.

This brought us to *Chapter 10*, and a more detailed treatise of the general properties of π-excessive heterocycles. Aromaticity amongst these species varies considerably, with furan (RE = 16 kcal/mol) and pyrrole (RE = 21 kcal/mol) being quite labile toward mineral acids. Thiophene (RE = 29 kcal/mol), on the other hand, has a reactivity profile more closely approximating benzene, from which it is only difficultly separable (much to the chagrin of a young Viktor Meyer). Simple mono-heteroatom members of this class are non-basic, while imidazole (pK_a 6.9) is far more basic than pyridine. Why?

Chapter 11 initiated our discussion of "de Novo" syntheses of π-excessive heterocycles, wherein most substituents are present from the beginning. Once again, aldol chemistry and imine formation play key roles, and the overall goal is to produce the aromatic ring directly in the proper oxidation state. Much of this chemistry was developed in the late nineteenth century, but in many cases it still represents the methodology of choice. And, of course, recourse might always be taken to introduction of new substituents (*Chapter 12*). Reflecting their electron-rich nature, nearly all π-excessive heterocycles undergo electrophilic substitution reactions (how different from their π-deficient cousins!). However, the reagents involved may require modification, except for with thiophene, where we can play "chemical hardball." α-Metalation reactions are also an effective means of functionalization.

Next, an entire chapter was devoted to ring transformations of π-excessive heterocycles, in particular Diels-Alder reactions (*Chapter 13*). Furan stands out as the most reactive of the parent ring systems, affording, for example, an essentially quantitative yield of *exo*-adduct with maleic anhydride. But such transformations are also readily reversible, and they are sensitive to steric hindrance. This brings to mind what is sometimes referred to as the most famous "failed" reaction in organic chemistry (think cantharidin). The oxazole ring is also noteworthy for its reactivity as a diene in 4π + 2π-cycloadditions, although the primary adducts are rarely isolated. In their stead are produced either pyridines (with alkene dienophiles) or furans (with alkynes). This was followed by an extensive discussion of heterocycles as synthons (*Chapter 14*), an overview of 1,3-dipolar cycloadditions (*Chapter 15*), and finally, we went back to basics (*Chapter 16*).

Have we answered the question of "Why heterocycles?" posed in the preface to this book? And have we provided some sense of the fascination of heterocyclic chemistry? Hopefully so, but in the end this will be left to the readers to decide. What can be said with certainty is that it has been an enjoyable journey for the author.

Index

Introductory Heterocyclic Chemistry, First Edition. Peter A. Jacobi.
© 2019 John Wiley & Sons Ltd. Published 2019 by John Wiley & Sons Ltd.